AF787814

Molecular and Cell Biology of Opportunistic Infections in AIDS

Molecular and Cell Biology of Human Diseases Series

Series Editors

D.J.M. Wright MD FRCPath
Reader in Medical Microbiology, Charing Cross and Westminster Medical School, London, UK

L.C. Archard PhD
Senior Lecturer in Biochemistry, Charing Cross and Westminster Medical School, London, UK

The continuing developments in molecular biology have made possible a new approach to a whole range of different diseases. The books in this series each concentrate on a disease or group of diseases where real progress is being made in understanding the pathogenesis, diagnosis and management. Experts on aspects of each area provide a text accessible to scientists and clinicians in a form which records advances and points to the potential application of these advances in a clinical setting.

Other titles in the series

1 *Molecular and Cell Biology of Sexually Transmitted Diseases*
Edited by David Wright and Leonard Archard

3 *Molecular and Cell Biology of Muscular Dystrophy*
Edited by Terry Partridge

4 *Molecular and Cell Biology of Neuropsychiatric Disease*
Edited by Frank Owen and Ruth Itzhaki

Molecular and Cell Biology of Opportunistic Infections in AIDS

EDITED BY

Steven Myint and Alan Cann

Department of Microbiology,
University of Leicester

SPRINGER-SCIENCE+BUSINESS MEDIA, B.V.

© 1993 Springer Science+Business Media Dordrecht
Originally published by Chapman & Hall in 1993

Typeset in 10/12 pt Sabon by
Columns Design and Production Services Ltd, Reading, UK
ISBN 978-94-010-4668-8 ISBN 978-94-011-1530-8 (eBook)
DOI 10.1007/978-94-011-1530-8

Apart from any fair dealing for the purposes of research or private study, or criticism or review, as permitted under the UK Copyright Designs and Patents Act, 1988, this publication may not be reproduced, stored, or transmitted, in any form or by any means, without the prior permission in writing of the publishers, or in the case of reprographic reproduction only in accordance with the terms of the licences issued by the Copyright Licensing Agency in the UK, or in accordance with the terms of licences issued by the appropriate Reproduction Rights Organization outside the UK. Enquiries concerning reproduction outside the terms stated here should be sent to the publishers at the London address printed on this page.

The publisher makes no respresentation, express or implied, with regard to the accuracy of the information contained in this book and cannot accept any legal responsibility or liability for any errors or omissions that may be made.

A catalogue record for this book is available from the British Library

Library of Congress Cataloging-in-Publication data available

Printed on permanent acid-free text paper, manufactured in accordance with the proposed ANSI/NISO Z 39.48-199X and ANSI Z 39.48-1984

Contents

Contributors

John P. Ackers
Department of Medical Parasitology
London School of Hygiene and Tropical Medicine
London WC1E 7HT
UK

Peter W. Andrew
Department of Microbiology
University of Leicester
Leicester LE1 9HN
UK

Alain Bonnin
Hopital du Bocage
2 Marechal de Lattre de Tassigny
BP 1542 F-21034, Dijon
France

Graham J. Boulnois
Department of Microbiology
University of Leicester
Leicester LE1 9HN
UK

Patrick Camerlynck
Hopital du Bocage
2 Marechal de Lattre de Tassigny
BP 1542 F-21034, Dijon
France

Alan J. Cann
Department of Microbiology
University of Leicester
Leicester LE1 9HN
UK

Jorge A. Cruz-Reyes
Department of Medical Parasitology
London School of Hygiene and Tropical Medicine
London WC1E 7HT
UK

Eduardo Dei-Cas
INSERM (U.42) 369 Rue J Guesde
F-59650 Villeneuve d'Ascq
France

Vincent C. Emery
University Division of Communicable Diseases
Royal Free Hospital and School of Medicine
London NW3 2QG
UK

Paul D. Griffiths
University Division of Communicable Diseases
Royal Free Hospital and School of Medicine
London NW3 2QG
UK

Julian M. Hopkin
Institute of Molecular Medicine
John Radcliffe Hospital
Oxford
UK

Wensi S. Hu
Department of Infectious and Parasitic Disease Pathology
American Registry of Pathology
Armed Forces Institute of Pathology
Washington DC 20306–6000
USA

Hitoshi Kotani
Department of Infectious and Parasitic Disease Pathology
American Registry of Pathology
Armed Forces Institute of Pathology
Washington DC 20306–6000
USA

Shyh-Ching Lo
Department of Infectious and Parasitic Disease Pathology
American Registry of Pathology
Armed Forces Institute of Pathology
Washington DC 20306–6000
USA

Ruth Matthews
Department of Medical Microbiology
University of Manchester
Manchester M13 9PT
UK

Steven H. Myint
Department of Microbiology
University of Leicester
Leicester LE1 9HN
UK

Nicola T. Ralphs
Department of Biological Sciences
Manchester Polytechnic
Manchester M1 5GD
UK

Joanne L. Rhoads
Department of Bacterial Immunology
Walter Reed Army Institute of Research
Washington DC 20307–5100
USA

Fran A. Rubin
Department of Bacterial Immunology
Walter Reed Army Institute of Research
Washington DC 20307–5100
USA

Demetris Savva
Department of Biochemistry and Physiology
University of Reading
Whiteknights PO Box 228
Reading RG6 2AJ
UK

William M. Spice
Department of Medical Parasitology
London School of Hygiene and Tropical Medicine
London WC1E 7HT
UK

David A.J. Tyrrell
PHLS Centre for Applied Microbiology and Research
Porton Down
Salisbury SP4 0JG
UK

Ann E. Wakefield
Institute of Molecular Medicine
John Radcliffe Hospital
Oxford
UK

Richard L. Warren
Department of Bacterial Immunology
Walter Reed Army Institute of Research
Washington DC 20307–5100
USA

Alison Webster
University Division of Communicable Diseases
Royal Free Hospital and School of Medicine
London NW3 2QG
UK

Martin Wiselka
Department of Infectious Diseases
Groby Road Hospital
Leicester LE3 9QE
UK

Preface

Human immunodeficiency virus and AIDS have received much attention in the last decade. As the major cause of death in young adults, this concern is merited. AIDS has also ushered onto the clinical agenda a number of unusual and interesting pathogens. These have had considerably less attention than the virus itself, but are arguably as important. This book attempts to put into one volume a state of the art review of the basic biology of these opportunistic organisms.

When compiling this book, the perceived difficulty was deciding how to restrict the number of chapters, as there is a large number of opportunistic organisms. It soon became clear, however, that the chapters would select themselves, as the work on many of these organisms was still in its infancy. We can be criticized for including *Salmonella*, which is not thought to be an opportunistic infection, and excluding *Cryptococcus*. Our defence is that the former has an interesting relationship with HIV and the latter is one of those organisms for which the basic biology is still in its early stages. The authors are all active researchers in their respective fields and we are grateful that they managed to review their subjects in the short turn-around time required for a book such as this not to be needlessly outdated on publication. We hope that, if nothing else, this book stimulates interest and more research on these agents: perhaps, in future editions, there will be much more to describe on their molecular biology.

Steven Myint and Alan Cann
Leicester, England.

Foreword

DAVID A.J. TYRRELL

I have spent my professional career being interested and involved in infectious diseases of man, and in the use of scientific methods to understand them, with, naturally, a particular interest in improved methods of diagnosis and treatment. At certain periods, new methods and approaches were introduced which quickly and radically altered what we were doing in all these areas. One such was the introduction of tissue cultures in the 1950s, which revealed a range of viruses which were important causes of disease of the CNS and respiratory tract. Another was the expansion in immunology in the 1960s and 70s, which gave us some clear concepts of cell mediated immunity and its role in protection against certain bacterial and viral infections, and the introduction of ELISA technology as a sensitive and efficient method of detecting a wide range of antibodies, and sometimes antigens, in clinical materials.

In spite of this, there was a widespread view, as the 1970s gave way to the 1980s, that infectious diseases were not 'interesting' clinically or scientifically, and this was reinforced by the perception that they were well on the way to being eliminated by the application of antibiotics, vaccination and hygiene.

I was stimulated to consider the field at that time and I contested the view of the day in a short monograph entitiled '*The Abolition of Infection?*'. In that review I indicated that hitherto unknown viruses could still enter the human population and might spread, and I referenced the first descriptions of AIDS. I certainly did not foresee, however, the way in which the virus would be detected, the remarkable speed with which diagnostic tests would be developed, the speed with which infection would become pandemic, or the mixture of success and failure which would attend the massive efforts to develop antiviral drugs and vaccines.

Looking back, it is clear that the world was indeed fortunate that at the time the disease was recognized, a new set of technologies and associated knowledge had become available, such as methods for culturing human lymphocytes, for cloning and sequencing nucleic acids, and also in understanding the genetics and biology of retroviruses which was largely derived from research in cancer and veterinary medicine. As a result, the causative virus was identified remarkably quickly and this lead to methods for screening blood products, identifying infected subjects, studying the abnormal immune system and monitoring the progress of disease and treatment. It is arguable that the striking success of the new methodologies, when previous methods would not work, helped to stimulate their application to other infectious diseases; this was, of course, enhanced by the fact that the major clinical problems in HIV-infected subjects were severe infections with a range of viruses, protozoa, fungi and bacteria, many of which were little known to clinical medicine before the AIDS epidemic.

Infectious disease workers, whether located in the clinic or the laboratory, are therefore faced with a need to understand and keep abreast of a cataract of publications and results obtained with methodologies which have been introduced or rapidly developed since they graduated, and in reference not only to the complex story of the HIV infections, but also the diverse and often bizarre infections and immune responses which contribute to the whole clinical panorama that AIDS presents. That panorama is so wide that it involves virtually all general and specialist branches of medicine and surgery.

For anyone who wishes to understand what this new knowledge is all about, I would recommend this book. The contributors have produced chapters which are accessible to anyone with an average level of undergraduate understanding in biomedical topics. They outline the main features of the biology of the opportunistic infections and they explain not only the principles of many molecular biological techniques and approaches, but what their application has taught us about these infections, which are at the same time both grave and fascinating. It is a book which deserves to be read through and also consulted and dipped into. Although it is remarkably up to date now, unfortunately it is bound to become out of date in detail in a few years, but I think it will remain useful for some years to come because of the attention it gives to broad principles, as well as to the wide range of abnormalities found in AIDS and other forms of immunodeficiency and reduced host resistance.

1

A clinical overview of opportunistic infections in patients with AIDS

MARTIN J. WISELKA

1.1 INTRODUCTION

The term 'opportunistic infection' does not yet have a precise definition. It is generally accepted, however, that these are infections that occur uncommonly in the immunocompetent, but relatively frequently in those with compromised host defences. 'Opportunist organisms' have three basic properties: they are usually organisms of low pathogenicity, they cause more severe disease in the immunocompromised, and they often have an atypical clinical presentation. The AIDS epidemic was first recognized in 1981 when the Center for Disease Control reported cases of *Pneumocystis carinii* pneumonia (PCP) and Kaposi's sarcoma in previously healthy homosexual men [18]. Since then over 20 opportunist agents have been identified. The human immunodeficiency virus (HIV) was identified in 1983 [4, 45] and the natural history of HIV infection has been well documented since.

The progression from asymptomatic HIV infection to AIDS is associated with a reduction in the number of CD4 positive T-lymphocytes (Figure 1.1) and the development of a wide range of bacterial, viral, fungal and protozoal infections (Table 1.1). Organisms which cause symptomatic infections in patients with AIDS include known pathogens (e.g. *Streptococcus pneumoniae*, *Mycobacterium tuberculosis*) and opportunistic agents which are not usually virulent in the normal host (e.g. *Pneumocystis carinii*, *Cryptosporidium*, *Cryptococcus neoformans*). This chapter discusses the clinical manifestations of these infections and highlights

Molecular and Cell Biology of Opportunistic Infections in AIDS
Edited by Steven Myint and Alan Cann
Published in 1992 by Chapman & Hall, London ISBN 0 412 45330 4

Table 1.1 Spectrum of infections which are frequently associated with the acquired immune deficiency syndrome

Organ system affected	*Bacteria*	*Viruses*	*Fungi*	*Protozoa*
Skin and mucosa	*Staphylococcus aureus* *Mycobacterium tuberculosis* *Mycobacterium avium-intracellulare*	Herpes simplex virus Varicella-zoster virus Molluscum contagiosum Human papilloma virus	Dermatophyte infections Mucocutaneous candida Cutaneous manifestations of disseminated mycoses	
Lung	*Streptococcus pneumoniae* *Haemophilus influenzae* *Branhamella catarrhalis* *Staphylococcus aureus* Group B streptococci *Mycobacterium tuberculosis* *Mycobacterium avium-intracellulare*	Cytomegalovirus	*Cryptococcus neoformans* *Histoplasma capsulatum* *Coccidioidomycosis*	*Pneumyocystis carinii*
Gastrointestinal tract and liver	*Salmonella* *Shigella* *Campylobacter* *Clostridium difficile* *Mycobacterium tuberculosis* *Mycobacterium avium-intracellulare*	Herpes simplex virus Cytomegalovirus	*Candida* Hepatic manifestations of disseminated mycoses	*Cryptosporidium* *Isospora belli* *Entamoeba histolytica* *Giardia lamblia* *Microsporidia*
Nervous system	*Mycobaterium tuberculosis* *Mycobacterium avium-intracellulare* *Treponema pallidum*	Papovavirus Herpes simplex virus Varicella-zoster virus Cytomegalovirus	*Candida* *Cryptococcus neoformans*	*Toxoplasma gondii*

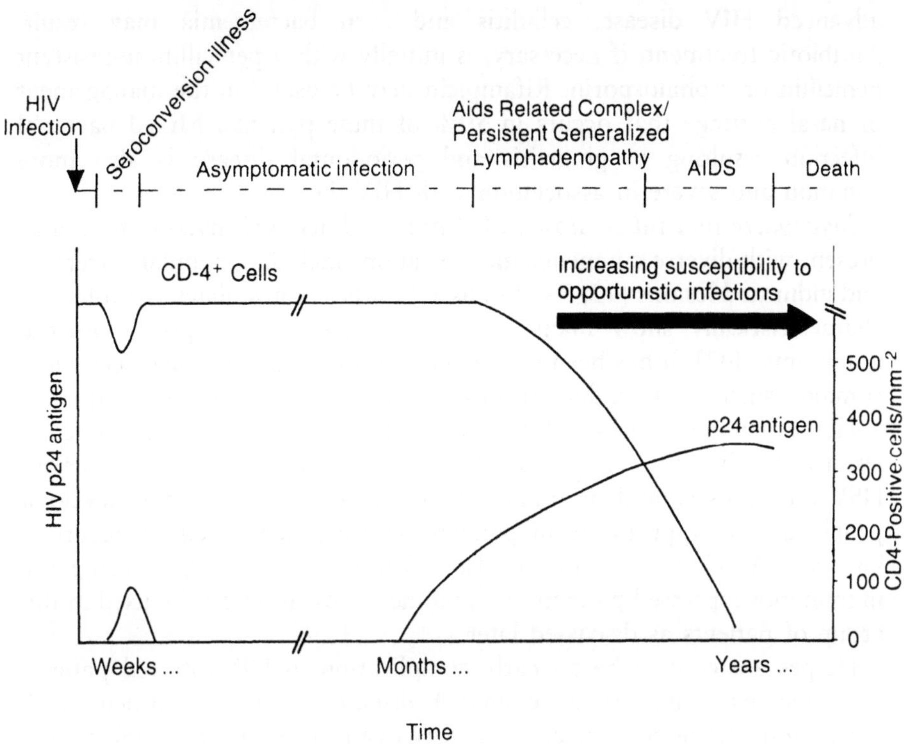

Figure 1.1 Time course of CD4+ cells.

conditions where current therapy is either inadequate or ineffective. The management of patients with AIDS poses particular problems, as the consequences of infection must be distinguished from those caused by HIV, drug therapy, and associated malignancies.

CLINICAL MANIFESTATIONS OF OPPORTUNISTIC INFECTIONS

1.2.1 Cutaneous and mucosal infections

Cutaneous and mucosal infections are extremely common and usually occur before the appearance of an AIDS-defining condition [71]. These infections may be caused by bacteria, viruses, fungi or protozoa [49, 70, 84, 93]. Superficial bacterial infections include impetigo and folliculitis, which are usually caused by *Staphylococcus aureus* [35, 105]. The most common sites are the trunk, inguinal region and face. The primary lesion tends to be a follicular pustule but may be urticarial. A common unusual feature in HIV-positive patients is intense pruritus. In patients with more advanced HIV disease, cellulitis and even bacteraemia may result. Antibiotic treatment, if necessary, is initially with a penicillinase-resistant penicillin or cephalosporin. Rifampicin may be useful in the management of nasal carriage that occurs in 50% of these patients. Mixed bacterial infection resulting in gingivitis and periodontal disease is also more common and severe in association with HIV disease.

Mycobacterium tuberculosis, *M. leprae* and atypical mycobacteria may present with diverse cutaneous manifestations including papular, ulcerated and indurated lesions [42, 74, 93]. Skin biopsies in mycobacterial infection characteristically show many acid-fast organisms with poorly formed granulomas [42]. It has been reported that these infections have been more common since the introduction of therapy with azidothymidine [3].

Herpes simplex virus (HSV) causes severe cutaneous and mucosal lesions, which may recur frequently in patients with AIDS [101]. Serious HSV infections should be treated with systemic acyclovir, and acyclovir prophylaxis is appropriate in patients who are debilitated by recurrent infection. Acyclovir resistance is also a feature which was first recognized in immunosuppressed patients. Vidarabine or foscarnet can be used in this group of patients as discussed later.

Herpes zoster may be an early complication in HIV-infected patients [21] and extensive or disseminated disease requires treatment with intravenous acyclovir. Epstein–Barr virus causes few overt problems but has been implicated in the aetiology of hairy leukoplakia, a warty-like growth usually occurring on the side of the tongue [56, 102]; and central nervous system lymphomas [83]. Molluscum contagiosum and human papilloma virus frequently cause florid and widespread lesions in patients with AIDS [49, 70, 84].

Superficial fungal infections are very common and include *Tinea cruris*, *T. pedis*, *Pityriasis versicolor* and *Candida* [49, 70]. Oral candidiasis occurs in nearly all patients and may be one of the first symptomatic manifestations of HIV infection [30]. It can take a number of forms. The most common presentation is its pseudomembranous form: white plaques which can be wiped off to reveal an erythematous base. An atrophic form presents as smooth red patches, and a hyperkeratotic form (candidal leucoplakia) manifests itself as white patches which cannot be removed easily. Fungal infections may initially respond to topical antifungal therapy but recurrences are frequent and systemic therapy is usually required.

Cutaneous lesions have been described in disseminated cryptococcal, histoplasmosis, coccidioidomycosis and sporotrichosis infections [9, 10, 32] and should be suspected in patients who have lived in endemic areas.

Skin manifestations of protozoal infections are seen rarely but cutaneous leishmaniasis and infection with *Pneumocystis carinii* have been reported [25].

1.2.2 Pulmonary infections

The lung is a frequent site of opportunistic infections in patients with AIDS. *Pneumocystis carinii*, either alone or co-existing with other infections, is the most common pathogen and occurs in approximately 80% of all patients [88, 89]. *Pneumocystis carinii* is a protozoan-like organism that infects the lungs of humans and many animals and has trophozoite and cystic forms. Transmission is thought to occur primarily by the respiratory route and extrapulmonary infections are rare. Symptoms of pneumocystis infection include unproductive cough, dyspnoea and fever and usually evolve over several weeks [72]. A minority of patients have a more acute onset, with rapid progression to respiratory failure [39]. The characteristic radiological appearance is a diffuse, bilateral, perihilar pulmonary infiltrate, although a wide range of radiographic appearances have been reported and about 5% of patients have a normal chest X-ray [5, 39]. The diagnosis of *Pneumocystis* pneumonia is established by methanamine silver staining of induced sputum, bronchial lavage specimens or bronchial biopsies [8, 44, 48].

Cytomegalovirus (CMV) is often isolated from sputum and bronchoalveolar lavage specimens in addition to *Pneumocystis carinii* or other lung pathogens. The clinical significance of CMV in this situation is unclear [88, 112], as studies have shown that the presence of concomitant CMV does not affect the outcome of patients presenting with *Pneumocystis* pneumonia [67]. Primary CMV pneumonitis is comparatively rare and is diagnosed if CMV is isolated from the lung and histology shows the presence of characteristic viral inclusion bodies. There are no distinctive clinical or radiological features [112].

Community-acquired bacterial pneumonias are frequently recognized in association with AIDS and the pathogens include *Streptococcus pneumoniae*, *Staphylococcus aureus*, *Mycoplasma pneumoniae*, *Branhamella catarrhalis*, *Haemophilus influenzae* and group B streptococci [88, 89, 99, 105]. Symptoms of bacterial pneumonia usually have an acute onset and chest radiographs show lobar, segmental or diffuse consolidation [99, 112]. The incidence of *Mycobacterium tuberculosis* infection is higher in patients with AIDS than in non-immunosuppressed patients [114]. In Africa, one third of HIV-positive patients have been found to have tuberculosis at presentation and it is clearly an important worldwide problem [58]. Tuberculosis is often very aggressive in patients with AIDS and there is a high incidence of extrapulmonary infection [114]. The radiographic features also differ from those seen in other groups of patients as cavitation is rare, infection frequently involves the middle and lower lobes of the lung and marked lymphadenopathy may be present [114]. The diagnosis of tuberculosis is established by culture or acid-fast staining of sputum or bronchoalveolar lavage specimens. Tuberculin skin tests are not usually helpful. Atypical mycobacterial infection occurs in up to half of all patients with AIDS [121]. Ninety-six percent of atypical mycobacteria are classified as *Mycobacterium avium-intracellulare* complex (MAC) [64]. Infection is generally widely disseminated and is associated with fever, night sweats, malaise, and weight loss, although many patients are asymptomatic [59]. The chest X-ray usually shows diffuse infiltrates if there is pulmonary infection [112]. Mycobacteria may be recovered from blood cultures, lung tissue, bone marrow, liver, lymph nodes or other biopsy specimens [59, 122].

A number of fungi cause pulmonary infections in patients with AIDS. *Cryptococcus neoformans* usually presents with meningitis or encephalitis, but may occasionally cause pneumonia with fever, cough and dyspnoea and diffuse infiltrates on the chest X-ray [118]. The diagnosis is established by histology of lung washings or biopsies which reveals organisms in the alveolar septae [43, 73]. Histoplasmosis occurs in endemic areas and pulmonary disease is often part of a disseminated infection [54]. Blood cultures are usually positive for *Histoplasma capsulatum* and organisms may be visualized in biopsy specimens from infected sites. Disseminated coccidioidomycosis occurs in patients from endemic areas [13], and usually results in a severe and progressive infection with large numbers of organisms in biopsy specimens [50]. Pulmonary aspergillosis and candidiasis have been reported in AIDS patients but are relatively uncommon [29, 31]. *Toxoplasma gondii* is a rare cause of pneumonitis [16] which presents with fever and dyspnoea and may resemble *Pneumocystis carinii* pneumonia. A maculopapular rash has been reported in toxoplasma infection [16].

Investigation of pulmonary infections

Initial investigation of patients with chest symptoms will include a chest X-ray, however the radiographic features are rarely diagnostic [5, 88, 89]. A number of other non-invasive tests have been used to determine whether there is evidence of early pulmonary infection, particularly in patients with a normal chest X-ray. These tests include gallium scanning [22], measurement of the transfer factor (TLCO) [107], arterial blood gas analysis and determination of oxygen saturation during rest and exercise [113].

If there is clinical evidence of pulmonary disease, further investigations include blood cultures and sputum microscopy and culture. Gram stain and Ziehl–Neelsen stains should be performed. *Pneumocystis carinii* has been identified in induced sputum specimens [8], but this technique is relatively time consuming and insensitive and most physicians will proceed to fibre-optic bronchoscopy, provided there are no clotting abnormalities or severe respiratory failure. Bronchoalveolar lavage provides the diagnosis in over 85% of cases [12, 44, 98, 112] and is generally preferable to transbronchial biopsy which is more invasive and seldom improves the diagnostic yield.

Treatment for pneumocystis pneumonia may be initiated several days before bronchoscopy without affecting the diagnostic sensitivity. Some investigators therefore recommend empirical treatment for pneumocystis pneumonia in patients with typical features of infection and perform bronchoscopy only in those patients who fail to respond. An open lung biopsy might be considered to establish the aetiology of a pulmonary infiltrate, or if the patient is not responding to treatment.

Treatment and prophylaxis of pulmonary infections

The conventional treatment for Pneumocystis pneumonia consists of high-dose oral or intravenous trimethoprim-sulphamethoxazole or intravenous pentamidine [41, 108, 119]. The frequency of drug reactions is increased in patients with AIDS [117] and 40–80% of those treated with trimethoprim-sulphamethoxazole develop toxicity, commonly fever, rash, hepatitis, leucopenia, or thrombocytopenia which usually appears between days 7 and 14 of treatment [72, 117]. Approximately 50% of patients given pentamidine develop adverse reactions, including marrow toxicity, uraemia, hepatitis, fever, skin rash, renal toxicity and hypoglycaemia [119]. Nebulized pentamidine has been used successfully in patients with early pneumocystitis pneumonia [86] but infection may recur. Seventy-five percent of patients with *Pneumocystis* pneumonia respond to initial therapy and improvement is often rapid, with a reduction in dyspnoea and cough after several days [108, 119]. The X-ray appearance, pulmonary function tests, and gallium scan all resolve more slowly. Poor prognostic features include, recurrent episodes of *Pneumocystis* pneumonia, a long history of symptoms preceding admission, tachypnoea, low pO_2 at presentation, and evidence of interstitial oedema in histological specimens [11, 69]. The possibility of concomitant infection with other pathogens should be considered in unresponsive patients. Salvage therapy in patients with *Pneumocystis* pneumonia who fail to improve on either trimethoprim-sulphamethoxazole or pentamidine includes the use of high dose methylprednisolone [82], eflornithine and trimetrexate [1]. Assisted ventilation may be required in some cases [37], but its role has yet to be defined.

Following successful treatment of *Pneumocystis* pneumonia, patients have a 40–60% risk of developing further episodes of infection [116], therefore secondary prophylaxis is essential. Oral trimethoprim-sulphamethoxazole [40] and nebulized pentamidine [47, 115] are both highly effective. Dapsone or sulphadoxine and pyrimethamine (Fansidar) have been used in patients who are unable to tolerate trimethoprim-sulphamethoxazole. Primary prophylaxis should be offered to patients with $CD4^+$ cell counts <200/mm^3 as they are at high risk of developing *Pneumocystis* infection within 6–12 months [116].

Cytomegalovirus pneumonitis has been treated with 9-(1,3-dihydroxy-2-propoxymethyl) guanine (DHPG or ganciclovir) [23] and foscarnet [104]; however, the response is poor. Both agents must be given intravenously and maintenance therapy must be continued indefinitely. Ganciclovir has dose-related and reversible bone marrow toxicity which is potentiated if the drug is given in combination with azidothymidine [23, 75, 95]. The most serious adverse effects of foscarnet are renal impairment, seizures, hypomagnesaemia and hypocalcaemia [92, 104]. There is therefore an urgent need for an effective and well tolerated oral agent to treat CMV infections.

Mycobacterium tuberculosis responds to standard antituberculous therapy [97]. *Mycobacterium avium-intracellulare* is resistant to most antimicrobial agents [59, 122] and treatment is very difficult. Regimens have included combinations of drugs including rifampicin, ethambutol, amikacin, rifabutin, ansamycin, clofazimine, cycloserine and ciprofloxacin, but none has been shown to substantially alter the clinical course of infection [59]. A recent uncontrolled study showed promising results with azithromycin which led to a reduction in mycobacteria cell counts and clinical improvement in approximately 75% of patients treated for at least 20 days [123]. Further clinical trials are clearly required.

Fungal lung infections in patients with AIDS generally respond poorly to therapy with amphotericin B [29, 31]. The newer imidazole agents, including itraconazole and fluconazole appear to be more effective and are well tolerated. Controlled clinical trials are awaited.

1.2.3 Gastrointestinal tract and liver infections

The most frequent infective cause of upper gastrointestinal symptoms are *Candida* species. Candidal oesophagitis occurs in approximately 14% of AIDS patients and is associated with dysphagia, odynophagia, retrosternal pain and oesophageal bleeding [31]. Concomitant oral involvement is usually found. Herpes simplex viruses types 1 and 2 are a cause of persistent oral ulceration and may involve the oesophagus and lower GI tract. Cytomegalovirus (CMV) infection can produce oesophagitis with linear or raised ulcers [120] which tend to be more extensive than those caused by herpes simplex. The diagnosis is confirmed by endoscopy and biopsies showing characteristic viral inclusion bodies.

Oral examination may reveal the presence of *Candida*, herpes simplex, or other lesions including Kaposi's sarcoma or hairy leukoplakia. If oesophageal *Candida* is suspected, a therapeutic trial of antifungal therapy may be given before arranging further investigations, including barium swallow or upper gastrointestinal endoscopy. Endoscopy is more sensitive than barium studies [24] and allows biopsies to be taken for confirmation of the diagnosis. Topical antifungal agents are sometimes effective for localized candidal infections, however more extensive disease requires systemic therapy with oral imidazole preparations including ketoconazole, fluconazole and itraconazole. Relapses are common and maintenance therapy is usually required. Fluconazole may be slightly more effective than ketoconazole [30] and has a lower incidence of hepatotoxicity, but is considerably more expensive.

Oesophagitis due to herpes simplex virus responds to treatment with oral acyclovir, however recurrences are common and acyclovir prophylaxis may be necessary. There is little clinical information on the use of ganciclovir or foscarnet in CMV oesophagitis.

A wide range of infecting organisms have been identified in the lower gastrointestinal tract in patients with AIDS. Protozoal agents are of particular importance and the incidence of *Entamoeba histolytica* and *Giardia lamblia* was known to be increased in homosexual men, even before the onset of the AIDS epidemic [100]. Another protozoal species, *Cryptosporidium*, is the most serious pathogen of the lower GI tract in AIDS patients [26]. *Cryptosporidium* is recognized as a cause of a mild, self-limiting, diarrhoeal illness in the immunocompetant host, but is more severe in AIDS patients, in whom persistent cryptosporidial infection may lead to relentless diarrhoea, malaise and weight loss. The diagnosis is made by the identification of cysts in the stools using a modified acid-fast stain [81]. Other important protozoal causes of diarrhoea and weight loss are *Isospora belli*, which is most common in the tropics [28], and *Enterocytozoon bienusi*, a microsporidial organism which has been associated with symptoms of diarrhoea and malabsorption [33, 36, 55]. *Encephalitozoon cunicule* is a more recently recognized AIDS-associated pathogen. These infections are diagnosed by light microscopy or electron microscopy of intestinal biopsies [36, 55].

Important bacterial infections of the gastrointestinal tract include *Salmonella*, *Shigella*, *Campylobacter* and *Clostridium difficile* [40, 94]. The incidence of salmonella bacteraemia and disseminated salmonella infections is increased in patients with AIDS compared to non-immunocompromised individuals [48].

Mycobacterium tuberculosis infection of the lower gastrointestinal tract is comparatively rare [114], but *Mycobacterium avium-intracellulare* is frequently isolated. Infection may be asymptomatic or give rise to diarrhoea, abdominal pain and pyrexia [53, 55]. Mycobacteria are identified by acid-fast staining of the stools or biopsy specimens which characteristically show foamy macrophages containing numerous organisms [53, 55].

CMV may infect any part of the gastrointestinal tract but is most common in the large bowel where it causes an acute colitis, giving clinical features of bloody diarrhoea, abdominal pain and fever [109]. CMV causes shallow ulceration and submucosal haemorrhage, and complications include toxic dilatation and bowel perforation. The diagnosis is confirmed by the histological appearance of characteristic viral inclusion bodies. Other viruses which have been implicated in the aetiology of diarrhoea in AIDS patients include adenoviruses, enteroviruses and rotaviruses [27].

Investigation and management of diarrhoea

Diarrhoea occurs in over half of patients with AIDS [2] and pathogens are isolated in up to 85% of episodes [109]. Initial investigations include microscopy and culture of stool specimens. Several samples should be sent as *Cryptosporidium* and other agents may be shed intermittently [109]. Stools should be tested for *Clostridium difficile* toxin if pseudomembranous colitis is suspected. Further investigations include rectal swabs for the detection of *Chlamydia* or gonorrhoea, sigmoidoscopy, rectal biopsy and upper gastrointestinal endoscopy. Duodenal aspiration and biopsy may reveal the presence of small intestinal pathogens including *Giardia lamblia*, *Enterocytozoon bienusi* and *Mycobacterium avium-intracellulare* complex [55].

The approach to treatment depends upon the pathogen(s). Giardiasis and amoebic infection respond to metronidazole, but there is no effective treatment for intestinal cryptosporidiosis. A number of agents have been used, including erythromycin and spiramycin, however the response has been poor [110]. As persistent cryptosporidiosis is one of the most distressing conditions in AIDS patients, further antimicrobial agents are urgently required. Paromomycin appears promising but controlled trials are awaited. *Isospora belli* responds to trimethoprim-sulphamethoxazole but may recur [28], and a clinical response to metronidazole was shown in patients with *Enterocytozoon bienusi*, although organisms were not eradicated [36].

Many agents have been used to treat *Mycobacterium avium-intracellulare*, (section 1.2.2) but infection is resistant to therapy and survival is usually less than 6 months [59, 64]. Intestinal CMV infection may be treated with either foscarnet of ganciclovir, however the response is poor compared to that of CMV retinitis [19, 65]. Human immunodeficiency virus may contribute to gastrointestinal symptoms in AIDS patients as HIV has been shown to infect the intestinal mucosa [91], causing partial villous atrophy [46], malabsorption and diarrhoea [85]. The ability of HIV to cause autonomic denervation in the GI tract [57] is another potential cause of diarrhoea.

Infections of the liver

Abnormalities of hepatic function may be secondary to a number of factors including opportunistic infections, drug therapy and malignant disease. A form of sclerosing cholangitis has been described in AIDS patients and usually presents with right upper quadrant pain associated with cholestatic liver function tests [17, 34]. Cryptosporidial and CMV infections of the gastrointestinal tract have been implicated in the aetiology of this condition. The diagnosis is confirmed by endoscopic retrograde cholangiography which shows irregular strictures of the biliary tract [17, 34]. Sphincterotomy may relieve symptoms due to a localized lesion.

The clinical features of hepatitis in patients with AIDS include fever, jaundice, right upper quadrant pain and hepatomegaly. A wide range of opportunistic infections may be responsible including mycobacteria, hepatitis viruses, CMV, histoplasmosis, coccidiomycosis and *Cryptococcus neoformans*. A liver biopsy may be needed to establish the diagnosis [15].

1.2.4 Infections of the nervous system

Infections causing space occupying lesions

Space occupying lesions may present acutely with fits and focal neurological signs or, more insidiously, with personality change, confusion and fever. The protozoon, *Toxoplasma gondii* is the most common opportunistic infection of the central nervous system in patients with AIDS [90]. The cat is the definitive host of *T. gondii* and infection is transmitted to man by the faeco-oral route or by eating undercooked meat. Reactivation of encysted organisms in the brain is believed to be the cause of toxoplasma encephalitis [63] which is associated with the presence of single or multiple cerebral lesions and usually presents with headache and focal neurological signs [78]. Latent infection is common, particularly in France and southern Europe where toxoplasma IgG antibodies are found in approximately 70% of the population [80]. It has been estimated that 12–30% of AIDS patients who are seropositive for toxoplasma IgG will develop cerebral toxoplasmosis [52].

Approximately 2–4% of patients with AIDS develop progressive multifocal leucoencephalopathy (PML) [103]. PML results from a papovavirus infection which causes lesions of the cerebral white matter. The condition presents insidiously with dysphasia, ataxia, or focal signs [7] and histological findings include focal demyelination, swollen astrocytes and oligodendrocytes containing eosinophilic inclusion bodies [103].

Investigation and management of patients with focal cerebral lesions

The differential diagnosis of focal neurological symptoms in patients with AIDS includes opportunistic infections, cerebral lymphoma, strokes, vasculitis and HIV encephalopathy. Initial investigations are likely to include a CT or magnetic resonance imaging (MRI) scan. MRI may be more sensitive for focal infections [78]. Toxoplasmosis gives characteristic mass lesions which show ring-enhancement after the injection of contrast material [78]. Lesions are usually multiple and occur predominantly in the basal ganglia or in the corticomedullary region of the occipital or frontoparietal lobes [78]. PML is associated with low density lesions in the periventricular white matter which do not enhance after contrast [78]. Examination of the cerebrospinal fluid (CSF) can be performed if there is no evidence of a focal mass or raised intracranial pressure, and may reveal the presence of opportunistic organisms or cerebral lymphoma. Serological tests are of limited value in the diagnosis of toxoplasmosis as an IgM response only occurs in approximately 20% of patients with encephalitis [80, 115].

Other causes of ring-enhancing lesions include *Mycobacterium tuberculosis*, *Candida albicans*, *Cryptococcus neoformans*, *Nocardia*, *Aspergillus*, CMV, Kaposi's sarcoma and cerebral lymphoma; however, these are much less common and most physicians would commence treatment for toxoplasmosis and investigate further only if there was no response. A brain biopsy is necessary in difficult cases.

Toxoplasma responds well to treatment with oral pyrimethamine and sulphadiazine [115]. Steroids may be beneficial in patients who have evidence of cerebral oedema or raised intracranial pressure and clindamycin is an effective alternative in patients who are unable to tolerate sulphonamides [76]. Treatment should be continued for 4–6 weeks and maintenance therapy should be given indefinitely as there is a high rate of relapse [115]. The role of primary prophylaxis in patients with serological evidence of previous *Toxoplasma gondii* infection has yet to be established, however it is possible that trimethoprim-sulphamethoxazole given for prophylaxis against *Pneumocystis carinii* might also prevent toxoplasmosis. No effective treatment is currently available for PML.

Infections causing meningoencephalitis

Cryptococcus neoformans is the most common cause of meningitis in patients with AIDS and is seen most frequently in African patients and injectable drug users [77]. The clinical presentation of cryptococcal meningitis is usually insidious with fever, lethargy and weight loss, followed by headache, nausea, neck stiffness and photophobia [20, 32]. The cerebrospinal fluid may be normal or show pleocytosis with a raised protein and lowered glucose [73]. The diagnosis is established by the detection of organisms on Indian ink staining, or cryptococcal antigen in CSF or blood [38]. Amphotencin and flucytosine have been used in cryptocloccal meningitis but the treatment of choice is fluconazole (200–400 mg daily for 6 weeks), a triazole antifungal agent which has good CSF penetration and may be given orally or intravenously [14, 111]. Maintenance therapy with fluconazole (100–200 mg daily) has been shown to prevent relapses [10] and should be continued indefinitely.

Encephalitis in patients with AIDS may present as an acute toxic confusional state associated with fever, or have a slower onset with cognitive impairment, lethargy, and generalized weakness. Physical signs include ataxia, weakness, hyper-reflexia and extensor plantar responses. The CT scan shows signs of cerebral atrophy with widened sulci and dilated ventricles.

HIV is neurotropic and is associated with a wide range of neurological manifestations including encephalopathy, polyneuropathy, peripheral neuropathy, mononeuritis multiplex and autonomic neuropathy [60, 62, 77]. HIV infection is the most common cause of subacute encephalopathy, but other possibilities which should be excluded are CMV, herpes simplex virus, varicella zoster virus, atypical mycobacterial infection and *Treponema pallidum* [68]. The choice of treatment depends on the causative organism.

Retinitis

CMV retinitis is a serious problem affecting up to 40% of AIDS patients. The characteristic findings are the appearance of florid perivascular haemorrhages and exudates which ultimately lead to retinal infarction and blindness if left untreated [65, 95]. Initial unilateral involvement progresses commonly to affect both eyes. Early retinitis responds well to treatment with ganciclovir or foscarnet, however maintenance therapy must be continued indefinitely as drug therapy is suppressive only [23, 61, 75, 92]. Despite maintenance, progression of the retinitis tends to occur.

Radiculopathy

Radiculopathy is a rare manifestation of CMV infection and often occurs in patients who have evidence of CMV retinitis [6, 87]. CMV radiculopathy starts in the lumbo-sacral area and nerve root involvement is rapidly progressive. Patients present with symptoms of a cauda equina syndrome with paraesthesia, sensory loss, and weakness of the buttocks and legs. Urinary retention may also occur [6, 87]. Investigation includes myelography or MRI scanning to exclude a mass lesion and examination of the CSF which usually shows raised protein, low glucose and polymorphonuclear leucocytosis [6, 87]. CMV is usually isolated from the CSF and radiculopathy responds to ganciclovir [87].

1.3 CONCLUSIONS

A wide spectrum of serious microbial infections are seen in patients with AIDS. Many of these infections are caused by opportunistic organisms which have low virulence in the normal host and were previously considered to be of little clinical importance. Effective therapy is currently available for several of the most important opportunistic pathogens including *Pneumocystis carinii*, *Toxoplasma gondii*, *Cryptococcus neoformans*, *Candida*, herpes simplex virus and CMV. Primary and secondary prophylaxis is now being offered for many of these conditions. The increasing use of antimicrobial prophylaxis and the introduction of antiretroviral agents have been responsible for a marked recent improvement in the prognosis of patients with AIDS [51, 96].

Outstanding problems in the management of opportunistic infections include drug toxicity, emergence of drug resistance [66, 106], and infections for which present therapy is ineffective [79]. Cryptosporidia and *Mycobacterium avium-intracellulare* continue to cause severe problems in patients with AIDS and deserve particular attention. An effective oral agent against CMV would also be of enormous benefit. Further research into the molecular biology of the opportunistic organisms associated with AIDS will undoubtedly lead to improved diagnostic techniques and a more rational approach to the development of novel antimicrobial agents.

REFERENCES

1. Allegra, C.J., Chabner, B.A., Tuazon, C.U., *et al.* (1987) Trimetrexate for the treatment of *Pneumocystis carinii* pneumonia in patients with the acquired immunodeficiency syndrome. *N. Engl. J. Med.*, **317**, 978–85.
2. Antony, M.A., Brandt, J., Klein, R.S., Bernstein, L.H. (1988) Infectious diarrhoea in patients with AIDS. *Dig. Dis. Sci.*, **33**, 1141–6.
3. Barbaro, D.J., Orcutt, V.L., Coldiron, B.M. (1989) Mycobacterium-avium-

Mycobacterium intracellulare infection limited to the skin and lymph nodes in patients with AIDS. *Rev. Infect. Dis.*, **1**, 625–8.

4. Barre-Sinoussi, F., Chermann, J.C., Rey, F., *et al.* (1983) Isolation of a T-lymphocyte retrovirus from a patient at risk for immunodeficiency syndrome (AIDS). *Science*, **220**, 868–71.
5. Barter, S.J., Rowland-Hill, C., Shaw, R., Pinching, A.J. (1987) The radiological features of the pulmonary complications of AIDS. *Thorax*, **42**, 205–9.
6. Behar, R., Wiley, C., McCutchan, J.A. (1987) Cytomegalovirus polyradiculopathy in acquired immune deficiency syndrome. *Neurology*, **37**, 557–61.
7. Berger, J.R., Kaszovitz, B., Donovan, P.J., Dickenson, G. (1987) Progressive multifocal leucoencephalopathy associated with human immunodeficiency syndrome. *Ann. Intern. Med.*, **107**, 78–87.
8. Bigby, T.D., Margolskee, D., Curtis, J.L., *et al.* (1986) The usefulness of induced sputum in the diagnosis of *Pneumocystis carinii* pneumonia in patients with the acquired immunodeficiency syndrome. *Am. Rev. Respir. Dis.*, **133**, 515–18.
9. Bonner, J.R., Alexander, W.J., Dismukes, W.E., *et al.* (1984) Disseminated histoplasmosis in patients with the acquired immune deficiency syndrome. *Arch. Intern. Med.*, **144**, 2178–81.
10. Bozzette, S.A., Larsen, R.A., Chiu, J. (1991) A placebo-controlled trial of maintenance therapy with fluconazole after treatment of cryptococcal meningitis in the acquired immunodeficiency syndrome. *N. Engl. J. Med.*, **324**, 580–4.
11. Brenner, M., Ognibene, F.P., Lack, E.E., *et al.* (1987) Prognostic factors and life expectancy of patients with acquired immunodeficiency syndrome and *Pneumocystis carinii* pneumonia. *Am. Rev. Respir. Dis.*, **136**, 1199–206.
12. Broaddus, C., Dake, M.D., Stulbarg, M.S., *et al.* (1985) Bronchoalveolar lavage and transbronchial biopsy for the diagnosis of pulmonary infections in the acquired immunodeficiency syndrome. *Ann. Intern. Med.*, **102**, 747–52.
13. Bronnimann, D.A., Adam, R.D., Galgiani, J.N., *et al.* (1987) Coccidioidomycosis in the acquired immunodeficiency syndrome. *Ann. Intern. Med.*, **106**, 372–9.
14. Byrne W.R. and Wajszczuk, C.P. (1988) Cryptococcal meningitis in the acquired immunodeficiency syndrome (AIDS): successful treatment with fluconazole after failure with amphotericin B. *Ann. Intern. Med.*, **108**, 384–5.
15. Cappell, M.S., Schartz, M.S., Biempica, L. (1990) Clinical utility of liver biopsy in patients with serum antibodies to the human immunodeficiency virus. *Am. J. Med.*, **88**, 123–30.
16. Catterall, J.M., Hofflin, J.M., Remington, J,S. (1986) Pulmonary toxoplasmosis. *Am. Rev. Respir. Dis.*, **133**, 704–5.
17. Cello, J.P. (1989) Acquired immunodeficiency syndrome cholangiopathy, spectrum of disease. *Am. J. Med.*, **86**, 539–46.
18. Centers for Disease Control. (1981) Kaposi's sarcoma and *Pneumocystis* pneumonia among homosexual men. *M.M.W.R.*, **30**, 305–8.

19. Chachoua, A., Dieterich, D., Krasinski, K., *et al.* (1987) 9-(1,3-dihydroxy-2-propoxymethyl) guanine (ganciclovir) in the treatment of cytomegalovirus gastrointestinal disease with the acquired immunodeficiency syndrome. *Ann. Intern. Med.*, **107**, 133–7.
20. Chuck, S.L., Sande, M.A. (1989) Infections with *Cryptococcus neoformans* in the acquired immunodeficiency syndrome. *N. Engl. J. Med.*, **321**, 794–9.
21. Cohen, P.R., Beltrani, V.P., Grossman, M.E. (1988) Disseminated herpes zoster in patients with human immunodeficiency virus infection. *Am. J. Med.*, **84**, 1076–80.
22. Coleman, D.L., Dodek, P.M., Golden, J.A. (1984) Correlation between gallium lung scans and fibreoptic bronchoscopy in patients with suspected *Pneumocystis carinii* pneumonia and the acquired immune deficiency syndrome. *Am. Rev. Respir. Dis.*, **130**, 1166–9.
23. Collaborative DHPG Treatment Study Group. (1986) Treatment of serious cytomegalovirus infections with 9-(1,3-dihydroxy-2-propoxymethyl) guanine in patients with AIDS and other immunodeficiencies. *N. Engl. J. Med.*, **314**, 801–5.
24. Connolly, G.M., Forbes, A., Gleeson, J.A., Gazzard, B.G. (1989) Investigation of upper gastrointestinal symptoms in patients with AIDS. *Aids*, **3**, 453–6.
25. Coulman, C.U., Greene, I., Archibald, R.W.R. (1987) Cutaneous pneumocystosis. *Ann. Intern. Med.*, **106**, 396–8.
26. Crawford, F.G., Vermund, S.H. (1988) Human cryptosporidiosis. *CRC Crit. Rev. Microbiol.*, **16**, 113–59.
27. Cunningham, A.L., Grohman, G.S., Harkness, J., *et al.* (1988) Gastrointestinal viral infections in homosexual men who were symptomatic and seropositive for human immunodeficiency virus. *J. Infect. Dis.*, **158**, 386–91.
28. DeHovitz, J.A., Pape, J.W., Boncy, M., Johnson, W.D. (1986) Clinical manifestations and therapy of *Isospora belli* infection in patients with the acquired immune deficiency syndrome. *New. Engl. J. Med.*, **315**, 87–90.
29. Denning, D.W., Follansbee, S.E., Scolaro, M., Norris, S., Edelstein, H., Stevens, D.A. (1991) Pulmonary aspergillosis in the acquired immunodeficiency syndrome. *N. Engl. J. Med.*, **324**, 654–62.
30. De Wit, S., Weerts, D., Goossens, H., Clumeck, N. (1989) Comparison of fluconazole and ketoconazole for oropharyngeal candidiasis in AIDS. *Lancet*, **i**, 746–8.
31. Diamond, R.D. (1991) The growing problem of mycoses in patients infected with the human immunodeficiency virus. *Rev. Inf. Dis.*, **13**, 480–6.
32. Dismukes, W.E. (1988) Cryptococcal meningitis in patients with AIDS. *J. Infect. Dis.*, **157**, 624–8.
33. Dobbins, W.O., Weinstein, W.M. (1985) Electron microscopy of the intestine and rectum in acquired immunodeficiency syndrome. *Gastroenterology*, **88**, 738–49.
34. Dowsett, J.F., Miller, R., Davidson, R., *et al.* (1988) Sclerosing cholangitis in acquired immunodeficiency syndrome. *Scand. J. Gastroenterol.*, **23**, 1267–74.

35. Duvic, M. (1987) Staphylococcal infections and the pruritis of AIDS-related complex. *Arch. Dermatol.*, **123**, 1599.
36. Eeftinck Schattenkerk, J.K.M., Van Gool, T., Van Ketel, R.J., *et al.* (1991) Clinical significance of small-intestinal microsporidiosis in HIV-1-infected individuals. *Lancet*, **337**, 895–8.
37. El-Sadr, W., Simberkoff, M.S. (1988) Survival and prognostic factors in severe *Pneumocystis carinii* pneumonia requiring mechanical ventilation. *Am. Rev. Respir. Dis.*, **137**, 1264–7.
38. Eng, R.H.K., Bishburg, E., Smith, S.M., Kapila, R. (1986) Cryptococcal infections in patients with the acquired immunodeficiency syndrome. *Am. J. Med.*, **81**, 19–23.
39. Engelberg, L.A., Lerne, C.W., Tapper, M.L. (1984) Clinical features of *Pneumocystis* pneumonia in the acquired immune deficiency syndrome. *Am. Rev. Respir. Dis.*, **130**, 689–94.
40. Fischl, M.A., Dickinson, G.M., Sinave, C., *et al.* (1986) *Salmonella* bacteraemia as manifestation of acquired immunodeficiency syndrome. *Arch. Intern. Med.*, **146**, 113–5.
41. Fischl, M.A., Dickenson, G.M., La Voie L. (1988) Safety and efficacy of sulphamethoxazole and trimethoprim chemoprophylaxis for *Pneumocystis carinii* pneumonia in AIDS. *J.A.M.A.*, **259**, 1185–9.
42. Freed, J.A., Pervez, N.K., Chen, V., *et al.* (1987) Cutaneous mycobacteriosis, Occurrence and significance in two patients with the acquired immunodeficiency syndrome. *Arch. Dermatol.*, **123**, 1601–3.
43. Gal. A.A., Koss, M.N., Hawkins, J., Evans, S., Einstein, H. (1986) The pathology of pulmonary cryptococcal infections in the acquired immunodeficiency syndrome. *Arch. Pathol. Lab. Med.*, **110**, 502–7.
44. Gal, A.A., Klatt, E.C., Koss, M.N., Strigle, S.M. (1987) The effectiveness of bronchoscopy in the diagnosis of *Pneumocystis carinii* and cytomegalovirus pulmonary infections in acquired immunodeficiency syndrome. *Arch. Pathol. Lab. Med.*, **111**, 238–41.
45. Gallo, R.C., Salahuddin, S.Z., Popovic, M., *et al.* (1984) Frequent detection and isolation of cytopathic retroviruses (HTLV-III) from patients with AIDS and at risk from AIDS. *Science*, **224**, 500–3.
46. Gillin, J.S., Shike, M., Alcock, N. (1985) Malabsorption and mucosal abnormalities of the small intestine in the acquired immunodeficiency syndrome. *Ann. Intern. Med.*, **102**, 619–22.
47. Girard, P.M., Landman, R., Lepetre, A.S., *et al.* (1989) Prevention of *Pneumocystis carinii* pneumonia by pentamidine aerosol in zidovudine-treated AIDS patients. *Lancet*, **i**, 1348–52.
48. Glaser, J.B., Morton-Kute, L., Berger, S.R., *et al.* (1985) Recurrent *Salmonella typhimurium* bacteraemia associated with the acquired immunodeficiency syndrome. *Ann. Intern. Med.*, **102**, 189–93.
49. Goodman, D., Teplitz, E.D., Wishner, A., *et al.* (1987) Prevalence of cutaneous disease in patients with acquired immunodeficiency syndrome (AIDS) or AIDS-related complex. *J. Am. Acad. Dermatol.*, **17**, 210–20.
50. Graham, A.R., Sobonya, R.E., Bronnimann, D.A., Galgiani, J.N. (1988)

Quantitative pathology of coccidioidomycosis in acquired immunodeficiency syndrome. *Hum. Pathol.*, **19**, 800–6.

51. Graham, N.M.H., Zeger, S.L., Park, L.P., *et al.* (1991) Effect of zidovudine and *Pneumocystis carinii* pneumonia prophylaxis on progression of HIV-1 infection to AIDS. *Lancet*, **338**, 265–9.
52. Grant, I.H., Gold, J.W., Rosenblum, M., Niedzwieki, D., Armstrong, D. (1990) *Toxoplasma gondii* serology in HIV infected patients, the development of central nervous system toxoplasmosis in AIDS. *Aids*, **4**, 519–21.
53. Gray, J.R., Rabenek, L. (1989) Atypical mycobacterial infection of the gastrointestinal tract in AIDS patients. *Am. J. Gastroenterol.*, **84**, 1521–4.
54. Graybill, J.R. (1988) Histoplasmosis and AIDS. *J. Infect. Dis.*, **158**, 623–6.
55. Greenson, J.K., Belitsos, P.C., Yardley, J.H., Bartlett, J.G. (1991) AIDS enteropathy: Occult enteric infections and duodenal mucosal alterations in chronic diarrhoea. *Ann. Intern. Med.*, **114**, 366–72.
56. Greenspan, J.S., Greenspan, D., Lennette, E.T., *et al.* (1985) Replication of Epstein–Barr virus within the epithelial cells of oral 'hairy' leucoplakia, an AIDS associated lesion. *N. Engl. J. Med.*, **313**, 1564–71.
57. Griffin, G.E., Miller, A., Batman, P., Pinching, A., Harris, J.R.W., Mathan, M. (1988) Damage to intrinsic jejunal autonomic nerves in HIV infection. *Aids*, **2**, 379–82.
58. Harries, A.D. (1991) Tuberculosis and human immunodeficiency virus infection in developing countries. *Lancet*, **335**, 387–90.
59. Hawkins, C.C., Gold, J.W.M., Whimbey, E., *et al.* (1986) *Mycobacterium avium* complex infections in patients with the acquired immunodeficiency syndrome. *Ann. Intern. Med.*, **105**, 184–8.
60. Ho, D.D., Rota, T.R., Schooley, R.T., *et al.* (1985) Isolation of HTLV-III from cerebrospinal fluid and neural tissues of patients with neurological syndromes related to the acquired immunodeficiency syndrome. *N. Engl. J. Med.*, **313**, 1493–7.
61. Holland, G.N., Sidikaro, Y., Kreiger, A.E., *et al.* (1987) Treatment of cytomegalovirus retinopathy with ganciclovir. *Ophthalmology*, **94**, 815–23.
62. Hollander, H., Levy, J.A. (1987) Neurological abnormalities and recovery of human immunodeficiency virus from cerebrospinal fluid. *Ann. Intern. Med.*, **106**, 692–5.
63. Holliman, R.E. (1988) Toxoplasmosis and the acquired immunodeficiency syndrome. *J. Infect.*, **16**, 121–8.
64. Horsburgh, C.R., Selik, R.M. (1989) The epidemiology of disseminated non-tuberculous mycobacterial infection in the acquired immunodeficiency syndrome (AIDS). *Am. Rev. Respir. Dis.*, **139**, 4–7.
65. Jacobson, M. A., Mills, J. (1988) Serious cytomegalovirus disease in the acquired immunodeficiency syndrome (AIDS). Clinical findings, diagnosis, and treatment. *Ann. Intern. Med.*, **108**, 585–94.
66. Jacobson, M.A., Berger, T.G., Fikrig, S., *et al.* (1990) Acyclovir-resistant varicella zoster virus infections after chronic acyclovir therapy in patients with the acquired immunodeficiency syndrome. *Ann. Intern. Med.*, **112**, 187–91.

67. Jacobson, M.A., Mills, J., Rush, J., *et al.* (1991) Morbidity and mortality of patients with AIDS and first-episode *Pneumocystis carinii* pneumonia unaffected by concomitant pulmonary cytomegalovirus infection. *Am. Rev. Respir. Dis.*, **144**, 6–9.
68. Johns, D.R., Tierney, M., Felsenstein, D. (1987) Alterations in the natural history of neurosyphilis by concurrent infection with the human immunodeficiency virus. *N. Engl. J. Med.*, **316**, 1569–72.
69. Kales, C.P., Murren, J.R., Torres, R.A., *et al.* (1987) Early predictors of in-hospital mortality for *Pneumocystis carinii* pneumonia in the acquired immunodeficiency syndrome. *Arch. Intern. Med.*, **147**, 1413–7.
70. Kaplan, M.H., Sadick, N., McNutt, S., *et al.* (1987) Dermatologic findings and manifestations of acquired immunodeficiency syndrome (AIDS). *J. Am. Acad. Dermatol.*, **16**, 485–506.
71. Klein, R.S., Harris, C.A., Small, C.B., *et al.* (1984) Oral candidiasis in high-risk patients as the initial manifestation of the acquired immune deficiency syndrome. *N. Engl. J. Med.*, **311**, 354–7.
72. Kovacs, J.A., Hiemenz, J.W., Macher, A.M. (1984) *Pneumocystis carinii* pneumonia, A comparison between patients with the acquired immunodeficiency syndrome and patients with other immunodeficiencies. *Ann. Intern. Med.*, **100**, 663–71.
73. Kovacs, J.A., Kovacs, A.A., Polis, M., *et al.* (1985) Cryptococcosis in the acquired immunodeficiency syndrome. *Ann. Intern. Med.*, **103**, 533–8.
74. Lamfers, E.J.P., Bastiaans, A.H., Mravunac, M., *et al.* (1987) Leprosy in the acquired immunodeficiency syndrome. *Ann. Intern. Med.*, **107**, 111–2.
75. Laskin, O.L., Stahl-Bayliss, C.M., Kalman, C.M., Rosecan, L.R. (1987) Use of ganciclovir to treat serious cytomegalovirus infections in patients with AIDS. *J. Infect. Dis.*, **155**, 323–7.
76. Leport, C., Bastuji-Garin, S., Peronne, C., *et al.* (1989) An open study of the pyrimethamine-clindamycin combination in AIDS patients with toxoplasmosis. *J. Inf. Dis.*, **160**, 557–8.
77. Levy, R.M., Janssen, R.S., Bush, T.J., Rosenblum, M.L. (1988) Neuro-epidemiology of acquired immunodeficiency syndrome. *J. AIDS.*, **1**, 31–40.
78. Levy, R.M., Mills, C.M., Posin, J.P., *et al.* (1990) The efficacy and clinical impact of brain imaging in neurologically symptomatic AIDS patients, a prospective CT/MRI study. *J. AIDS*, **3**, 461–71.
79. Laughon, B.E., Allaudeen, H.S., Becker, J.M., *et al.* (1991) Summary of the workshop on future directions in discovery and development of therapeutic agents for opportunistic infections associated with AIDS. *J. Infect. Dis.*, **164**, 244–51.
80. Luft, B.J., Remington, J.S. (1988) Toxoplasmosis encephalitis. *J. Infect. Dis.*, **157**, 1–4.
81. Ma, P., Soave. R. (1983) Three step stool examination for cryptosporidiosis in 10 homosexual men with protracted watery diarrhoea. *J. Infect. Dis.*, **147**, 824–8.
82. McFadden, D.K., Edelson, J.D., Hyland, R.H., Rodriquez, C.H., Inouye, T., Rebuck, A.S. (1987) Corticosteroids as adjunctive therapy in treatment of

Pneumocystis carinii pneumonia in patients with the acquired immunodeficiency syndrome. *Lancet*, **ii**, 1477–9.

83. Macmahon, E.M.E., Glass, J.D., Hayward, S.D. (1991) Epstein–Barr virus in AIDS-related primary central nervous system lymphoma. *Lancet*, **338**, 969–73.
84. Matis, W.L., Triana, A., Shapiro, R., *et al.* (1987) Dermatologic findings associated with human immunodeficiency virus infection. *J. Am. Acad. Dermatol.*, **17**, 746–51.
85. Miller, A., Griffin, G.E., Batman, P., *et al.* (1988) Jejunal mucosal architecture and fat absorption in male homosexuals infected with human immunodeficiency virus. *Q. J. Med.*, **69**, 1009–19.
86. Miller, R.F., Godfrey-Fausett, P., Semple, S.J.G., *et al.* (1989) Nebulised pentamidine as treatment for *Pneumocystis carinii* pneumonia in the acquired immune deficiency syndrome. *Thorax*, **44**, 565–9.
87. Miller, R.G., Storey, J.R., Greco, C.M. (1990) Ganciclovir in the treatment of progressive AIDS-related polyradiculopathy. *Neurology*, **40**, 569–74.
88. Murray, J.F., Felton, C.P., Garay, S.M., *et al.* (1984) Pulmonary complications of the acquired immune deficiency syndrome. *N. Engl. J. Med.*, **310**, 1682–8.
89. Murray, J.F., Garay, S.M., Hopewell, P.C., Mills, J., Snider, G.L., Stover, D.E. (1987) Pulmonary complications of the acquired immune deficiency syndrome, an update. *Am. Rev. Respir. Dis.*, **135**, 504–9.
90. Navia, B.A., Petito, C.K., Gold, J.W., *et al.* (1986) Cerebral toxoplasmosis complicating the acquired immune deficiency syndrome; clinical and neuropathological findings in 27 patients. *Ann. Neurol.*, **20**, 224–38.
91. Nelson, J.A., Wiley, C.A., Reynolds-Kohler, C., Reese, C.E., Margaretten, W., Levy, J.A. (1988) Human immunodeficiency virus detected in bowel epithelium from patients with gastrointestinal symptoms. *Lancet*, **i**, 259–62.
92. Palestine, A.G, Polis, M.A., De Smet, M.D., *et al.* (1991) A randomised, controlled trial of foscarnet in the treatment of cytomegalovirus retinitis in patients with AIDS. *Ann. Intern. Med.*, **115**, 665–73.
93. Penneys, N.S., Hicks, B. (1985) Unusual cutaneous signs of acquired immunodeficiency syndrome. *J. Am. Acad. Dermatol.*, **13**, 845–52.
94. Perlman, D.M., Ampel, N.M., Schifman, R.B., *et al.* (1988) Persistent *Campylobacter jejuni* infections in patients infected with human immunodeficiency virus. *Ann. Intern. Med.*, **108**, 540–6.
95. Peters, B.S., Beck, E.J., Anderson, S., *et al.* (1991) Cytomegalovirus infection in AIDS. Patterns of disease, response to therapy and trends in survival. *J. Infect.*, **23**, 129–37.
96. Peters, B.S., Beck, E.J., Coleman, D.G. (1991) Changing disease patterns in patients with AIDS in a referral centre in the United Kingdom, the changing face of AIDS. *Br. Med. J.*, **302**, 203–7.
97. Pitchenik, A.E., Rubinson, H.A. (1985) The radiographic appearance of tuberculosis in patients with the acquired immune deficiency syndrome (AIDS) and pre-AIDS. *Am. Rev. Respir. Dis.*, **131**, 393–6.
98. Pitchenik, A.E., Ganjei, P., Torres, A., Evans, D.A., Rubin, E., Baier, H.

(1986) Sputum examination for the diagnosis of *Pneumocystis carinii* pneumonia in the acquired immunodeficiency syndrome. *Am. Rev. Respir. Dis.*, **133**, 226–9.

99. Polsky, B., Gold, J.W.M., Whimbey, E., *et al.* (1986) Bacterial pneumonia in patients with the acquired immunodeficiency syndrome. *Ann. Intern. Med.*, **104**, 38–41.
100. Quinn, T.C., Stamm, W.E., Goodell, S.E., *et al.* (1983) The polymicrobial nature of intestinal infections in homosexual men. *N. Engl. J. Med.*, **309**, 576–82.
101. Quinnan, G.V., Masur, H., Rook, A.H., *et al.* (1984) Herpesvirus infections in the acquired immune deficiency syndrome. *J.A.M.A.*, **252**, 72–7.
102. Resnick, L., Herbst, J., Ablashi, D.V., *et al.* (1988) Regression of oral hairy leucoplakia after orally administered acyclovir. *J.A.M.A.*, **259**, 384–8.
103. Richardson, E.P. (1988) Progressive multifocal leucoencephalopathy 30 years later. *N. Engl. J. Med.*, **318**, 315–6.
104. Ringden, O., Wilczek, H., Lunnqvist, B., Gahrton, G., Wahren, B., Lernestedt, J-O. (1985) Foscarnet for cytomegalovirus infections. *Lancet*, **i**, 1503–4.
105. Rolston, K.V.I., Uribe-Botero, G., Mansell, P.W.A. (1987) Bacterial infections in adult patients with the acquired immune deficiency syndrome (AIDS) and AIDS-related complex. *Am. J. Med.*, **83**, 604–5.
106. Safrin, S., Berger, T.G., Gilson, I., *et al.* (1991) Foscarnet therapy in five patients with AIDS and acyclovir-resistant varicella-zoster virus infection. *Ann. Intern. Med.*, **115**, 19–21.
107. Shaw, R.J., Roussak, C., Forster, S.M., Harris, J.W., Pinching, A.J., Mitchell, D.M. (1988) Lung function abnormalities in patients infected with the human immunodeficiency virus with and without overt pneumonitis. *Thorax*, **43**, 436–40.
108. Shelhamer, J.H., Ognibene, F.P., Macher, A.M., *et al.* (1984) Persistence of *Pneumocystis carinii* in lung tissue of acquired immunodeficiency syndrome patients treated for *Pneumocystis* pneumonia. *Am. Rev. Respir. Dis.*, **130**, 1161–5.
109. Smith, P.D., Lane, H.C., Gill, V.J., *et al.* (1988) Intestinal infections in patients with the acquired immunodeficiency syndrome. *Ann. Intern. Med.*, **108**, 328–33.
110. Soave, R., Armstrong, D. (1986) Cryptosporidium and cryptosporidiosis. *Rev. Infect. Dis.*, **8**, 1012–23.
111. Stern, J.J., Hartman, B.J., Sharkey, P., *et al.* (1988) Oral fluconazole therapy for patients with acquired immunodeficiency syndrome and cryptococcosis, experience with 22 patients. *Am J. Med.*, **85**, 477–80.
112. Stover, D.E., White, D.A., Romano, P.A., Gellene, R.A., Robeson, W.A. (1985) Spectrum of pulmonary diseases associated with the acquired immune deficiency syndrome. *Am. J. Med.*, **78**, 429–37.
113. Stover, D.E., Greeno, R.A., Gagliardi, A.J. (1989) The use of a simple exercise test in the diagnosis of *Pneumocystis carinii* pneumonia in patients with AIDS. *Am. Rev. Respir. Dis.*, **139**, 1343–6.

114. Sunderam, G., McDonald, R.J., Maniatis, T., Oleske, J., Kapila, R., Reichman, L.B. (1986) Tuberculosis as a manifestation of the acquired immunodeficiency syndrome. *J.A.M.A.*, **256**, 362–6.
115. Thomas, S., O'Doherty, M., Bateman, N. (1990) *Pneumocystis carinii* pneumonia. *Br. Med. J.*, **300,** 211–2.
116. Tuazon, C.U. (1989) Toxoplasmosis in AIDS patients. *J. Antimicrob. Chemother.*, **23** (Suppl. A); 77–82.
117. Van der Ven, A.J.A.M., Koopmans, P.P., Vree, T.B., Van der Meer, J.W.M. (1991) Adverse reactions to co-trimoxazole in HIV infection. *Lancet*, **338**, 431–3.
118. Wasser, L., Talavera, W. (1987) Pulmonary cryptococcosis in AIDS. *Chest*, **92**, 692–5.
119. Wharton, M., Coleman, D.L., Wofsy, C., *et al.* (1986) Prospective randomised trial of trimethoprim-sulphamethoxazole versus pentamidine for *Pneumocystis carinii* pneumonia in the acquired immune deficiency syndrome. *Ann. Intern. Med.*, **105**, 37–44.
120. Wilcox, C.M., Diehl, D.L., Cello, J.P., Margaretten, W., Jacobson, M.A. (1990) Cytomegalovirus esophagitis in patients with AIDS. A clinical endoscopic and pathologic correlation. *Ann. Intern. Med.*, **113**, 589–93.
121. Wong, B., Edwards, F.F., Kiehn, T.E., *et al.* (1985) Continuous high grade *Mycobacterium avium intracellulare* infections in patients with the acquired immunodeficiency syndrome. *Am. J. Med.*, **78**, 35–40.
122. Young, L.S., Inderlied, C.B., Berlin, O.G., Gottlieb, M.S. (1986) Mycobacterial infections in AIDS patients, with an emphasis on the *Mycobacterium avium* complex. *Rev. Infect. Dis.*, **8**, 1024–33.
123. Young, L.S., Wiviott, L., Wu, M., Kolonoski, P., Bolan, R., Inderlied, C.B. (1991) Azithromycin for treatment of *Mycobacterium avium-intracellulare* complex infection in patients with AIDS. *Lancet*, **338**, 1107–9.

2

The application of molecular biology to the diagnosis of infectious disease

STEVEN H. MYINT

2.1 INTRODUCTION

The techniques of molecular biology have enabled rapid advances in the study of infectious human diseases. Advances in our knowledge of the biology and pathogenic mechanisms of disease-causing organisms have recently arisen from the use of such techniques and this will be evident in subsequent chapters. This chapter will review the general principles behind another application that is described in subsequent chapters, that of gene detection.

2.2 FUNDAMENTALS

All replicative organisms, whether prokaryotes or eukaryotes, contain either deoxyribonucleic acid (DNA), ribonucleic acid (RNA), or both. The basic building bricks of these nucleic acids are the four nucleosides: adenine (A), cytidine (C), guanine (G) and either thymidine (T) in DNA or uridine (U) in RNA. RNA is, with few exceptions, made up of a single chain of phosphorylated nucleosides (nucleotides). DNA usually exists as a duplex, with one chain of nucleotides being 'complementary' to the other. This complementarity arises from the specific binding that exists between A and T (or U) and C and G. Gene probes consist of nucleotide (nt) sequences, which have been labelled to enable their detection, that are complementary to a sought nucleotide sequence. Hybrids can be formed

Molecular and Cell Biology of Opportunistic Infections in AIDS
Edited by Steven Myint and Alan Cann
Published in 1992 by Chapman & Hall, London ISBN 0 412 45330 4

that consist of two strands of DNA (DNA : DNA), of RNA (RNA : RNA), or one of each (RNA : DNA). The temperature at which only half of the strands are in the form of a duplex is known as the melting temperature (T_m). The T_m of a DNA : DNA hybrid is dependent on the base composition of the hybrid, the ionic strength of the hybridization solution and the concentration of agents such as formamide [16, 62]. This is represented in the equation:

$$T_m = \frac{81.5 + 16.6 \log M + 0.41(\% G + C) - 500}{n - 0.61 \ (\% \text{ formamide})}$$

where M is the ionic strength in moles per litre and n is the length of the shorter chain of a duplex. A simplified formula which more accurately reflects hybrids where n is shorter than 50 bases is:

$$T_m = 4(G + C) + 2(A + T)$$

The T_m of an RNA : RNA hybrid is 9–12 °C higher than the equivalent DNA : DNA duplex [9]; RNA : DNA hybrids may have a higher [10, 17] or lower T_m [16].

There are five principal variants in any hybridization reaction: the probe, the conditions of hybridization, the hybridization format, the labelling or detection system, and the target sequence(s). These will be discussed in turn.

2.2.1 Probes and probe selection

In principle, probes are designed from known nucleic acid sequences. They may be many hundreds of nucleotides in length or short sequences of only tens of nucleotides. The former tend to be derived from cloned nucleic acid, the latter synthesized chemically. 'Long' probes are generally used in direct probe techniques and short oligonucleotide sequences used for gene amplification methods. In either case, it is a prerequisite that the target sequence is known and unique to the agent sought.

Long or 'full-length' probes have greater specificity because of their length and complexity. They can also be labelled more extensively, thus increasing the level of sensitivity. Synthetic oligonucleotides, however, may be easier and less costly to generate; they can also be synthesized in an already modified form, for example with a detector molecule(s) incorporated, and have a far faster rate of hybridization. Most useful oligonucleotide probes are 18–30 bases in length, have 40–60% G + C, are without intrastrand complementary regions and have less than 70% homology with non-target regions. Computer software is now available to aid the design of these oligonucleotide probes.

2.2.2 Hybridization conditions

These determine both the kinetics and stringency of hybridization:

1. Temperature: Standard hybridization for full length probes takes place most readily at $T_m - 25\,°C$; the calculation of T_m can be made from the formulae given above. In general, this will usually be between 60 and 75 °C. Temperatures closer to T_m can be used if greater stringency is required. There is less flexibility when employing oligonucleotide probes, and typically hybridization would be allowed to take place at $T_m - 5\,°C$.
2. Salt concentration: This predominantly affects the rate of hybridization [54]. Typically a sodium concentration of between 0.4 and 1.0 molar is used.
3. Formamide concentration: The use of formamide reduces the T_m of the duplex [42] and is particularly useful for RNA : RNA hybridizations. Dimethylsulphoxide (DMSO) can be used in place of formamide, particularly in the polymerase chain reaction since Taq polymerase is inhibited by formamide.
4. Hybridization accelerators: A number of inert substances have been employed to speed up hybridization reactions with large probes [60]. The most commonly used is dextran sulphate in final concentrations of 5–10%. Similar final concentrations are used of polyethylene glycol (PEG) [4]. These substances are believed to act by decreasing the volume that nucleic acids can occupy, thus speeding up the interaction between probe and target sequences. Polyacrylic acid (2–4% final concentration) [45], phenol [35] and 4M guanidine thiocyanate [58] are also used under appropriate circumstances. The last of these is particularly useful in RNA : RNA hybridization reactions. These hybridization accelerators are not effective with probes less than 150 base pairs (bp) long, but are generally not needed with these short probes.

2.2.3 Hybridization kinetics

The formation of hybrids is a dynamic process. If the two complementary sequences are present in equimolar amounts, the process of re-association is subject to the rules of second order kinetics and can be expressed mathematically by the following equation [62]:

$$\frac{1}{\text{fss}} = \frac{K_2\text{Cot}}{2} + 1$$

where fss is the fraction of single stranded sequences, K_2 is the second order rate constant, Cot the initial molar concentration of single stranded nucleotide sequence, and t the reaction time in seconds.

From this equation, and under standard experimental conditions, it is possible to determine the $Cot_{1/2}$ value for a given DNA duplex (i.e. the Cot value where 50% of the duplex has re-associated). This reflects the sequence complexity of the DNA. Other major factors influencing the rate of hybridization are the temperature at which annealing is allowed to occur, the monovalent cation concentration and the viscosity of the milieu.

In practice, particularly with oligonucleotide probes, the target DNA concentration is not known and excess probe is used, and the kinetics involved more closely mirror first order reactions. Meinkoth and Wall [43] have suggested that the time required to hybridize half the probe can be calculated from the following formula:

$$t_{1/2} = \frac{\ln 2}{kC}$$

where C is the probe concentration in moles per litre, k is the rate constant which is dependent on probe length (L), complexity (N) and the nucleation constant (k_n) as defined by the following equation:

$$k = \frac{k_n L^{0.5}}{N}$$

The nucleation constant is itself determined by the ionic strength, pH and temperature at which hybridization is allowed to proceed. Under the conditions commonly used (pH 5–9, Na^+ concentration 0.4 – 1M, hybridization temperature T_m – 25 °C) for full length probes $t_{1/2}$ can be calculated from:

$$t_{1/2} = \frac{N \ln_2}{3.5 \times 10^5 (L^{0.5})C}$$

Hybridization should be allowed to take place for up to $3 \times t_{1/2}$.

2.2.4 Hybridization format

Essentially, there are two formats in common use. Single phase hybridization implies a liquid environment, and two phase (or mixed phase) hybridization has either the probe or target bound to a solid matrix. Liquid phase hybridization is used in the technologies employing gene or probe amplification. Most non-amplification based probe methods use a two phase system.

There are several advantages in fixing either the probe or target to a solid support [43]. The time for hybridization can be reduced, multiple targets or probes can be incorporated into a single hybridization and the solid phase can be recovered to enable detection of hybrids. Also, bound strands of nucleic acid will not self-anneal. The main problem is that of non-specific binding to the solid phase producing so-called 'background signal'. This can be reduced by prehybridizing the solid phase with components such as non-target nucleic acid (e.g. salmon sperm). Additionally, the hybridization mix may include blocking agents such as Denhardt's solution (0.2% w/v each of bovine serum albumen, Ficoll, polyvinylpyrrolidone) [19] or detergents such as sodium dodecyl sulphate (SDS, usually 0.1–0.5%). A washing step with a detergent is also usually incorporated post-hybridization to remove non-specifically bound components. The stringency of this wash can be adjusted by varying both the temperature of the wash and the amount of detergent.

2.2.5 Types of solid phase support

Nucleic acid can be fixed to a solid phase either covalently or non-covalently. The most commonly used solid supports are nitrocellulose [13, 25] and nylon [15, 49]. Single stranded (denatured) DNA can be non-covalently bound by its sugar–phosphate backbone to nitrocellulose by heat applied under a vacuum. Nitrocellulose membranes have been used reliably for many years and provide an inexpensive way of immobilizing nucleic acids. They cannot, however, bind RNA or short DNA strands (less than 300 bases) effectively. Nitrocellulose membranes also have several other limitations: the formation of complete duplexes results in loss of the DNA from the filter, prolonged hybridization reactions at high temperature tend to lead to loss of bound nucleic acid, there is often batch-to-batch variation in the binding properties, and the filters are physically fragile and can only be reprobed once or twice before breakage.

Alternatively, nucleic acids can be cross-linked covalently to nylon membranes with or without the use of ultraviolet light. Nylon has all the advantages of nitrocellulose, apart from its low cost, and overcomes the disadvantages stated above. Moreover, the binding capacity of a nylon membrane is claimed to be 3–5 times that of the same size nitrocellulose membrane. The main problem with the use of nylon membranes is non-specific binding of proteins, which is particularly troublesome when some non-radioactively labelled probes are used. There are a number of commercially available nylon based membranes which differ slightly in their precise characteristics.

Other solid supports which are being used increasingly are magnetic beads [59] and polystyrene microtitre plates [29, 34, 48]. The use of magnetic beads allows easy separation of hybridized products. The microtitre plate methodologies, when coupled with fluorimetric or colorimetric detection methods enable the processing of large numbers of specimens [23, 48, 57].

The now classical method of filter hybridization used was that developed by Southern [56]. The target DNA is separated by gel electrophoresis and transferred to a membrane by capillary action. The membrane can then be hybridized to probes in solution. Northern blotting uses the same methodology for detecting target RNA species [3]. Reverse hybridization is an adaptation of these techniques where the probe is bound to the filter and the target sequence in solution.

The target nucleic acid can be directly applied to a filter either as a 'dot' or a 'slot' for subsequent probing. The slot–blot method has the added advantage of being semi-quantitative if a densitometer is used. Several manufacturers produce manifolds that allow nucleic acids in solution to be applied under vacuum to a membrane, thus ensuring uniformity of size of dots or slots.

In situ hybridization allows the localization of nucleic acids within cells and tissues [26, 31, 47]. Intact cells are fixed on a solid phase, either a membrane or microscope slide, probed and examined under a microscope. Non-isotopic methods allow direct visualization of target sites [2, 14]. Sandwich hybridization uses two probes, one that is unlabelled to capture target sequences and the other, labelled, to detect the captured target. This method can easily be adapted to the microtitre plate format [23, 34, 48].

Gene and probe amplification methodologies are discussed in section 2.3.

2.2.6 Labels

Probes can be labelled with isotopic (radioactive) or non-isotopic (non-radioactive) markers. For research purposes, isotopes such as ^{32}P, ^{35}S, or ^{125}I are used to label probes and allow sensitive detection by autoradiography. For safety reasons, the routine use of radioactivity to detect infectious agents is being challenged by a number of non-radioactive detection methods. The first of these alternative systems utilized the biotin–avidin interaction [1, 14, 39, 40, 51]. Biotin is a small water soluble vitamin of 244 D. It has a very high affinity for the much larger glycoprotein, avidin. Typically, biotinylated nucleotides are incorporated into the probe and avidin coupled to an enzyme, such as alkaline phosphatase, which is used as a secondary detector molecule. After the addition of an appropriate substrate for the enzyme, a colorimetric or chemiluminescent reaction can be detected. The sensitivity is near to that of using ^{32}P or other radioisotopes and less than a picogram of target DNA can be detected. The main technical difficulty with this system tends to be a high level of background signal, particularly for tissue *in situ* hybridization. Alternatively, the second detector molecule can be an antibody labelled with such an enzyme. Digoxigeninin can be substituted for biotin in such a system [20]. It is also possible to directly label the nucleotide probe with an enzyme [30] or use labelled antibodies directed against nucleic acid duplexes [11, 45]. These are all modifications on a theme and have roughly equivalent sensitivity. Chemiluminescent detection has the added advantage that the results can be conveniently recorded on an X-ray film [32, 33].

2.2.7 Labelling methods

Oligonucleotide probes are easily labelled at the time of synthesis by the incorporation of modified nucleotides [1, 51]. Long probes can be labelled by a variety of techniques.

Nick translation is a well tested and relatively inexpensive method of generating DNA probes of high specific activity [50]. The method relies on DNase I to introduce random single stranded breaks (nicks) in both strands of a DNA duplex and then for DNA polymerase I to remove one or more bases at the exposed 5′ phosphoryl end to replace them with radiolabelled substitutes at the corresponding 3′ terminus. The nick would have been shifted ('translated') in a 3′ direction. In this way probes can be produced with an activity of 10^6–10^9 cpm/μg DNA.

Greater specific activities can be achieved using the random priming method [24]. Short oligonucleotides with random nucleotide sequences are used to act as primers for the synthesis of new radiolabelled copies of probe sequences, using the Klenow fragment of DNA polymerase I. Less than 200 ng of input DNA is required and the probe can be used without the removal of unincorporated nucleotides (unlike nick translated) probes. The method also has the advantages, compared to nick translation, of being applicable to single stranded DNA templates and DNA templates of less than 500 bp.

It is also possible to synthesize probes from molecularly cloned templates. Specific single stranded DNA probes can be made by cloning and inserting the sequence of interest into the multiple cloning site (MCS) of a vector such as M13mp19 [28]. Then, with a known M13 primer and DNA polymerase, it is possible to generate strand specific probes. A modification of this approach involves the use of the polymerase chain reaction (section 2.3).

Similarly, single stranded RNA probes can be generated from riboprobe vectors such as pGEM(R) [44]. These incorporate promoters either side of an MCS for SP6, T7 or T3 RNA polymerases so that both complementary and anti-complementary single stranded RNA probes can be synthesized. The advantages of using RNA probes are that RNA : RNA and RNA : DNA hybrids tend to be more stable than DNA : DNA hybrids, they can be used without denaturation, they do not self-anneal, and unhybridized probe can be removed by RNase digestion. The major problem with this type of probe is that an RNase-free environment is required or the probe will be destroyed.

2.2.8 Target sequences

Sample preparation is as equally important a part of hybridization procedures as the other components. For most biological samples, some degree of processing is required to enable the extraction and concentration of target nucleic acid. The possible exception to this is *in situ* hybridization, where the nucleic acid target may already be concentrated within cells. There are a number of potential nucleic acid species that can be targets. In viruses, there is a simple choice between genomic nucleic acid and messenger RNA. In higher organisms, to these can be added others targets such as mitochondrial and ribosomal RNA. Ribosomal RNA is particularly attractive as there may be several thousand copies of a sequence in a single bacterium.

DNA preparation

Samples containing cellular material may have to be lysed to release the nucleic acid constituents. A combination of physical methods (shearing, ultrasound) and detergent is usually sufficient for this purpose. However, in these crude preparations, proteins and other materials my interfere with hybridization of the probe to the target sequence. Proteins may be partially digested with proteases and then separated from the nucleic acids. The commonest procedure used to deproteinize samples depends on the differential solubility of nucleic acids and proteins in organic solvents such as phenol. Phenol extraction of proteins results in partition of nucleic acids into the aqueous component, whilst the organic phenol phase extracts the lipids and some hydrophobic proteins. The interphase will also contain some proteinaceous material. Nucleic acid DNA can then be recovered by precipitating in salt and ethanol (or isopropanol) [53].

RNA preparation

Preparation of RNA is technically more demanding than that of DNA [18]. This is because of the ubiquitous nature of RNases; many cell types contain high levels of RNase activity. Thus an RNase-free environment must be attained. This involves treating glassware and solutions with diethylpyrocarbonate (DEPC) and extensive autoclaving of materials. It is often necessary to incorporate inhibitors of RNase activity into the extraction mix. Guanidium isothyocyanate rapidly reduces RNase activity and many standard procedures utilize this compound [12, 53]. These procedures are effective but labour intensive. Alternatively, more specific inhibitors such as vanadyl–ribonucleoside complex (VRC) [8] and human placental or recombinant RNase inhibitor (RNasin) can be used. These latter methods are technically less demanding, but have other inherent disadvantages, for example, VRC also inhibits a number of other enzymes and will therefore interfere with subsequent enzymatic manipulations.

2.3 GENE AND PROBE AMPLIFICATION

2.3.1 Gene amplification

Gene amplification technologies, principally the polymerase chain reaction (PCR) have found widespread application in the few years since their introduction [52, 21, 22]. The basis of this method is shown in Figure 2.1. Two oligonucleotide primers are annealed to regions flanking the target region in duplex DNA and then extended by a DNA polymerase. Using a thermostable DNA polymerase, it is possible to perform many temperature cycles without the addition of new enzyme at each cycle and obtain enormous amplification of known target sequences. Typically, 30 to 50 cycles are used and the entire process is usually automated. The products of a PCR reaction can then be visualized by gel electrophoresis. The length of the product is that of the two primers plus the intervening region. It is usual to confirm the identity of the products as being the specific target region by probing with an oligonucleotide probe. 'Nested PCR' involves two rounds of amplification, the second round utilizing primers complementary to sequences within the product of the first round. This adaptation increases the degree of amplification achievable and may increase specificity. Amplification of RNA target sequences requires the prior step of making cDNA copies. This reverse transcription can be done with a separate enzyme such as AMV reverse transcriptase or with one of the newly available thermostable enzymes, such as rTth polymerase, which has both reverse transcriptase and DNA polymerase properties.

There are a number of potential problems in the application of these methods. Currently, thermostable polymerases have a higher misincorporation rate than, for example, the thermolabile T4 DNA polymerase. This is because of the absence of 3′ to 5′ exonuclease ('proofreading') activity in the former; Taq polymerase is calculated to have a misincorporation rate of the order of 10^{-5} nucleotides per cycle. A newer thermostable enzyme, VENT DNA polymerase isolated from *Thermococcus litoralis*, has proofreading activity and therefore a likely lower misincorporation rate, but has drawbacks for sequence specific priming reactions.

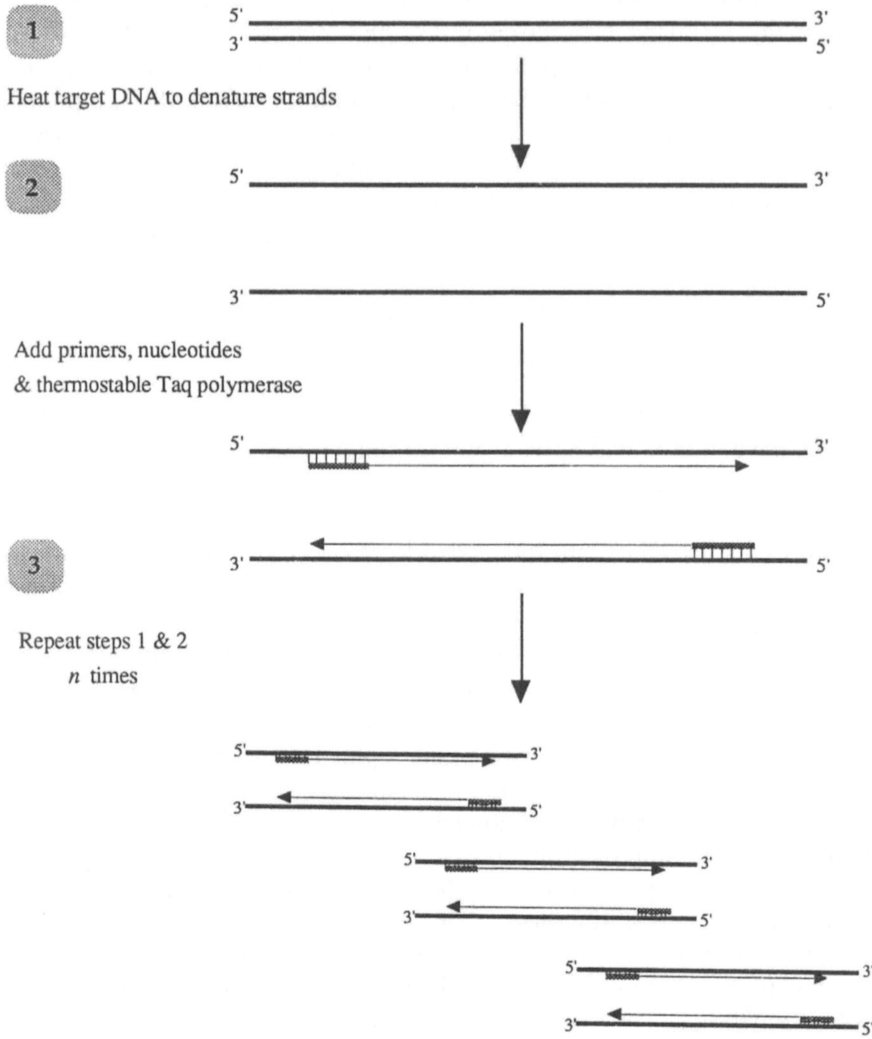

Figure 2.1 Principles of the polymerase chain reaction (see text).

The specificity of priming has also been a problem with priming of non-target sequences resulting in primer artefacts. Recent developments such as the addition of the thermostable polymerase at high temperature only [22], the so-called 'hot start', and the addition of DNA binding proteins may be the solution to this. The hot start also seems to reduce the incidence of 'primer-dimers', a product which consists essentially of primer and complementary sequences only.

However, the major problem with such an exquisitely sensitive technique is that of contamination. This can result from previous PCR reactions (product carry over) or stray molecularly cloned DNA. The most important measure to prevent this is attention to strict handling procedures. These include separation of preparation areas from those for analysis of reaction products and clinical material or plasmid DNA, the use of dedicated and positive-displacement pipettes and pre-aliquoting of reagents [37] and most important of all, the incorporation of appropriate negative controls into all amplification protocols. In addition to these precautions, short wave ultraviolet irradiation of the PCR reaction mix prior to addition of the template can be used to reduce or eliminate contaminating nucleic acids. Another strategy, which is commercially available, uses substituted nucleotides. Deoxyuridine triphosphate (dUTP) is used in amplification mixtures in place of deoxythymidine triphosphate (dTTP). In subsequent PCR reactions, the enzyme uracil-*N* glycosylase (UNG) enables the excision of uracil from any contaminating DNA containing uracil. The resultant abasic polynucleotides cannot function as templates and are susceptible to alkaline hydrolysis at the high temperatures used in a PCR reaction. The enzyme does not excise uracil from RNA which can still be used as template.

The PCR can also be used to produce probes for hybridization. This method is akin to the M13 template methodology already described. An excess of one primer and labelled nucleotides are used in a standard reaction to produce an excess of one strand ('asymmetric PCR') which is labelled.

There have been a number of modifications of the basic PCR protocol, including quantitation by blotting and probing products, adaptation of the procedure to microtitre plates and modification of primers so that colorimetric detection is possible [5, 29, 34, 46]. It is also likely that commercial kits for PCR diagnosis will be available in the near future. Perhaps the biggest obstacles that still need to be overcome are the cost and labour intensiveness of these methods; some of these modifications may prove to be of help to overcome this.

An alternative gene amplification technique is transcription based [36]: TAS (transcription-based amplification system), which is applicable to RNA targets. A cDNA copy is made of the RNA target using an appropriate primer and reverse transcriptase. This primer is sequence specific and contains a polymerase binding site. This partial duplex is then denatured and a second primer and additional reverse transcriptase used for second cDNA strand synthesis. The resultant double stranded DNA is then used as a template for T7 RNA polymerase with the result that a further 10–100 RNA transcripts are produced. These RNA transcripts can then be used for a second cycle, and so on. The sensitivity of the method is of the same order as that of PCR but, currently, is proving difficult to automate.

2.3.2 Probe amplification

There are two systems that show promise. The first of these is based on the ability of an RNA-dependent RNA polymerase, Qβ replicase, to synthesize 10^6–10^9 copies of a template sequence [41]. The principle of this method is illustrated in Figure 2.2. The probe sequences are inserted into the MCS of Qβ phage-derived plasmid containing a T7 promoter. T7 RNA polymerase is then added to produce RNA transcripts containing the probe sequence. These are hybridized to their target, and after elution, amplified by the addition of Qβ replicase. This process is several times more rapid than the thermal cycling of a PCR reaction. The other major advantage is that, because the amount of amplified RNA depends on the amount hybridized, the method is quantifiable. The major difficulty here is non-specific hybridization. Once this difficulty is overcome, the Qβ replicase system may offer a potential alternative to PCR.

Perhaps the most exciting development since automated PCR is the ligase chain reaction (LCR). This technique has been developed consequent to the cloning of the thermostable enzyme, Taq ligase [6]. The method (Figure 2.3) relies on complete complementarity at the junction of two contiguous oligonucleotides hybridized to a target sequence [7]. A single base mismatch means that ligation will not occur. The use of four oligonucleotides, two complementary to each target strand, and repeated cycles of denaturation, hybridization and ligation results in multiple copies of ligated oligonucleotides. It has been shown that sub-picomolar quantities of lambda phage DNA can be detected by this method [7]. Developments have paralleled those of PCR, but have been even more rapid. Labelling of one oligonucleotide with biotin, and the other with a reporter molecule allows non-isotopic detection without electrophoresis [38]. 'Nested LCR' utilizes, as in its PCR equivalent, two rounds of amplification. This may improve the signal to background ratio. Perhaps the most significant development for the detection of infectious disease is a combination of PCR and LCR, 'PLCR'. Primary amplification of DNA or RNA is achieved by PCR/RT-PCR followed by subsequent LCR. The exquisite sensitivity of PCR is combined with the specificity of LCR. Automation of this procedure has much potential for the detection of infectious agents.

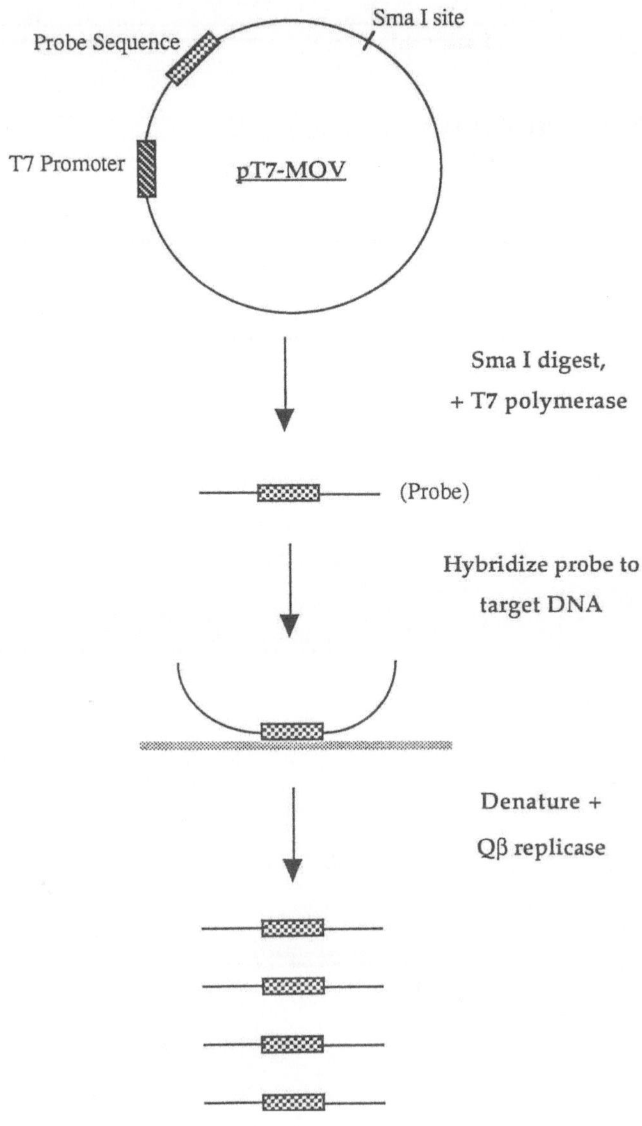

Figure 2.2 Principles of the Qβ replicase probe amplification system (see text).

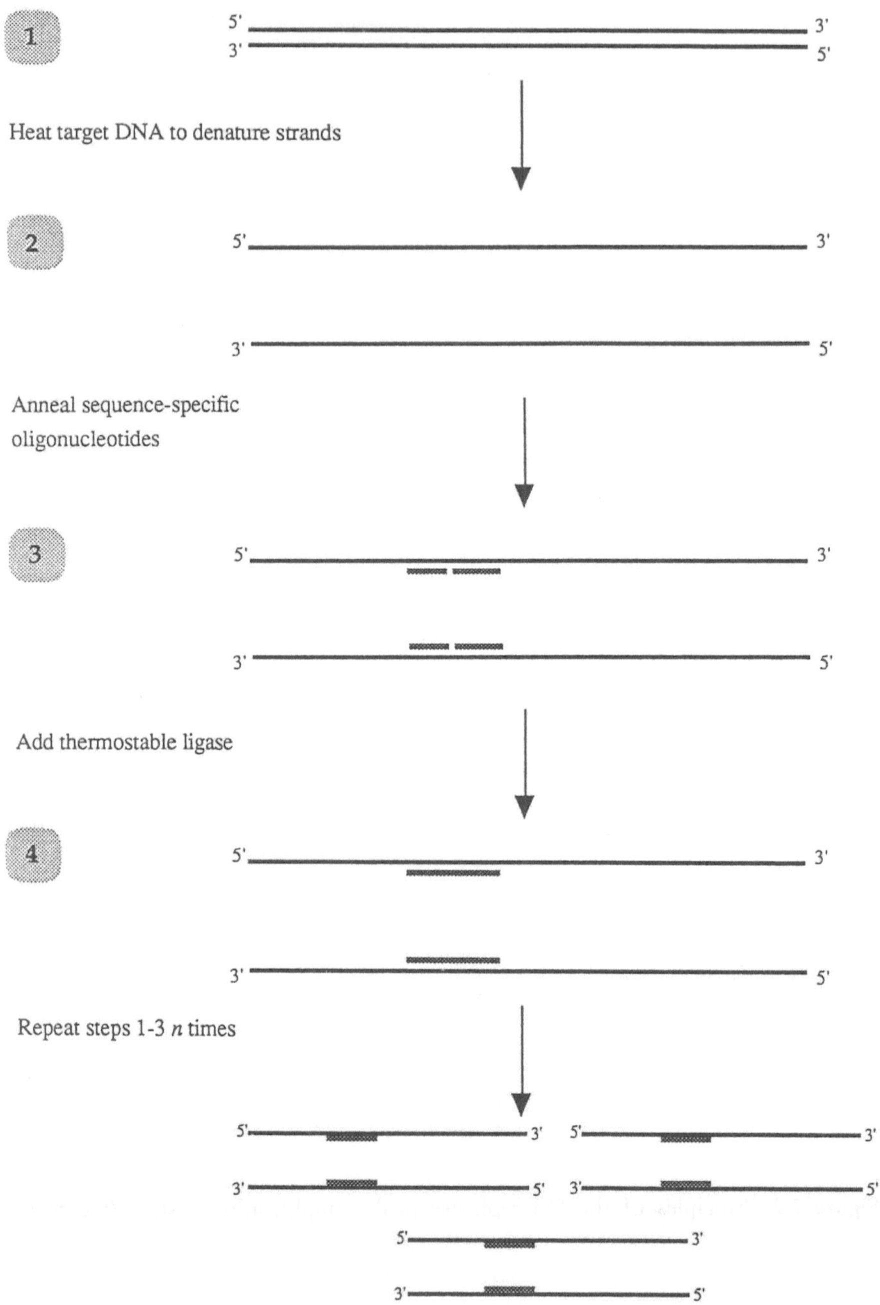

Figure 2.3 Principles of the ligase chain reaction (see text). Only one strand detection shown.

REFERENCES

1. Al-Hakim, H., Hull, R. (1986) Studies towards the development of chemically synthesised nonradioactive biotinylated nucleic acid hybridization probes. *Nucleic Acids Res.*, **14**, 9965–72.
2. Allan, G.M., Todd, D., Smyth, J.A. (1989) *In situ* hybridization: an optimised protocol for a biotinylated DNA probe renders it more sensitive than a comparable ^{35}S-labelled probe. *J. Virol. Methods*, **24**, 181–90.
3. Alwine, J.C., Kemp, D.J., Stark, G.R. (1977) Method for the detection of specific RNAs in agarose gels by transfer to diazobenzoxylmethyl paper and hybridization with DNA probes. *Proc. Natl. Acad Sci. USA*, **74**, 5350–4.
4. Amasino, R.M. (1986) Acceleration of nucleic acid hybridization rate by polyethylene glycol. *Anal. Biochem.*, **152**, 304–7.
5. Balauger, P., Terouanne, B., Boussioux, A.M., Nicolas, J.C. (1991) Quantification of DNA sequences obtained by polymerase chain reaction using a bioluminescent adsorbent. *Anal. Biochem.*, **195**, 105–10.
6. Barany, F. (1991) Genetic disease detection and DNA amplification using cloned thermostable ligase. *Proc. Natl. Acad. Sci. USA*, **88**, 189–93.
7. Barany, F. (1991) The ligase chain reaction (LCR) in a PCR world. PCR, Methods and applications. **1**, 5–17.
8. Berger, S., Birkenmeir C. (1979) Inhibition of intractable nucleases with ribonucleoside-vanadyl complexes, isolation of messenger ribonucleic acid from resting lymphocytes. *Biochemistry*, **18**, 5143–9.
9. Billeter, M.A., Weissman, C., Warner, R.C. (1966) Properties of double stranded RNA from *Escherichia coli* infected with bacteriophage MS2. *J. Mol. Biol.*, **17**, 145–73.
10. Bishop, J.A. (1972) Molecular hybridization of ribonucleic acid in a large excess of deoxyribonucleic acid. *Biochem. J.*, **126**, 171–85.
11. Boguslawski, S.J., Smith, D.E., Michalak, M.A., *et al.* (1986) Characterisation of a monoclonal antibody to DNA:RNA and its application to the immunodetection of hybrids. *J. Immunol Methods*, **89**, 123–30.
12. Brawerman, G., Mendecki, J., Lee, S. (1972) A procedure for the isolation of mammalian messenger ribonucleic acid. *Biochemistry*, **11**, 637–40.
13. Bresser, J., Gillespie, D. (1983) Quantitative binding of covalently closed circular DNA to nitrocellulose in NaI. *Anal. Biochem.*, **129**, 357–64.
14. Brigatti, D.J., Myerson, D., Leary, J.J., *et al.* (1983) Detection of viral genomes in cultured cells and paraffin-embedded tissue sections using biotin-labelled hybridization probes. *Virology*, **126**, 32–50.
15. Cannon, G., Heinhorst, S., Weissbach, A. (1975) Quantitative molecular hybridization on nylon membranes. *Anal. Biochem.*, **149**, 229–37.
16. Casey, J., Davidson, M. (1977) Rate of formation and thermal stabilities of RNA:DNA and DNA:DNA duplexes at high concentrations of formamide. *Nucleic Acids Res.*, **4**, 1539–52.
17. Chamberlin, M., Berg, P. (1964) Mechanism of action of RNA polymerase, formation of DNA:RNA hybrids with single-stranded templates. *J. Mol. Biol.* **8**, 297–313.
18. Chirgwin, J., Przybyla, A., MacDonald, R., Rutter, W. (1979) Isolation of

biologically active ribonucleic acid from sources enriched in ribonuclease. *Biochemistry*, **18**, 5284–90.

19. Denhardt, D. (1966) A membrane filter technique for the detection of complementary DNA. *Biochem. Biophys. Res. Commun.*, **23**, 641–6.
20. Dooley, S., Radtke, J., Blin, N., Unteregger, G. (1988) Rapid detection of DNA-binding factors using protein blotting and digoxigenin-dUTP marked probes. *Nucleic Acids Res.*, **16**, 11839.
21. Ehrlich, H. (1989) *PCR technology, Principles and applications for DNA amplification.* Stockton Press, New York.
22. Ehrlich, H.A., Gelfand, D., Sninsky, J. (1991. Recent advances in the polymerase chain reaction. *Science*, **252**, 1643–51.
23. Ezaki, T., Hashimoto, Y., Yamamoto, H., *et al.* (1990) Evaluation of the microplate hybridization method for rapid identification of the species of *Legionella. Eur. J. Clin. Microbiol.*, **9**, 213.
24. Feinberg, A.P., Vogelstein, B. (1983) A technique for radiolabelling DNA restriction endonuclease fragments to high specific activity. *Anal. Biochem*, **132**, 6–13.
25. Flavell, R.A., Birfelder, E.J., Sanders, P.M., Borst, P. (1974) DNA:DNA hybridization on nitrocellulose filters. 1. General considerations and non-ideal kinetics. *Eur. J. Biochem.*, **47**, 535–43.
26. Gall, J.G., Pardue, M.L. (1969) Formation and detection of RNA:DNA hybrid molecules in cytological preparations. *Proc. Natl. Acad. Sci. USA*, **63**, 378–83.
27. Gillespie, D., Spiegelman, S. (1965) A quantitative assay for DNA–RNA hybrids immobilised on a membrane *J. Mol. Biol.*, **12**, 829–42.
28. Hu, N., Messing, J. (1982) The making of strand specific M13 probes. *Gene*, **17**, 271–7.
29. Inouye, S., Hondo, R. (1991) Microplate hybridization of amplified viral DNA segment. *J. Clin. Microbiol.*, **28**, 1469–72.
30. Jablonski, E., Moomaw, E.W., Tullis, R.H., Ruth, J.L. (1986) Preparation of olidodeoxynucleotide-alkaline phosphatase conjugates and their use as hybridization probes. *Nucleic Acids Res.*, **14**, 6115–28.
31. John, H.A., Birnstiel, M.L., Jones, K.W. (1969) RNA:DNA hybrids at the cytological level. *Nature*, **223**, 582–7.
32. Keller, G.H., Cumming, C.U., Huang, D.P., Manak, M.M., Ting. R. (1988) Chemical method for introducing haptens onto DNA probes. *Anal. Biochem.*, **170**, 441–50.
33. Keller, G.H., Huang, D.P., Manak, M.M. (1989) Labelling of DNA probes with a photoactivatable hapten. *Anal. Biochem.*, **177**, 392–5.
34. Keller, G.H., Huang, D.P., Manak, M.M. (1991) Detection of human immunodeficiency virus type 1 DNA by polymerase chain amplification and capture hybridization in microtitre wells. *J. Clin. Microbiol.*, **29**, 638–41.
35. Kohne, D.E., Levinson, S.A., Byers, M.J. (1977) Room temperature method for increasing the rate of DNA reassociation by many thousandfold, the phenol re-emulsion technique. *Biochemistry*, **16**, 5329–41.
36. Kwoh, D.Y., Davis, G.R., Whitfield, K., Chapelle, H., DiMichele, L., Gingeras, T.R. (1988) Transcription based amplification system and detection of amplified HIV-1 sequences in infected cells using a bead-based sandwich

hybridization format. *Practical Aspects of Molecular Probes*. Third San Diego Conference. Abstract.
37. Kwok, S., Higuchi, R. (1989) Avoiding false positives with PCR. *Nature*, **339**, 237–8.
38. Landegren, U., Kaiser, R., Sanders, J., Hood, L (1988) A ligase mediated gene detection technique. *Science*, **241**, 1077–80.
39. Langer, P.R., Waldrop, A.A., Ward, D.C. (1981) Enzymatic synthesis of biotin-labelled polynucleotides. *Proc. Natl. Acad. Sci. USA*, **78**, 6633–7.
40. Leary, J.J., Brigati, D.J., Ward, D.C. (1983) Rapid and sensitive colorimetric method for visualising biotin-labelled probes hybridised to DNA or RNA immobilised on nitrocellulose bio-blots. *Proc. Natl. Acad. Sci. USA*, **80**, 4045–9.
41. Lizardi, P.M., Guerra, C.E., Lomeli, H., Tussie-Luna, I., Kramer, F.R. (1988) Exponential amplification of recombinant RNA hybridization probes. *Biotechnology*, **6**, 1197–202.
42. McConaughy, B.L., Laird, C.L., McCarthy, B.J. (1969) Nucleic acid reassociation in formamide. *Biochemistry*, **8**, 3289–95.
43. Meinkoth, J., , Wahl, G.M. (1984) Hybridization of nucleic acids immobilised on solid supports. *Anal. Biochem.*, **138**, 267–84.
44. Melton, D.A., Krieg, P.A., Rebagliati, M.R., Maniatis, T., Zinn, K., Green, M.R. (1984) Efficient *in-vitro* synthesis of biologically active RNA and RNA hybridization probes from plasmids containing a bacteriophage SP6 promoter. *Nucleic Acids Res.*, **12**, 7035–56.
45. Miller, C.A., Patterson, W.L., Johnson, P.K., *et al.* (1988) Detection of bacteria by hybridization of rRNA with DNA latex and immunodetection of hybrids. *J. Clin. Micro.*, **26**, 1271–6.
46. Olson, J.D., Panfili, P.R., Zuk, R.F., Sheldon, E.L. (1991) Quantitation of DNA hybridization in a silicon sensor-based system, application to PCR. *Mol. Cell. Probes*, **5**, 351–8.
47. Polak, J.M., McGee, J.O.D. (eds) (1990) In Situ *Hybridization: Principles and Practice*. Blackwell, Oxford.
48. Polsky-Cynkin, R., Parsons, G.H., Allerdt, L., Landes, G., Davis, G., Rashtchian, A. (1985) Use of DNA immobilised on plastic and agarose supports to detect DNA by sandwich hybridization. *Clin. Chem.*, **31**, 1438–43.
49. Reed, K.C., Mann, D.A. (1985) Rapid transfer of DNA from agarose gels to nylon membranes. *Nucleic Acids Res.*, **13**, 7207–21.
50. Rigby, P.W.S., Dieckman, M., Rhodes, C., Berg, P. (1977) Labelling deoxyribonucleic acid to high specific activity *in vitro* by nick translation with DNA polymerase I. *J. Mol. Biol.*, **113**, 237–51.
51. Ruth, J.L. (1984) Chemical synthesis of non-radioactively labelled probes. *DNA*, **3**, 123.
52. Saiki, R.K., Gelfand, D.H., Stoffel, S., *et al.* (1988) Primer directed enzymatic amplification of DNA with a thermostable polymerase. *Science*, **239**, 487–94.
53. Sambrook, J., Fritsch, E.F., Maniatis, T. (1989) *Molecular Cloning, A Laboratory Manual*. Cold Spring Harbour, New York.

54. Schildkraut, C., Lifson, S. (1965) Dependence of the melting temperature of DNA on salt concentration. *Biopolymers*, **3**, 195–208.
55. Smith, M.J., Britten, R.J., Davidson, E.H. (1975) Studies on nucleic acid reassociation kinetics, reactivity of single strand tails in DNA:DNA renaturation. *Proc. Natl. Acad. Sci. USA*, **72**, 4805–9.
56. Southern, E.M. (1975) Detection of specific sequences among DNA fragments separated by gel electrophoresis. *J. Mol. Biol.*, **98**, 503–17.
57. Syvanen, A.C., Tchen, P., Ranki, M., Soderlund, H. (1986) Time-resolved fluorimetry, a sensitive method to quantify DNA hybrids. *Nucleic Acids Res.*, **14**, 1017–28.
58. Thompson, J., Gillespie, D. (1987) Molecular hybridization with RNA probes in concentrated solutions of guanidine thiocyanate. *Anal. Biochem.*, **163**, 281–91.
59. Uhlen, M. (1989) Magnetic separation of DNA. *Nature*, **340**, 733–4.
60. Wahl, G.M., Stern, M., Stark, G.R. (1979) Efficient transfer of large DNA fragments from agarose gels to diazobenzyloxymethyl paper and rapid hybridization using dextran sulphate. *Proc. Natl. Acad. Sci. USA*, **76**, 3683–7.
61. Wallace, R.B., Shaffer, J., Murphy, R.F., Bonner, J., Hirose, T., Itabura, K. (1979) Hybridization of synthetic oligo deoxyribonycleotides to ϕ174 DNA: the effect of single base pair mismatch. *Nucleic Acids Res.*, **6**, 3543–56.
62. Wetmur, J.G., Davidson, N. (1968) Kinetics of renaturation of DNA. *J. Mol. Biol.*, **31**, 349–70.

3

Salmonella, *Shigella* and *Campylobacter*

FRAN A. RUBIN, JOANNE L. RHOADS and RICHARD L. WARREN

3.1 THE SCOPE OF THE PROBLEM

Infection with the human immunodeficiency virus (HIV) causes profound derangements in humoral as well as cell-mediated immunity. The resulting spectrum of clinical illness includes asymptomatic infection, chronic generalized lymphadenopathy, mucocutaneous disease (as persistent T-helper cell deficiency develops) and opportunistic infection and neoplasia (when absolute T-helper cell depletion occurs) [53]. Early in the AIDS epidemic, clinical attention was focused on unusual infections with the opportunistic protozoa, fungi and viruses. However, serious infections with common bacterial pathogens (disseminated atypical mycobacteria, *Streptococcus pneumoniae*, *Haemophilus influenzae* and *Salmonella* spp.) were soon recognized as additional complications of HIV disease [49, 64, 68, 133, 144, 155]. When compared with the general population, HIV infected subjects have a higher incidence of serious infection with the enteric organisms *Salmonella*, *Shigella* and *Campylobacter*. Those with late stage HIV disease often have an atypical presentation of infection with these organisms, marked by bacteraemia and frequent relapse despite appropriate therapy.

In this chapter, we will discuss mechanisms of pathogenesis, epidemiology, clinical features, treatment and potential use of molecular techniques for diagnosis and management of infections with enteric bacteria in those with HIV disease. Because of the predominance of salmonella infections in HIV patients, this chapter will emphasize *Salmonella* and include brief summaries of infections caused by *Shigella* and *Campylobacter*.

Molecular and Cell Biology of Opportunistic Infections in AIDS
Edited by Steven Myint and Alan Cann
Published in 1992 by Chapman & Hall, London ISBN 0 412 45330 4

3.1.2 Infections with *Salmonella, Shigella* and *Campylobacter*

Salmonella are motile, Gram negative rods of the family *Enterobacteriaceae.* Although more than 2000 species have been identified using the Kauffman and White serologic classification, over 95% of serotypes isolated from humans belong to groups A, B, C, D, and E. In addition to group specific somatic O antigens, *Salmonella* synthesize monophasic or diphasic flagellar antigens; a few serotypes synthesize a capsular Vi antigen (*S. typhi* and *S. paratyphi* C). *Salmonella* are primarily pathogens of lower animals and have been isolated from almost all animal species. Humans typically acquire *Salmonella* through the ingestion of contaminated food or water, although transmission by fomites or direct faecal–oral spread can occur [82].

Human salmonellosis is an important medical problem throughout the world. The majority of non-typhoid salmonella infections in man present with gastroenteritis and fever. Several other clinical syndromes including enteric fever, bacteraemia, localized infection, and the chronic carrier state may occur and overlap [151].

Prior to HIV disease, bacteraemia occurred in 5–10% of known non-typhoid salmonella infections. In several studies, however, an increased risk of bacteraemia was strongly associated with the presence of underlying diseases which result in defective cell mediated immunity, such as leukaemia, lymphoma, or systemic lupus erythematosis [1, 28, 183]. For example, in a 1971 study from Memorial Hospital in New York, 35% of immunocompromised cancer patients with salmonella infections presented with bacteraemia [183].

S. typhimurium was the predominant serotype in 412 cases of bacteraemia representing 7% of the total cases of non-typhoid salmonella reported to the New York Department of Health from 1962–1971, and accounted for a greater proportion of septicaemia than its overall proportion in total isolates. The majority of patients with bacteraemia in this study either had an underlying disease associated with defective cellular immune function or were under the age of five, with the majority of these being under the age of one, another group known to be at high risk for salmonella bacteraemia [28].

As expected, subjects with HIV disease are at increased risk of serious salmonella infections, particularly septicaemia, and *S. typhimurium* appears to be the predominant serotype responsible for bacteraemia in this population.

Bacteraemia with non-typhoid *Salmonella* in AIDS patients has been reported in the United States, Europe and Africa. However, the true incidence of salmonella bacteraemia in HIV infected persons is difficult to determine. In the US, it is estimated that only 1–5% of salmonella infections are reported to the Centers for Disease Control (CDC) [25] making comparisons between the incidence and clinical features of AIDS-related salmonellosis with that of the general population impossible to interpret. Celum *et al.* [19] addressed these issues by using a population based surveillance system for AIDS and *Salmonella* to analyse the incidence, clinical features, and serotypes of salmonella infections in men with and without AIDS in San Francisco, California. The average annual incidence of salmonellosis among AIDS patients in San Francisco was 20 times higher than the incidence in the age and sex matched control population, with a 5-fold increased incidence of bacteraemia in the AIDS group. The predominant serotype isolated from the bacteraemic AIDS patients was *S. typhimurium.*

Table 3.1 summarizes the clinical and microbiological features of HIV patients with salmonella bacteraemia as reported in studies from the US, Europe, Africa and South America. All of the studies address salmonella bacteraemia in late stage HIV disease (AIDS or advanced ARC); for many subjects in these studies, recurrent salmonella bacteraemia was the index AIDS diagnosis, according to the revised CDC surveillance case definition for AIDS [22]. Profeta *et al.* [134] describe a minimum salmonella infection rate of 2% in their Newark, New Jersey AIDS patients, 100-fold higher than the general population of Newark. Smith [161] reports a 4.2% prevalence of non-typhoid *Salmonella* in the stools of AIDS patients seen and cultured at the National Institutes of Health, Bethesda, Maryland, also confirming a high incidence of salmonella infection in advanced HIV disease. The incidence of salmonella infections in early HIV disease is unknown and may well be under reported.

Although the data is incomplete, several studies report the occurrence of bacteraemia without accompanying complaints referable to the gastrointestinal tract and with negative stool cultures, analogous to earlier experiences with salmonella bacteraemia in immune compromised hosts [28]. Patients typically present with non-specific symptoms of fever, chills, night sweats and headache, usually without septic shock unless other infections coexist. Among patients with gastrointestinal symptoms and bacteraemia, no presentation is typical, with symptoms ranging from mild abdominal cramping to severe and bloody diarrhoea, and from acute onset to a prolonged course over a month or more.

The geographic prevalence of salmonella serotypes may help to determine which of these cause disease in HIV infected subjects. Before 1986, in the US and Africa, the most common serotype isolated from the blood of those with AIDS was *S. typhimurium* and *S. enteritidis*, consistent with the two most common blood isolates reported prior to the AIDS epidemic. Recent data from the US Centers for Disease Control show more *S. enteritidis* than *S. typhimurium* in blood isolates reported in 25 to 45-year-old men in states with high and middle AIDS incidences during

Table 3.1 Salmonella bacteraemia in patients with HIV

				Number of patients with:					
Year	*Country*	*Total no. patients*	*Salmonella isolated*	*Positive blood culture*	*Gastrointestinal symptoms*	*Stool culture positive*	*Relapse*	*HIV stage*[a]	*Reference no.*
1982–3	Africa	8	*S. typhimurium*	6	0	1	2	5 to 6	37
			S. dublin	1	0	0	1	5 to 6	
			Group D	1	0	1	1	5 to 6	
1985	USA	16	*S. typhimurium*	16	15	7	13	5 to 6	64, 85, 161
1985	USA	6	*S. typhimurium*	3	3	1	3	6	117
			S. enteritidis	3	1	NS	1	6	
1986	USA	6	3 group B	3	1	1	5 of 6	6	49
			S. cholerae-suis	2	2	1		6	
			1 group D	1	0	0		6	
1987	USA	15	*S. typhimurium*	7	NS	NS	NS	6	19
			S. dublin	5	NS	NS	NS	6	
			other	3	NS	NS	NS	6	
1987	Spain	10	*S. enteritidis*	9	4 of 10	5	NS	5 to 6	96
			S. typhimurium	1		1	NS	5 to 6	
1990	Peru	8	*S. typhi*	3	NS	NS	no	2 to 3	67
			S. paratyphi	1	NS	1	no	2 to 3	
			S. typhi	3	3	2	1	6	
			S. paratyphi	1	1	1	1	6	
1990	Africa	12	*S. typhimurium*	10	NS	NS	2	NS	63
			S. typhi	2	NS	NS	0	NS	

NS, not stated.
[a] Authors' interpretation of patient stage according to the Walter Reed Staging Classification for HIV infection [136].

1983–1987; this parallels the sharp overall increase in *S. enteritidis* infections in the northeastern US [102]. In Lima, Peru where *S. typhi* and *S. paratyphi* are endemic, eight cases of typhoid or paratyphoid fever were reported among 117 HIV seropositives, a rate approximately 60 times that of the general population, and 25 times that of the 25 to 35-year-old age group of Lima, indicating that those with HIV infection are at risk for infection with the organisms prevalent in their environment [67].

This concept is further supported by Celum's San Francisco study where less common serotypes of salmonella (*S. dublin* and *S. newport*) were found to be more prevalent in AIDS patients. Since *S. dublin* outbreaks have been associated with consumption of raw (unpasteurized) milk, the authors propose that the prevalence of these organisms may be related to the use of 'natural' food products in AIDS patients in San Francisco. The ingestion of rattlesnake preparations resulting in *S. arizoniae* bacteraemia in HIV infected individuals also emphasizes the association between salmonella infection and the ingestion of specific vehicles [5, 18, 122, 142].

Salmonella bacteraemia in HIV disease is complicated by focal infections, including brain and muscle abscesses, and a high rate of relapse without an identifiable focus, in spite of appropriate antibiotic therapy. Although data are incomplete, more than half of those with advanced HIV disease and salmonella bacteraemia are reported to have recurrent bacteraemia once antibiotic therapy is discontinued.

It appears more likely that the individuals do not clear their infection rather than that they become re-infected. Cases appear to be sporadic and not clustered, and isolation of the same salmonella serotype is common in recurrent bacteraemias. In one case, DNA plasmid fingerprints of a stool isolate were identical to that of a blood isolate obtained more than 4 months later, indicating that the initial gastrointestinal infection was not adequately cleared [109].

Prior to the AIDS epidemic, the high prevalence of enteric pathogens in homosexual men was well documented. The frequent detection of enteric pathogens in both symptomatic and asymptomatic homosexual subjects was thought to be secondary to promiscuity, oral–anal sexual practices and frequent untreated infections [6]. In contrast to *Salmonella* spp. which were not detected, the incidence of shigella and campylobacter enteritis in homosexual men was increased when compared with the general population [137]. Since homosexual men represent the majority of HIV infected persons in industrialized countries, it is to be expected that the same enteric pathogens previously isolated from this population are also found as major causes of enteritis in those with HIV disease. It is also expected that the frequency of faecal–oral infection with these agents is enhanced by their low infective dose, as few as 500 organisms for *Campylobacter* [10] and less than 200 organisms for *Shigella* [40].

Shigella are non-motile, Gram negative rods of the family *Enterobacteriaceae.* Four species based on biochemical profiles and antigenic properties have been described, *S. dysenteriae*, *S. flexneri*, *S. boydii* and *S. sonnei*. Further subtyping is based on specific somatic antigens. *Shigella* species have long been recognized as a cause of infectious enteritis, most prevalent in children and homosexual men. Extra-intestinal shigella infections are rare. Prior to AIDS, most shigella bacteraemias were caused by *S. flexneri* and occurred in malnourished children [108]. Analogous to the experience with AIDS-associated salmonella infections, shigellosis in AIDS patients is associated with an unusual degree of bacteraemia, and relapse following therapy. Whereas *S. sonnei* is the species most commonly isolated in the USA [21], *S. flexneri* is more often found as a cause of bacteraemia in AIDS. In contrast to the experience with *Salmonella* in end stage HIV disease, which presents as a febrile illness without gastrointestinal symptoms in almost half of the reported cases, shigella bacteraemia is always associated with diarrhoea and fever. Diarrhoea is usually severe, often bloody and accompanied by cramping and nausea; faecal leucocytes are usually present in the stools [7, 65, 107].

Campylobacter are motile, curved Gram negative rods of the family *Spirillaceae* that are generally microaerophilic and unable to ferment or oxidize the usual carbohydrate substrates available in the diagnostic laboratory. Detection and identification of these bacteria is complicated by the variety of different conditions including incubation temperature, atmospheric conditions, length of incubation, and the appropriate media which they require. In addition to somatic antigens, capsular and flagellar antigens have been recognized. Although this genus currently includes 14 species, only some of the species have been associated with human disease [128]. Individual *Campylobacter* spp. are metabolically and genetically so diverse that reclassification may be necessary. This has already occurred with the species *C. pylori*, now designated *Helicobacter pylori.*

Campylobacter spp. are recognized as a major cause of human diarrhoeal disease in the immunocompetent host in many countries [10], particularly in children [23, 124]. Although *C. jejuni* is the most common species isolated from adults and children, this may be a reflection of the relative ease of isolating this species as compared to other *Campylobacter.* Outbreaks have been associated with fresh water, milk and contaminated poultry [23]. Studies of homosexual men with diarrhoea and proctitis reveal a 2–6% recovery rate for *Campylobacter* spp., indicating that venereal transmission may occur [136]. Specific species, *C. cinaedi* and *C. fennelliae*, have been isolated almost exclusively from homosexual men [170].

In AIDS patients, campylobacter infection usually presents as diarrhoeal illness. Bacteraemia [9], cholecystitis [181], refractory diarrhoea [41], and relapse in spite of therapy [129] have all been reported. Bacteraemia has been reported with *C. cinaedi* [30] and *C. fennelliae* [119] as well as *C. fetus* [31] previously associated with bacteraemia in the immune compromised host.

3.1.3 Treatment

In the immunocompetent host, acute salmonella gastroenteritis is usually not treated, since antibiotic treatment has been shown to prolong faecal excretion and favour the development of resistant organisms [4]. Due to the frequency of severe complications and relapse, all HIV patients with salmonella enteritis or bacteraemia should receive antibiotic therapy, determined by susceptibility testing of the organisms. For susceptible organisms, ampicillin, chloramphenicol, and trimethoprim-sulphamethoxazole (TMP–SMZ) have all been used successfully, with improvement in symptoms and clearing of bacteraemia. However, more than 20% of *S. typhimurium* strains are resistant to ampicillin [28], chloramphenicol does not cure the gastrointestinal carrier state [143], and TMP–SMZ has a high rate of side effects for HIV infected patients [66]. The fluoroquinolones, norfloxacin and ciprofloxacin, are effective against *Salmonella* and other enteric pathogens, and early data from their use is encouraging. Because of the high rate of bacteraemic relapse with *Salmonella*, many physicians recommend the use of chronic suppressive therapy after an initial 10–14-day course. Although there are no controlled trials, reports of bacteraemic breakthroughs while on suppressive doses of ampicillin, combined with the frequency of patient intolerance to TMP–SMZ, have led several investigators to recommend ciprofloxacin for chronic suppressive therapy, as well as for initial treatment [86, 91].

Antibiotic therapy of shigella infections in normal hosts has been shown to decrease the duration of symptoms and the shedding of *Shigella* in stools [76]. Because of the risk of severe disease and the potential for person-to-person spread, all HIV subjects with shigella infections, even if they are asymptomatic, should receive antibiotic treatment with cure documented by follow-up culture. Campylobacter enteritis in immunocompetent hosts is not uniformly treated with antimicrobials, but it also merits therapy in those with HIV disease. Those with shigellaemia or with *Campylobacter* spp. isolated from the blood or other normally sterile sites, require treatment with parenteral antibiotics for 7–14 days, with cure documented by culture.

Since an increasing number of shigella isolates are resistant to ampicillin and TMP–SMZ [7, 107, 156] and there have been reports of campylobacter isolates resistant to multiple antibiotics [41], antimicrobial susceptibility testing is indicated, especially in those with prolonged enteritis or systemic disease. The quinolone antibitoics, norfloxacin and ciprofloxacin, have excellent activity against bacterial enteric pathogens and are appropriate choices for empirical therapy of suspected infection with *Salmonella*, *Shigella* or *Campylobacter*.

3.1.4 Prevention

HIV infected patients predominantly acquire *Salmonella* through the ingestion of contaminated food or water. Salmonella infection in AIDS patients has resulted from the ingestion of raw milk [19], snake powders [5, 18, 122, 142], and contact with pet turtles [166]. Salmonella infections in the general population have also been associated with the ingestion of raw milk [20, 159], pasteurized milk [148], hamburger [162], poultry [83] and eggs [24, 42, 163]. While transmission of *Shigella* and *Campylobacter* within segments of the HIV infected population may be facilitated by sexual practices, these organisms are water and food borne pathogens as well. Outbreaks of shigellosis have resulted from contaminated lettuce [35] and many outbreaks of campylobacteriosis have been associated with ingestion of contaminated water [125, 177], raw milk [32], and undercooked poultry [77]. It is therefore important to caution HIV infected subjects against the consumption of raw or 'natural' products, and undercooked meat and poultry and eggs. Griffen and Tauxe [69] from the US CDC specifically recommend that all immunocompromised patients be counselled to cook raw foods of animal origin well to avoid cross-contaminating cooked foods with raw foods, and to avoid eating raw shellfish, raw milk, raw meat and raw eggs. In addition, the use of folk remedies (e.g. rattlesnake capsules, and other unauthorized treatments) by HIV infected individuals needs to be evaluated for their potential to transmit disease. HIV infected patients should be informed regarding the potential transmission of enteric pathogens that may result from specific sexual practices. Unsanitary water supplies should be avoided and infection control measures used in the treatment of those with enteric infections.

Environmental risk factors are important sources of opportunistic infections in AIDS patients by *Salmonella*, *Shigella* and *Campylobacter*. Issues to be addressed include:

1. Better control of sanitation in the food industry;
2. Identification and treatment of typhoid carriers in typhoid endemic countries with programmes to screen food handlers;
3. Import regulations to control salmonellosis associated with turtles;
4. Counselling.

3.2 MECHANISMS OF PATHOGENESIS

3.2.1 *Salmonella*, *Shigella* and *Campylobacter*

Our understanding of the pathogenesis of *Salmonella* sp. results in a large part from studies with a *S. typhimurium* mouse model. This model has defined some of the properties of *Salmonella* spp. that contribute to pathogenesis. Investigators have manipulated both the genetics of *Salmonella* and of the mouse in these studies.

Since the primary route of infection by *Salmonella* occurs by ingestion of contaminated food or water, it must be able to survive in a wide range of growth conditions with varying pH, nutritional supply and temperature. Once ingested, *Salmonella* must survive the low pH of the stomach, then attach to intestinal epithelial cells. *Salmonella* express receptors that regulate chemotaxis which results in attraction or repulsion of the bacteria in response to changes in the local concentrations of nutrients. Neither chemotaxis nor motility seem to contribute to the virulence of *Salmonella* [17, 105].

Next *Salmonella* must attach to epithelial or macrophage cells. This attachment can be mediated by bacterial products such as fimbriae, and by host products such as complement receptors or Fc receptors. Although some salmonellae express type 1 fimbriae, Jones showed that they play no significant role in attachment [87] to HeLa cells *in vivo*. In contrast, Lindquist showed that type 1 fimbriae enhanced attachment of *Salmonella* to enterocytes [103].

Once attached to host cells *Salmonella* are subject to phagocytosis. Some of the genes contributing to the survival of *Salmonella* in macrophages have been defined by experiments with transposon insertions. With this technique a transposon is introduced into specific sites in the bacterial DNA; mutant phenotypes are produced with a reduced capacity to invade and survive within macrophages [15, 16, 47, 58, 70, 103]. Of 23 highly attenuated transposon mutants, all but one invaded and replicated with the same efficiency as the parent strain in a human adenocarcinoma cell line Caco-2 [58] and survived less efficiently than the parent in macrophage cell line J774. One mutant attached and invaded both cell lines at a lower frequency than the parent strain. Insertions mapping to the PhoP/PhoQ operon increased the sensitivity of *Salmonella* to macrophage antibiotic peptides, defencins [47, 70, 71, 112, 113, 114]. PhoP is a positive regulator of phosphate regulated genes in *Salmonella* [70]. Using a similar approach Miller *et al.* [111] isolated TnPhoA insertions that reduced survival of *Salmonella* in macrophages. TnPhoA is a genetic tool used to define bacterial gene products that are transported to the periplasmic or outer membrane of Gram negative bacteria. Nine of 15

TnPhoA mutants were defective in LPS biosynthesis, while the other insertions mapped to the chromosome and caused defects in undefined secreted proteins. Using recombinant DNA techniques, Elsinghorst *et al.* isolated a 33 kb DNA fragment from *S. typhi* that confers on *Escherichia coli* HB101 the capacity to invade Henle 407 cells (human embryonic intestine) [44]. Transposon mutagenesis showed that the 33 kb fragment encoded four genes necessary for invasion, inv A–D. Homologous DNA isolated from *S. typhimurium* failed to confer the capacity to invade Henle cells on HB101, suggesting that *S. typhimurium* had a different invasion mechanism. Ishibashi [84] showed that opsonized *S. typhimurium* binds to complement receptor C1, while *S. typhi* binds to complement receptor C3 of mouse macrophages. Phagocytosis followed by phagolysosome fusion was observed for *S. typhi.* Phagocytosis without phagolysosome fusion was observed for *S. typhimurium.* Antibody to C3 receptor blocked binding of *S. typhi* and phagolysosome fusion. The relationship of the genetically defined attachment and invasion systems of the bacteria and host receptors has not been defined. However, these studies suggest that survival of *S. typhimurium* in mouse macrophages is associated with attachment to the C1 receptor rather than the C3 receptor.

Growth and spread to the tissues of animals and man of *Salmonella* is controlled by genes regulating the metabolism of the bacteria in the host. These mutations fall into two major classes; mutations that make the bacteria auxotrophic and mutations in global regulons. Mutations in aromatic biosynthetic pathway, AroA and AroD, [14, 46, 95, 115, 132, 154, 168, 184, 186] attenuate *Salmonella* and have been used as live vaccine carriers for heterologous proteins. Uridine 5′-diphosphate (UDP)-glucose-4-epimerase, GalE, mutants grown on glucose-free galactose containing medium synthesize minimal LPS and are lysed by the toxic effects of accumulated galactose-1-phosphate [61, 62]. GalE mutants of salmonella are attenuated and have been used as vaccines [52, 80, 81, 120, 185]. Mutations in either the cyclic AMP receptor protein (crp) or in adenyl cyclase (cya) have global effects on the expression of bacterial genes controlled by cyclic AMP. Both cya and crp mutants attenuate *Salmonella* [33]. Mutations in the PhoP/PhoQ operon affect synthesis of phosphate regulated genes and attenuate *Salmonella* [33, 59, 70, 71, 112, 113, 114, 135]. OmpR is a transcriptional activator of outer membrane proteins and its expression is controlled by an environmental sensor gene product of EnvZ. Mutations of OmpR are highly attenuated, while OmpC and OmpD insertion mutants, outer membrane gene under the control of OmpP, are only slightly attenuated, suggesting other OmpR regulated operons contributed to virulence [27].

Mutations that affect the cell wall, LPS and outer membranes of salmonella also reduce the capacity of the bacteria to survive within the infected host. Asd mutants of salmonella undergo autolysis when grown in the absence of diaminopimelic acid [34, 60], an essential component in cell wall biosynthesis. LPS defective strains are less virulent and less invasive than strains with complete LPS [45, 88, 111, 112]. Also, outer membrane proteins OmpC and OmpD contribute to virulence of salmonella [27].

Once salmonellae invade the Peyer's patches they are subject to both non-specific and specific host immune responses. O side chains of LPS protect *Salmonella* by steric inhibiting complement deposition on the outer membrane. Antibodies to common Gram negative antigens as well as salmonella-specific antibodies bind to *Salmonella*, increasing complement deposition, antibody-dependent cell-mediated cytotoxicity and phagocytosis and killing [27]. *Salmonella* evades the effects of serum antibody and complement by invading and surviving within macrophages [15, 47, 48, 58, 70, 71, 100, 150, 176] and histocytes. In addition to chromosomal genes, *Salmonella* carries a virulence associated plasmid (except *S. typhi*) [72]. Loss of this plasmid is associated with decreased virulence. A 35 kDa region of this plasmid was initially thought to carry the virulence-associated genes [47, 60, 73, 74, 78, 110, 121, 157]. At least two other genetically distinct plasmid regions have been shown to contribute to virulence [157]. It is unclear if these plasmid genes promote the spread of the bacteria from the Peyer's patch to mesenteric lymph nodes [73, 74] or if they promote survival of the bacteria in the tissues [78].

Salmonella can also evade antibodies by antigenic variation and by down-regulating the immune response. Suppression of specific and non-specific immune responses by *Salmonella* appears to be mediated by macrophages [2, 36, 43, 78, 89, 101, 118] The capacity of *S. typhimurium* to depress specific as well as non-specific immunity in the mouse is associated with a large virulence plasmid [78]. Transposon insertions mapping outside of as well as within the 35 kDa virulence-associated region knock out the capacity of the infecting strain to suppress the immune response of the host (Sizemore and Warren, unpublished data).

In summary, survival of *Salmonella* within the host is a complex interaction of bacterial as well as host factors. A better understanding of the details of this host–parasite interaction in both normal and immunocompromised individuals is required in order to develop rational treatment and prevention strategies.

Shigella spp. has a number of characteristics that contribute to its capacity to cause diarrhoea in healthy adults. These include, but are not limited to, attachment to and invasion of colonic cells, multiplication within epithelial cells, and production of shiga toxin. The invasive phenotype is mediated by genes carried by 180–240-kb plasmids. In addition, the expression of at least eight chromosomal genes is necessary for full virulence (for review, [75]). All of these properties contribute to the initial colonization of the colon and diarrhoea.

Campylobacter spp. are a major cause of acute diarrhoeal disease in humans; these pathogens can also produce bacteraemia as well as local infections. Recently, *H. pylori* (formerly *C. pylori*) has been implicated in gastritis and peptic ulcerations (for review, Peterson [130]). There is controversy about the pathogenesis of *Campylobacter* because of conflicting reports about invasiveness and toxin production [12, 178]. Recently, a transient depression of intestinal brush border has been described using weanling ferrets as an animal model for studying campylobacteriosis [8]. We expect pathogenic mechanisms will be elucidated from intensive investigations in several laboratories.

3.2.2 HIV and bacterial enteric pathogens

Human immunodeficiency virus (HIV) selectively results in the premature death of T-helper/inducer lymphocytes, resulting in progressive destruction of the cellular immune system [90]. Although the severity of clinical disease caused by HIV appears to correlate with the degree of T-helper cell depletion [140], infection with HIV results in widespread immunosuppression involving multiple aspects of humoral as well as cell mediated immunity. Subjects with HIV infection suffer quantitative loss and impaired clonal proliferation of $CD4^+$ cells, as well as decreased production of gamma interferon by these cells [13, 98]. These defects, combined with diminished phagocytic activity by monocytes and macrophages contribute to the prevalence of infections with intracellular organisms like *Salmonella* which require an intact cell mediated response for host protection.

In vitro functional studies of lymphocytes from patients with HIV infection have demonstrated intrinsic B-cell defects and defective T-cell control of B-cell function. The number of B-cells spontaneously secreting immunoglobulins is increased (polyclonal B-cell activation) but these B-cells have decreased proliferative responses to both T-cell dependent and T-cell independent antigens [99]. Clinically, HIV subjects demonstrate impaired antibody responses following both immunization and natural infection [3, 141, 155, 182]. HIV patients are subject to serious infections with organisms such as *Streptococcus pneumoniae*, *Haemophilus influenzae*, *Shigella* and *Campylobacter*, where humoral immunity plays a major role in host defence.

The prevalence of serious infection with enteric bacteria in AIDS patients most likely also results from failure of the local gut immune response. Mucosal integrity may be disrupted by chronic diarrhoea which is reported to occur in 30–90% of those with AIDS. There is a reduced number of $CD4^+$ T-helper cells and reduced macrophage function in the intestinal mucosa of AIDS patients. These cellular immune defects, combined with a defective IgA mediated immune response in the gut of patients with HIV disease, enhance both the incidence of localized enteric infection as well as the frequency of systemic spread [94].

Although there is no direct evidence that *Salmonella* and HIV interact, Tagliabue *et al.* [165] reported that natural antibacterial activity against *Salmonella* is reduced as well as the antibody response to LPS. This depressed natural antibacterial activity and reduced anti-LPS titres may contribute to the increased sensitivity to salmonella infections. LPS-treated HIV-infected macrophages are non-permissive for viral replication [92]. In contrast, IL2 and IL6 enhance the production of virus from HIV infected macrophages [171]. Since salmonella antigens are potent activators of lymphokines and monokines [39, 172] and since *Salmonella* is capable of depressing both specific and non-specific immune responses, it is possible that salmonella infections increase HIV load rather than depress viral replication.

3.3 MOLECULAR APPROACHES TO DIAGNOSIS

3.3.1 Nucleic acid probes for detection and identification of *Salmonella*, *Shigella* and *Campylobacter*

Culture methods have been the gold standard for the diagnosis of infections caused by *Salmonella*, *Shigella* and *Campylobacter*. These bacteria are isolated from clinical specimens with differential and selective media. Identification of a pathogen is based on biochemical reactions and serological tests. Plasmid profiles and DNA fingerprinting analysis have been useful for epidemiological studies [79, 123].

Recently, nucleic acid probes have been used for diagnosis and epidemiological studies. The hallmark of a nucleic acid probe is specificity since a probe is based on a unique nucleic acid sequence used in a hybridization reaction to detect homologous regions in samples. Probes have been developed from random sequences of nucleic acid as well as sequences encoding antigens or virulence associated traits; sequences encoding ribosomal RNA (rRNA) have been used, taking advantage of a natural amplification of target, since rRNA is present in 5000–10 000 copies per cell. A positive result with a probe assay is independent of antigen expression, can be obtained in the presence of other micro-organisms, and does not require pathogen isolation. Therefore, nonviable, as well as difficult to grow micro-organisms, can be identified with probes. Large numbers of clinical and/or environmental samples can be processed more rapidly than with conventional methods. High sensitivity with probe assays can be obtained if sufficient target is present in the specimen. Although growth amplification has been used successfully in probe assays, the polymerase chain reaction (PCR) is currently being used to achieve a more rapid assay with high sensitivity since nucleic acid sequences in a sample can be amplified over a million times in a few hours using PCR [149].

This section summarizes molecular approaches including probe assays for the detection and identification of *Salmonella*, *Shigella* and *Campylobacter.*

As discussed above, salmonella serotypes that have been isolated with increased frequency from AIDS patients include *S. typhimurium* [19, 37, 63, 64, 85, 96, 102, 117, 161], *S. enteritidis* [96, 102, 117], *S. dublin* [19, 102], *S. arizoniae* [5, 18, 122, 142] and *S. typhi* [63, 67]. DNA fingerprint analysis has been useful in early and current studies. Chromosomal probe fingerprinting, using cloned random sequences of *S. enteritidis*, was used to analyse clinical isolates of *S. enteritidis*, *S. typhimurium* and *S. dublin*; both species specific and genus specific patterns were observed [169]. Recurrent infection, rather than infection with a different strain of *S. typhimurium* in an HIV patient, was demonstrated using plasmid DNA fingerprinting [109].

Identification of salmonellae using colonies from clinical specimens [152] and growth enriched food samples [50], have been reported using probes from genomic DNA of *S. enteritidis* and *S. typhimurium*, respectively. A second generation assay for detection of salmonellae in food has been developed with a probe encoding salmonella rRNA sequences using an enzymatic detection method and a colorimetric endpoint [26].

Probes for the detection and identification of *S. typhi* are based on sequences encoding Vi and flagellin antigens. An 8.6 kb Vi DNA probe has been used to detect growth amplified *S. typhi* from the blood of typhoid patients [146]. Modifications of the protocol using an alkaline phosphatase conjugated oligonucleotide derived from the 8.6 kb fragment and chemiluminescent detection provides a positive test for *S. typhi* within 24 hours after specimen collection [147], compared to 2–5 days or longer with conventional blood culture. Hybridizing the oligonucleotide probe to PCR products from blood should reduce the time required for a positive result to 6–8 hours (Rubin, unpublished results). Preparation of stool samples for DNA probe assays has been a challenge because of as yet undefined inhibitors in stools, and low numbers of pathogens compared to faecal flora and non-specific background during detection. A preliminary study to identify chronic carriers of *S. typhi* in Lima, Peru, demonstrated promising results using a protocol of sequential hybridization of Vi and flagellar probes to colony lifts of dilutions of stool samples (Gotuzzo,

Rubin and Frankel, unpublished results). PCR protocols are currently being developed for use with stool samples for the diagnosis of acute cases of typhoid fever as well as the identification of chronic carriers using Vi and flagellar primers and probes

Shigella flexneri [7, 65, 107, 156] and *Shigella sonnei* [38] infections have been reported in patients with AIDS. Previous studies using DNA fingerprint analysis with nucleic acid probes has been useful in epidemiologic studies [104, 164]; this type of approach could be useful in following outbreaks as well as carriage of *Shigella*. Table 3.2 summarizes nucleic acid probes for the identification of *Shigella*. These probes have been cloned from virulence associated sequences [11, 158, 174, 175]. Methods using PCR and probes have been developed for identifying Shigella in stools [57] and food [97].

Campylobacter spp. associated with AIDS infections include *C. jejuni* [41, 129, 160], *C. pylori* [54], *C. fetus* [31, 138], *C. cinaedi* and *C. fennelliae* [119]. In a recent evaluation of ten different methods to

Table 3.2 Nucleic acid probes for the identification of *Salmonella*, *Shigella* and *Campylobacter*

Specificity	*DNA sequence*	*Reference no.*
Salmonella spp.	Random fragments[a]	50, 152, 169
	rRNA[a]	26[c]
S. typhi	Flagellar antigen[a]	56
	Vi antigen[a]	146, 147[c,d]
Shigella spp.	Virulence associated[b]	11, 153[c]
	Invasion associated[b] (ail locus)	57[c,d], 158
	Invasion associated outer membrane proteins (ipa B, ipa C, ipa D and ipa H)[b]	174, 175, Venkatesan (unpublished results)[c,d]
Campylobacter spp.	rRNA genes[a]	139[c]
	Random sequences[a]	29[c]
C. jejuni	Membrane antigens[a]	93
	Random sequences[a]	131
C. cinaedi/*C. fennelliae*	Random sequences[a]	170
C. pylori (*Helicobacter pylori*)	Random sequences[a]	173, 180[c]
	rRNA genes[a]	116[d]

[a] chromosomal.
[b] plasmid.
[c] nonisotopic.
[d] oligonucleotide.

distinguish epidemic-associated campylobacter strains, DNA fingerprint analysis including ribotyping (hybridization of restriction digests of chromosomal DNA hybridized with *Escherichia coli* 16S and 23S rRNA) was shown to be one of the most sensitive methods [126]. Genus specific probes cloned from rRNA as well as species specific probes for the identification of *Campylobacter* have been reviewed recently [179] and are summarized in Table 3.2; the number of species specific probes is most likely a reflection of the differences between strains within this genus. A PCR protocol using gastric biopsy and aspirate specimens has been reported which is rapid, accurate and sensitive for the detection of *H. pylori* [173].

3.3.2 Prospects for the future

Since pathogen specific nucleic acid probes are based on unique sequences of DNA, these probe assays have high specificity. In early studies, radiolabelled nucleic acid probes were used for diagnosis and epidemiology. These protocols are not practical for clinical microbiology laboratories because of the hazards to personnel, problems with waste disposal and short half-life (14 days for ^{32}P). To circumvent the problems with radioactive assays, non-isotopic techniques with biotin [153, 180], digoxigenin [167], sulphone [180], acetylaminofluorene [29] and alkaline phosphase [57, 147] have been developed recently for labelling probes for enteric pathogens. Hybridization of these probes with homologous bacterial DNA result in colorimetric or chemiluminescent [147] signals that can be measured directly or indirectly with an intermediate step to bind the detector moeity.

The use of oligonucleotide probes represent a major improvement in probe assays. These probes are approximately 20 base pairs in length; this short oligonucleotide length used at high molar concentrations decrease the hybridization time and thus reduces the time required for a positive result. Other advantages include stability of the reagent; there is less batch to batch variation because large amounts of probe with a long shelf-life are made synthetically. In addition, the problem of vector contamination is eliminated.

Sample preparation has a direct effect on the yield of target bacterial DNA and the subsequent sensitivity of the assay. The sensitivity of probe assays has varied requiring the development of methods to obtain sufficient target DNA from the sample to be detected with the probe. A number of protocols have emerged that are determined by the nature of the sample. For example, a concentration and growth amplification step has been used to detect *S. typhi* from blood samples because the average level of bacteraemia in typhoid patients is 15 organisms/ml of blood. The food industry uses a growth enrichment for the detection of *Salmonella* spp. However, the greatest advance in this area has been the discovery of the polymerase chain reaction, since one copy of target sequence from a pathogen can be amplified and detected with a nucleic acid probe, resulting in a high level of sensitivity.

It would be especially helpful in the AIDS population to have a probe assay to identify diarrhoeal pathogens. Simultaneous amplification of three virulence genes in diarrhoeal stools has been reported in which *Escherichia coli* producing heat-stable enterotoxin, *Escherichia coli* producing heat-labile enterotoxin, and *Shigella* were detected [55]. A format is envisaged for a probe assay in which samples would be detected at the genus level with one set of DNA probes with subsequent serotype identification with another set of probes. However, as shown in Table 3.2, probes have been reported to differentiate subspecies of *Campylobacter*, but not serotypes of *Shigella* or *Salmonella*, with the exception of *S. typhi*. Since the characteristics that differentiate these serotypes can be cloned and probes developed, it is only a matter of time before these probes become available.

Methods using monoclonal antibodies (Mab) as probes have been developed [127] for use in clinical microbiology; they can be used with isolated colonies but are frequently not sensitive enough to use directly with clinical samples. However, Mab specific for *Salmonella* [145] may be valuable for capturing bacteria from clinical or environmental samples using magnetic beads coated with Mab prior to PCR amplification. An immunomagnetic assay using anti-lipopolysaccharide Mab has been reported for the detection of *Salmonella* (O : 6,7) which had an increased sensitivity when compared to latex agglutination, but was used with broth enriched cultures of blood and stool samples [106]. The addition of PCR would both increase the sensitivity and reduce the time required for a positive result. A magnetic immuno-PCR assay for the detection and identification of *Salmonella* with results available in 5 hours has recently been reported [51], [181a] and will provide the basis for the development of tests for other pathogens.

In the AIDS population, it is important to make the diagnosis and initiate treatment as soon as possible. There are likely to be cost savings in the use of gene probe technology, but this will depend on the number of tests requested and the workflow practices in the diagnostic laboratory. We believe that a combination of nucleic acids probes and PCR can reduce the time required for the detection and identification of pathogens when compared to conventional culture methods. In addition, PCR of stool samples using a cocktail of primers would be less expensive. Molecular methods using DNA probes have been used to analyse community outbreaks and could be used to monitor reservoirs and carriage of bacterial enteric pathogens. We expect rapid, sensitive and specific DNA probe assays, with formats for use in the clinical laboratory, will have an impact on the diagnosis and management of infections with *Salmonella*, *Shigella* and *Campylobacter*.

ACKNOWLEDGEMENTS

We thank Fredrick Koster for helpful comments on the manuscript, Malabi Venkatesan for discussions relating to the manuscript, and Bryan Frain and Sybil Goode for their help in preparing the manuscript.

The views of the authors do not purport to reflect the position of the Department of the Army or the Department of the Defense.

REFERENCES

1. Abramson, S., Kramer, S. B., Radin, A., *et al.* (1985) *Salmonella* bacteraemia in systemic lupus erythematosis, eight year experience at a municipal hospital. *Arthritis Rheum.*, **28**, 75–9.
2. Al-Ramadi, B.K., Brodkin, M.A., Mosser, D.M., *et al.* (1991) Immunosuppression induced by attenuated *Salmonella*. Evidence for mediation by macrophage precursors. *J. Immunol.*, **146**, 2737–46.
3. Ammann, A.J., Schiffman, G., Abrams, D., *et al.* (1984) B-cell immunodeficiency in acquired immunodeficiency syndrome. *JAMA*, **251**, 1447–9.
4. Aserkoff, B., Bennett, J.V. (1969) Effect of antibiotic therapy in acute salmonellosis on the fecal excretion of salmonellae. *N. Engl. J. Med.*, **281**, 636–40.
5. Babu, K., Sonnenberg, M., Kathpalia, S., *et al.* (1990) Isolation of salmonellae from dried rattlesnake preparations. *J. Clin. Microbiol.*, **28**, 361–2.
6. Bartlett, J.G., Laughon, B., Quinn, T.C. (1988) Gastrointestinal complications of AIDS. in *AIDS: Etiology, Diagnosis, Treatment, and Prevention*, (eds V.T. DeVita *et al.*), Lippincott, Philadelphia, pp. 227–44.
7. Baskin, D.H., Lax, J.D., Barenberg, D. (1987) Shigella bacteraemia in patients with the acquired immune deficiency syndrome. *Am. J. Gastroenterol.*, **82**, 338–41.
8. Bell, J.A., Manning, D.D. (1990) Role of temporary intestinal brush border dysfunction in *Campylobacter jejuni* diarrhoea. *Curr. Microbiol.*, **21**, 355–9.
9. Bernard, E., Roger, P.M., Carles, D., *et al.* (1989) Diarrhoea and *Campylobacter* infections in patients infected with the human immunodeficiency virus (letter). *J. Infect. Dis.*, **159**, 143–4.
10. Blaser, M.J., Reller, L.B. (1981) Campylobacter enteritis. *N. Engl. J. Med.*, **305**, 1444–52.
11. Boileau, C.R., d'Hauteville, H.M., Sansonetti, P.J. (1984) DNA hybridization technique to detect *Shigella* species and enteroinvasive *Escherichia coli*. *J. Clin. Microbiol.*, **20**, 959–61.
12. Bok, H.E., Greeff, A.S., Crewe-Brown, H.H. (1991) Incidence of toxigenic *Campylobacter* strains in South Africa. *J. Clin. Microbiol.*, **29**, 1262–4.

13. Bowen, D.L., Lane, H.C., Fauci, A.S. (1985) Immunopathogenesis of the acquired immunodeficiency syndrome. *Ann. Intern. Med.*, **103**, 704–9.
14. Brown, A., Hormaeche, C.E., Demarco, *et al.* (1987) An attenuated aroA *Salmonella typhimurium* vaccine elicits humoral and cellular immunity to cloned beta-galactosidase in mice. *J. Infect. Dis.*, **155**, 86–92.
15. Buchmeier, N.A., Heffron, F. (1989) Intracellular survival of wild-type *Salmonella typhimurium* and macrophage-sensitive mutants in diverse populations of macrophages. *Infect. Immun.*, **57**, 1–7.
16. Buchmeier, N.A., Heffron, F. (1990) Induction of *Salmonella* stress proteins upon infection of macrophages. *Science*, **248**, 730–2.
17. Carsiotis, M., Stocker, B.A.D., Holder, I.A. (1989) *Salmonella typhimurium* virulence in a burned-mouse model. *Infect. Immun.*, 57, 2842–6.
18. Casner, P.R. and Zuckerman, M.J. (1990) *Salmonella arizonae* in patients with AIDS along the U.S.–Mexican border. *N. Engl. J. Med.*, **323**, 198–9.
19. Celum, C.L., Chaisson, R.E., Rutherford, G.W., *et al.* (1987) Incidence of salmonellosis in patients with AIDS. *J. Infect. Dis.*, **156**, 998–1002.
20. Centers for Disease Control. (1984) *Salmonella dublin* and raw milk consumption–California. *MMWR*, **33**, 196–8.
21. Centers for Disease Control. (1985) Shigellosis–U.S.A., 1984. *MMWR*, **34**, 600–2.
22. Centers for Disease Control. (1987) Revision of the CDC Surveillance Case Definition for acquired immunodeficiency syndrome. *MMWR*, **36** (Suppl), 3S–15S.
23. Centers for Disease Control. (1988) *Campylobacter* isolates in the United States, 1982–1986. *MMWR*, **37**, 1–13.
24. Centers for Disease Control. (1990) Update: *Salmonella enteritidis* infections and shell eggs–United States, 1990. *MMWR*, **39**, 909–12.
25. Chalker, R.B., Blaser, M.J. (1988) A review of human salmonellosis, III. Magnitude of salmonella infection in the United States. *Rev. Infect. Dis.*, **10**, 111–24.
26. Chan, S.W., Wilson, S.G., Vera-Garcia, M., *et al.* (1990) Comparative study of colourimetric DNA hybridization method and conventional culture procedure for detection of *Salmonella* in foods. *J. Assoc. Off. Anal. Chem.*, 73, 419–24.
27. Chatfield, S.N., Dorman, C.J., Hayward, C., *et al.* (1991) Role of ompR-dependent genes in *Salmonella typhimurium* virulence: mutants deficient in both ompC and ompF are attenuated *in vivo*. *Infect. Immun.*, **59**, 449–52.
28. Cherubin, C.E., Neu, H.C., Imperato, P.J., *et al.* (1974) Septicemia with non-typhoid *Salmonella*. *Medicine*, **53**, 365–76.
29. Chevrier, D., Larzul, D., Megraud, F., *et al.* (1989) Identification and classification of *Campylobacter* strains by using nonradioactive DNA probes. *J. Clin. Microbiol.*, **27**, 321–6.
30. Cimolai, N., Gill, M.J., Jones, A., *et al.* (1987) *Campylobacter cinaedi* bacteraemia: case report and laboratory findings. *J. Clin. Microbiol.*, **25**, 942–3.
31. Costel, E.E., Wheeler, A.P., Gregg, C.R. (1984) *Campylobacter fetus* spp.

fetus cholecystitis and relapsing bacteraemia in a patient with acquired immunodeficiency syndrome. *Southern Med. Journal*, 77, 927–8.
32. Cunningham, G.C. (1982) The hazard of consuming raw milk. *West. J. Med.*, **136**, 51–2.
33. Curtiss, R., III, Goldschmidt, R.M., Fletchall, N.B., *et al.* (1988) Avirulent *Salmonella typhimurium* delta cya delta crp oral vaccine strains expressing a streptococcal colonization and virulence antigen. *Vaccine*, **6**, 155–60.
34. Curtiss, R., III, Nakayama, K., Kelly, S.M. (1989) Recombinant avirulent salmonella vaccine strains with stable maintenance and high level expression of cloned genes *in vivo*. *Immunol. Invest.*, **18**, 583–96.
35. Davis, H., Taylor, J.P., Perdue, J.N., *et al.* (1988) A shigellosis outbreak traced to commercially distributed shredded lettuce. *Am. J. Epidemiol.*, **128**, 1312–21.
36. Deschenes, M.M.G., Nauciel, C. (1989) Suppression of primary antibody response in genetically susceptible mice infected with *Salmonella typhimurium*, restoration by catalase. *Res. Immunol.*, **140**, 55–65.
37. DeWit, S., Taelman, H., Perre, P.V.d., *et al.* (1988) Salmonella bacteraemia in African patients with human immunodeficiency virus infection. *Eur. J. Clin. Microbiol. Infect. Dis.*, **7**, 45–7.
38. Dryden, M.S., Shanson, D.C. (1988) The microbial causes of diarrhoea in patients infected with the human immunodeficiency virus. *J. Infect.*, **17**, 107–14.
39. Duncan, L.M., Meegan, L.S., Unanue, E.R. (1991) IL-1 gene expression in lymphoid tissues. *J. Immunol.*, **146**, 565–71.
40. DuPont, H.L., Levine, M.M., Hornick, R.B., *et al.* (1989) Inoculum size in shigellosis and implications for expected mode of transmission. *J. Infect. Dis.*, **159**, 1126–8.
41. Dworkin, B., Wormser, G.P., Abdoo, R.A., *et al.* (1986) Persistence of multiply antibiotic-resistant *Campylobacter jejuni* in a patient with the acquired immune deficiency syndrome. *Am. J. Med.*, **80**, 965–70.
42. Editorial. (1988) *Salmonella enteritidis* phage type 4, chicken and egg. *Lancet*, ii, 720–1.
43. Eisenstein, T.K., Killar, L.M., Stocker, B.A.D., *et al.* (1984) Cellular immunity induced by avirulent *Salmonella* in LPS-defective C3H/HeH mice. *J. Immunol.*, **133**, 958–61.
44. Elsinghorst, E.A., Baron, L.S., Kopecko, D.J. (1989) Penetration of human intestinal epithelial cells by *Salmonella*: molecular cloning and expression of *Salmonella typhi* invasion determinants in *Escherichia coli*. *Proc. Natl. Acad. Sci. U.S.A.*, **86**, 5173–7.
45. Euteneuer, B., Storkel, S., Loos, M. (1985) Differences in attachment and phagocytosis of *Salmonella minnesota* strains (Sform, Re mutant) by mouse peritoneal macrophages: participation of endogenous C1q and bacterial surface components (LPS, porins). *Curr. Top. Microbiol. Immunol.*, **121**, 85–97.
46. Fairweather, N.F., Chatfield, S.N., Makoff, A.J., *et al.* (1990) Oral vaccination of mice against tetanus by use of a live attenuated salmonella carrier. *Infect. Immun.*, **58**, 1323–6.

47. Fields, P.I., Groisman, E.A., Heffron, F. (1989) A salmonella locus that controls resistance to microbicidal proteins from phagocytic cells. *Science*, **243**, 1059–62.
48. Fields, P.I., Swanson, R.V., Haidaris, C.G., *et al.* (1986) Mutants of *Salmonella typhimurium* that cannot survive within the macrophage are avirulent. *Proc. Natl. Acad. Sci. U.S.A.*, **83**, 5189–93.
49. Fischl, M.A., Dickinson, G.M., Sinave, C., *et al.* (1986) Salmonella bacteraemia as manifestation of acquired immunodeficiency syndrome. *Arch. Intern. Med.*, **146**, 113–5.
50. Fitts, R., Diamond, M., Hamilton, C., *et al.* (1983) DNA–DNA hybridization assay for detection of *Salmonella* spp. in foods. *Appl. Environ. Microbiol.*, **46**, 1146–51.
51. Fluit, C., Torensma, R., Widjojoatmodjo, M., *et al.* (1991) The magnetic immuno-PCR assay, a rapid and sensitive detection and identification system for bacteria. *Abstr. Ann. Meeting Microbiol.*, 362.
52. Formal, S.B., Baron, L.S., Kopecko, D.J., *et al.* (1981) Construction of a potential bivalent vaccine strain, introduction of *Shigella sonnei* form I antigen genes into the galE *Salmonella typhi* Ty21, a typhoid vaccine strain. *Infect. Immun.*, **34**, 746–50.
53. Francis, D.P., Jaffe, H.W., Fully, P.N., *et al.* (1985) The natural history of infection with the lymphadenopathy-associated virus human T-lymphotropic virus Type III. *Ann. Intern. Med.*, **103**, 719–22.
54. Francis, N.D., Logan, R.P.H., Walker, M.M., *et al.* (1990) *Campylobacter pylori* in the upper gastrointestinal tract of patients with HIV-1 infection. *J. Clin. Pathol.*, **43**, 60–2.
55. Frankel, G., Giron, J.A., Valmassoi, J., *et al.* (1989) Multi-gene amplification, simultaneous detection of three virulence genes in diarrhoeal stool. *Mol. Microbiol.*, **3**, 1729–34.
56. Frankel, G., Newton, S.M. C., Schoolnik, G.K., *et al.* (1989) Unique sequences in region VI of the flagellin gene of *Salmonella typhi*. *Mol. Microbiol.*, **3**, 1379–83.
57. Frankel, G., Riley, L., Giron, J.A., *et al.* (1990) Detection of *Shigella* in feces using DNA amplification. *J. Infect. Dis.*, **161**, 1252–6.
58. Gahring, L.C., Heffron, F., Finlay, B.B., *et al.* (1990) Invasion and replication of *Salmonella typhimurium* in animal cells. *Infect. Immun.*, **58**, 443–8.
59. Galan, J.E., Curtiss, R., III (1989) Virulence and vaccine potential of phoP mutants of *Salmonella typhimurium*. *Microb. Pathog.*, **6**, 433–43.
60. Galan, J.E., Nakayama, K., Curtiss, R., III (1990) Cloning and characterization of the asd gene of *Salmonella typhimurium*: use in stable maintenance of recombinant plasmids in salmonella vaccine strains. *Gene*, **94**, 29–35.
61. Germanier, R. (1976) Vaccination against typhoid fever with a live oral vaccine. *Dev. Biol. Stand.*, **33**, 85–8.
62. Germanier, R., Furer, E. (1971) Isolation and characterization of gal E mutant Ty21a of *Salmonella typhi*: a candidate strain for a live, oral typhoid vaccine. *J. Infect. Dis.*, **131**, 553–8.
63. Gilks, C.F., Brindle, R.J., Otieno, L.S., *et al.* (1990) Life-threatening

bacteraemia in HIV-1 seropositive adults admitted to hospital in Nairobi, Kenya. *Lancet*, **336**, 545–9.
64. Glaser, J.B., Morton-Kute, L., Berger, S.R., *et al.* (1985) Recurrent *Salmonella typhimurium* bacteraemia associated with the acquired immunodeficiency syndrome. *Ann. Intern. Med.*, **102**, 189–93.
65. Glupczynski, Y., Hansen, W., Jonas, C., *et al.* (1985) *Shigella flexneri* bacteraemia in a patient with acquired immune deficiency syndrome. *Acta Clin. Belg.*, **40**, 388–90.
66. Gordin, F.A., Simon, G.L., Wofsky, C.B., *et al.* (1984) Adverse reactions to trimethoprim-sulfamethoxazole in patients with the acquired immunodeficiency syndrome. *Ann. Intern. Med.*, **100**, 495–9.
67. Gotuzzo, E., Frisancho, O., Sanchez, J., *et al.* (1991) Association between the acquired immunodeficiency syndrome and infection with *Salmonella typhi* or *Salmonella paratyphi* in an endemic typhoid area. *Arch. Intern. Med.*, **151**, 381–2.
68. Greene, J.B., Sidhu, S.S., Lewin, S., *et al.* (1982) *Mycobacterium avium-intracellulare*: A cause of disseminated life-threatening infection in homosexuals and drug abusers. *Ann. Intern. Med.*, **97**, 539–46.
69. Griffin, P.M., Tauxe, R.V. (1988) Food counseling for patients with AIDS. *J. Infect. Dis.*, **158**, 668.
70. Groisman, E.A., Chiao, E., Lipps, C.J., *et al.* (1989) *Salmonella typhimurium* phoP virulence gene is a transcriptional regulator. *Proc. Natl. Acad. Sci. U.S.A.*, **86**, 7077–81.
71. Groisman, E.A., Saier, M.J. (1990) *Salmonella* virulence: new clues to intramacrophage survival. *Trends Biochem. Sci.*, **15**, 30–3.
72. Gulig, P.A. (1990) Virulence plasmids of *Salmonella typhimurium* and other *salmonellae*. *Microb. Pathog.*, **8**, 3–11.
73. Gulig, P.A., Chiodo, V.A. (1990) Genetic and DNA sequence analysis of the *Salmonella typhimurium* virulence plasmid gene encoding the 28 000–molecular weight protein. *Infect. Immun.*, **58**, 2651–8.
74. Gulig, P.A., Curtiss, R., III. (1988) Cloning and transposon insertion mutagenesis of virulence genes of the 100–kilobase plasmid of *Salmonella typhimurium*. *Infect. Immun.*, **56**, 3262–71.
75. Hale, T.L. (1991) Genetic basis of virulence in *Shigella* species. *Microbiol. Rev.*, **55**, 193–205.
76. Haltalin, K., Nelson, J.R., Ring, I., *et al.* (1967) Double blind treatment study of shigellosis comparing ampicillin, sulfadiazone and placebo. *J. Pediatr.*, **70**, 970.
77. Harris, N.V., Weiss, N.S., Nolan, C.M. (1988) The role of poultry and meats in the etiology of campylobacter jejuni/colienteritis. *Am. J. Public Health*, **76**, 407–11.
78. Hoertt, B.E., Ou, J., Kopecko, D.J., *et al.* (1989) Novel virulence properties of the *Salmonella typhimurium* virulence-associated plasmid: immune suppression and stimulation of splenomegaly. *Plasmid*, **21**, 48–58.
79. Holmberg, S.D., Wachsmuth, K. (1989) Plasmid and chromosomal DNA analyses in the epidemiology of bacterial disease, in *Nucleic Acid and*

Monoclonal Antibody Probes (ed. Prakash), Marcel Dekker, New York, pp. 105–29.

80. Hone, D., Attridge, S., Van Den Bosch, L., *et al.* (1988) A chromosomal integration system for stabilization of heterologous genes in salmonella based vaccine strains. *Microb. Pathog.*, **5**, 407–18.
81. Hone, D.M., Attridge, S.R., Forrest, B., *et al.* (1988) A gal E via (Vi antigen-negative). mutant of *Salmonella typhi* Ty2 retains virulence in humans. *Infect. Immun.*, **56**, 1326–33.
82. Hook, E.W. (1990) *Salmonella* species, in *Principles and Practice of Infectious Diseases*, (ed. G.L. Mandell), Churchill Livingstone, New York, pp. 1700–16.
83. Humphrey, T.J., Mead, G.C., Rowe, B. (1988) Poultry meat as a source of human salmonellosis in England and Wales. *Epidemiol. Infect.*, **100**, 175–84.
84. Ishibashi, Y., Arai, T. (1990) Specific inhibition of phagosome-lysosome fusion in murine macrophages mediated by *Salmonella typhimurium* infection. *FEMS Microbiol. Immunol.*, **2**, 35–43.
85. Jacobs, J.L., Gold, J.W.M., Murray, H.W., *et al.* (1985) Salmonella infections in patients with the acquired immunodeficiency syndrome. *Ann. Intern. Med.*, **102**, 186–8.
86. Jacobson, M.A., Hahn, S.M., Gerberding, J.L., *et al.* (1989) Ciprofloxacin for salmonella bacteraemia in the acquired immunodeficiency syndrome (AIDS). *Ann. Intern. Med.*, **110**, 1027–9.
87. Jones, G.W., Richardson, L.A., Uhlman, D. (1981) The attachment to and invasion of HeLa cells by *Salmonella typhimurium*: the contribution of mannose-sensitive and mannose-resistant haemagglutinating activities. *J. Gen. Microbiol.*, **127**, 351–60.
88. Kawahara, K., Hamaoka, T., Suzuki, S., *et al.* (1989) Lipopolysaccharide alteration mediated by the virulence plasmid of *Salmonella. Microb. Pathog.*, **7**, 195–202.
89. Killar, L.M., Eisenstein, T.K. (1984) Differences in delayed-type hypersensitivity responses in various mouse strains in the C3H lineage infected with *Salmonella typhimurium*, strain SL3235. *J. Immunol.*, **133**, 1190–6.
90. Klatzmann, D., Barre-Sinoussi, F., Nugeyre, M.T., *et al.* (1984) Selective tropism of lymphadenopathy associated virus (LAV) for helper-inducer T lymphocytes. *Science*, **225**, 59–63.
91. Koehler, J.E., Chaisson, R.E. (1989) Bacterial infections in acquired immunodeficiency syndrome. *Infect. Dis. Ther.*, **3**, 231–51.
92. Kornbluth, R.S., Oh, P.S., Munis, J.R., *et al.* (1990) The role of interferons in the control of HIV replication in macrophages. *Clin. Immunol. Immunopathol.*, **54**, 200–19.
93. Korolik, V., Coloe, P.J., Krishnapillai, V. (1988) A specific DNA probe for the identification of *Campylobacter jejuni. J. Gen. Microbiol.*, **134**, 521–9.
94. Kotler, D.P., Scholes, J.V., Tierney, A.R. (1987) Intestinal plasma cell alterations in acquired immunodeficiency syndrome. *Dig. Dis. Sci.*, **32**, 129–38.
95. Kumar, S., Gorden, J., Flynn, J.L., *et al.* (1990) Immunization of mice against

Plasmodium vinckei with a combination of attenuated *Salmonella typhimurium* and malarial antigen. *Infect. Immun.*, **58**, 3425–9.

96. Lahoz, J.M.G., Aguado, C.G., Fernandez, R.M., *et al.* (1989) Bacteraemia caused by *Salmonella* non-typhi in patients infected with HIV. *Rev. Clin. Esp.*, **185**, 6–8.
97. Lampel, K.A., Jagow, J.A., Trucksess, M., *et al.* (1990) Polymerase chain reaction for detection of invasive *Shigella flexneri* in food. *Appl. Environ. Microbiol.*, **56**, 1536–40.
98. Lane, H.C., Depper, J.M., Green, W.C., *et al.* (1985) Qualitative analysis of immune function in patients with the acquired immunodeficiency syndrome: Evidence for a selective defect in soluble antigen recognition. *N. Engl. J. Med.*, **313**, 79–84.
99. Lane, M.C., Masur, H., Edgar, L.C., *et al.* (1983) Abnormalities of B cell activation and immunoregulation in patients with the acquired immunodeficiency syndrome. *N. Engl. J. Med.*, **309**, 453–8.
100. Langermans, J.A.M., Van Der Hulst, M., Nibbering, P.H., Van Furth, R. (1990) Activation of mouse peritoneal macrophages during infection with *Salmonella typhimurium* does not result in enhanced intracellular killing. *J. Immunol.*, **144**, 4340–6.
101. Lee, J.C., Gibson, C.W., Eisenstein, T.K. (1985) Macrophage-mediated mitogenic suppression induced in mice of the C3H lineage by a vaccine strain of *Salmonella typhimurium*. *Cell. Immunol.*, **91**, 75–91.
102. Levine, W.C., Buehler, J.W., Bean, N.H., *et al.* (1991) Epidemiology of nontyphoidal salmonella bacteraemia during the human immunodeficiency virus epidemic. *J. Infect. Dis.*, **164**, 81–7.
103. Lindquist, B.L., Lebenthal, E., Lee, P.C., *et al.* (1987) Adherence of *Salmonella typhimurium* to small-intestinal enterocytes of the rat. *Infect. Immun.*, **55**, 3044–50.
104. Litwin, C.M., Storm, A.L., Chipowsky, S., *et al.* (1991) Molecular epidemiology of shigella infections: Plasmid profiles, serotype correlation, and restriction endonuclease analysis. *J. Clin. Microbiol.*, **29**, 104–8.
105. Lockman, H.A. and Curtiss, R., III (1990) *Salmonella typhimurium* mutants lacking flagella or motility remain virulent in BALB/c mice. *Infect. Immun.*, **58**, 137–43.
106. Luk, J.M.C., Lindberg, A.A. (1991) Rapid and sensitive detection of *Salmonella* (O : 6,7) by immunomagnetic monoclonal antibody-based assays. *J. Immunol. Meth.*, **137**, 1–8.
107. Mandell, W., Neu, H.C. (1986) Shigella bacteraemia in adults. *JAMA*, **255**, 3116–7.
108. Martin, T., Habbick, B.F., Nyssen, J. (1983) Shigellosis with bacteraemia: a report of two cases and a review of the literature. *Pediatr. Infect. Dis.*, **2**, 21–6.
109. Mayer, K.H., Hanson, E. (1986) Recurrent salmonella infection with a single strain in the acquired immunodeficiency syndrome. Confirmation by plasmid fingerprinting. *Diagn. Microbiol. Infect. Dis.*, **4**, 71–6.
110. Michiels, T., Popoff, M.Y., Durviaux, S., *et al.* (1987) A new method for the

physical and genetic mapping of large determinants on the 90 kb plasmid of *Salmonella typhimurium*. *Microb. Path.*, **3**, 109–16.

111. Miller, I., Maskell, D., Hormaeche, C., *et al.* (1989) Isolation of orally attenuated *Salmonella typhimurium* following TnphoA mutagenesis. *Infect Immun.*, **57**, 2758–63.
112. Miller, S.I., Kukral, A.M., Mekalanos, J.J. (1989) A two-component regulatory system (phoP phoQ) controls *Salmonella typhimurium* virulence. *Proc. Natl. Acad. Sci. U.S.A.*, **86**, 5054–8.
113. Miller, S.I., Mekalanos, J.J. (1990) Constitutive expression of the phoP regulon attenuates salmonella virulence and survival within macrophages *J. Bacteriol.*, **172**, 2485–90.
114. Miller, S.I., Pulkkinen, W.S., Selsted, M.E., *et al.* (1990) Characterization of defensin resistance phenotypes associated with mutations in the phoP virulence regulon of *Salmonella typhimurium*. *Infect. Immun.*, **58**, 3706–10.
115. Molina, N.C., Parker, C.D. (1990) Murine antibody response to oral infection with live aroA recombinant *Salmonella dublin* vaccine strains expressing filamentous hemagglutinin antigen from *Bordetella pertussis*. *Infect. Immun.*, **58**, 2523–8.
116. Morotomi, M., Hoshina, S., Green, P., *et al.* (1989) Oligonucleotide probe for detection and identification of *Campylobacter pylori*. *J. Clin. Microbiol.*, **27**, 2652–5.
117. Nadelman, R.B., Mathur-Wagh, U., Yancovitz, S.R., *et al.* (1985) *Salmonella* bacteraemia associated with the acquired immunodeficiency syndrome (AIDS). *Arch. Intern. Med.*, **145**, 1968–71.
118. Nauciel, C., Ronco, E., Guenet, J.L. (1988) Genetic control of *Salmonella typhimurium*-induced depression of delayed-type hypersensitivity to sheep erythrocytes in mice. *Infect. Immun.*, **56**, 310–3.
119. Ng, V.L., Hadley, W.K., Fennell, C.L., *et al.* (1987) Successive bacteraemia with *Campylobacter cinaedi* and *Campylobacter fennelliae* in a bisexual male. *J. Clin. Microbiol.*, **25**, 2008–9.
120. Nnalue, N.A., Stocker, B.A.D. (1987) Test of the virulence and live-vaccine efficacy of auxotrophic and galE derivatives of *Salmonella choleraesuis*. *Infect. Immun.*, **55**, 955–62.
121. Norel, F., Coynault, C., Miras, I., *et al.* (1989) Cloning and expression of plasmid DNA sequences involved in salmonella serotype typhimurium virulence. *Mol. Microbiol.*, **3**, 733–43.
122. Noskin, G.A., Clarke, J.T. (1990) *Salmonella arizonae* bacteraemia as the presenting manifestation of human immunodeficiency virus infection following rattlesnake meat ingestion. *Rev. Infect. Dis.*, **12**, 514–7.
123. Owen, R.J. (1989) Chromosomal DNA fingerprinting – a new method of species and strain identification applicable to microbial pathogens. *J. Med. Microbiol.*, **30**, 89–99.
124. Pai, G.H., Sorger, S., Lackman, K., *et al.* (1979) Campylobacter gastroenteritis in children. *J. Pediatr.*, **94**, 589–91.
125. Palmer, S.R., Gully, P.R., White, J.M., *et al.* (1983) Water-borne outbreak of campylobacter gastroenteritis. *Lancet*, i, 287–90.

126. Patton, C.M., Wachsmuth, I.K., Evins, G.M., *et al.* (1991) Evaluation of 10 methods to distinguish epidemic-associated campylobacter strains. *J. Clin. Microbiol.*, **29**, 680–8.
127. Payne, W.J., Marshall, D.L., Shockley, R.K., *et al.* (1988) Clinical laboratory applications of monoclonal antibodies. *Clin. Microbiol. Rev.*, **1**, 313–29.
128. Penner, J.L. (1988) The genus *Campylobacter*: A decade of progress. *Clin. Microbiol. Rev.*, **1**, 157–72.
129. Perlman, D.M., Ampel, N.M., Schifman, R.B., *et al.* (1988) Persistent *Campylobacter jejuni* infections in patients infected with human immunodeficiency virus. *Ann. Intern. Med.*, **108**, 540–6.
130. Peterson, W.L. (1991) *Helicobacter pylori* and peptic ulcer disease. *N. Engl. J. Med.*, **324**, 1043–8.
131. Picken, R.N., Wang, Z., Yang, H.L. (1987) Molecular cloning of a species-specific DNA probe for *Campylobacter jejuni*. *Mol. Cell Probes*, **1**, 245–59.
132. Poirier, T.P., Kehoe, M.A., Beachey, E.H. (1988) Protective immunity evoked by oral administration of attenuated aroA *Salmonella typhimurium* expressing cloned streptococcal M protein. *J. Exp. Med.*, **168**, 25–32.
133. Polsky, B., Gold, J.W.M., Whimbey, E., *et al.* (1986) Bacterial pneumonia in patients with the acquired immune deficiency syndrome. *Ann. Intern. Med.*, **104**, 38–41.
134. Profeta, S., Forrester, C., Eng, R.H.K., *et al.* (1985) *Salmonella* infections in patients with acquired immunodeficiency syndrome. *Arch. Intern. Med.*, **145**, 670–2.
135. Pulkkinen, W.S., Miller, S.I. (1991) A *Salmonella typhimurium* virulence protein is similar to a *Yersinia enterocolitica* invasion protein and a bacteriophage lambda outer membrane protein. *J. Bacteriol.*, **173**, 86–93.
136. Quinn, T.C., Goodell, S.E., Fennell, C., *et al.* (1984) Infections with *Campylobacter jejuni* and campylobacter-like organisms in homosexual men. *Ann. Intern. Med.*, **101**, 187–92.
137. Quinn, T.C., Stamm, W.E., Goodell, S.E., *et al.* (1983) The polymicrobial origin of intestinal infections in homosexual men. *N. Engl. J. Med.*, **309**, 576–82.
138. Rao, G.G., Karim, Q.N., Maddocks, A., *et al.* (1990) *Campylobacter fetus* infections in two patients with AIDS. *J. Infect.*, **20**, 170–2.
139. Rashtchian, A., Eldredge, J., Ottaviani, M., *et al.* (1987) Immunological capture of nucleic acid hybrids and application to nonradioactive DNA probe assay. *Clin. Chem.*, **33**, 1526–30.
140. Redfield, R.R., Wright, D.C., Tramont, E.C. (1986) The Walter Reed Staging Classification for HTLV-III/LAV Infection. *N. Engl. J. Med.*, **314**, 131–2.
141. Rhoads, J.L., Brix, D.L., Wright, C., *et al.* (1991) Safety and immunogenicity of multiple conventional immunizations administered during early HIV infection. *J. Acquir. Immune. Defic. Syndr.*, **4**, 724–31.
142. Riley, K.B., Antoniskis, D., Maris, R., *et al.* (1988) Rattlesnake capsule-associated *Salmonella arizona* infections. *Arch. Intern. Med.*, **148**, 1207–10.
143. Robertson, R.P., Wahab, M.F., Raasch, F.O. (1968) Evaluation of chloramphenicol and ampicillin in salmonella enteric fever. *N. Engl. J. Med.*, **278**, 171–6.

144. Rolston, K.V., Uribe-Botero, G., Mansell, P.W. (1987) Bacterial infection in adult patients with the acquired deficiency syndrome (AIDS) and AIDS-related complex. *Am. J. Med.*, **83**, 604–5.
145. Rubin, F.A. (1990) Probes for the identification of *Salmonella*, in *Gene Probes for Bacteria* (eds. A.J.L. Macario, E.C. de Macario), Academic Press, New York, pp. 323–52.
146. Rubin, F.A., McWhirter, P.D., Punjabi, N.H., *et al.* (1989) Use of a DNA probe to detect *Salmonella typhi* in the blood of patients with typhoid fever. *J. Clin. Microbiol.*, **27**, 1112–4.
147. Rubin, F.A., Voyta, J.C., Murphy, O.J., *et al.* (1990) Rapid identification of *Salmonella typhi* using a Vi DNA probe and chemiluminescent detection. *Abstr. Annu. Meeting Am. Soc. Microbiol.*, p. 88.
148. Ryan, C.A., Nickels, M.K., Hargrett-Bean, N.T., *et al.* (1987) Massive outbreak of antimicrobial resistant salmonellosis traced to pasteurized milk. *JAMA*, **258**, 3269–74.
149. Saiki, R.K., Scharf, S., Faloona, F., *et al.* (1985) Enzymatic amplification of beta-globin genomic sequences and restriction site analysis for diagnosis of sickle cell anemia. *Science*, **230**, 1350–54.
150. Sant'Anna, O.A., Massa, S., Mouton, D., *et al.* (1989) *Salmonella typhimurium* infection in high and low antibody responder mice: inverse correlation between antibody responsiveness and resistance to infection. *FEMS Microbiol. Immunol.*, **1**, 465–71.
151. Saphra, I., Winter, J.W. (1957) Clinical manifestations of salmonellosis in man: an evaluation of 7779 human infections identified at the New York Salmonella Center. *N. Engl. J. Med.*, **256**, 1128–34.
152. Scholl, D., Kaufmann, C., Jollick, J. D., *et al.* (1990) Clinical application of novel sample processing technology for the identification of *Salmonellae* by using DNA probes. *J. Clin. Microbiol.*, **28**, 237–41.
153. Sethabutr, O., Hanchalay, S., Echeverria, P., *et al.* (1985) A non-radioactive DNA probe to identify *Shigella* and enteroinvasive *Echerichia coli* in stools of children with diarrhoea. *Lancet*, **ii**, 1095–7.
154. Sigwart, D.F., Stocker, B.A., Clements, J.D. (1989) Effect of a purA mutation on efficacy of salmonella live-vaccine vectors. *Infect. Immun.*, **57**, 1858–61.
155. Simberkoff, M.S., Sadr, W.E., Schiffman, G., *et al.* (1984) *Streptococcus pneumoniae* infections and bacteraemia in patients with acquired immune deficiency syndrome. *Am. Rev. Respir. Dis.*, **130**, 1174–6.
156. Simor, A.E., Poon, R., Borczyk, A. (1989) Chronic *Shigella flexneri* infection preceding development of acquired immunodeficiency syndrome. *J. Clin. Microbiol.*, **27**, 353–5.
157. Sizemore, D.R., Fink, P.S., Ou, J.T., *et al.* (1991) Tn5 mutagenesis of the *Salmonella typhimurium* 100 kb plasmid: Definition of new virulence regions. *Microb. Pathog.*, **10**, 493–9.
158. Small, P.L.C., Falkow, S. (1986) Development of a DNA probe for the virulence plasmid of *Shigella* spp. and enteroinvasive *Escherichia coli*, in *Microbiology–1986*, (ed. Lieve), American Society for Microbiology, Washington, D.C., pp. 121–4.

159. Small, R.G., Sharp, J.C.M. (1979) A milk-borne outbreak due to *Salmonella dublin*. *J. Hyg.*, **82**, 95–100.
160. Smith, P.D., Lane, H.C., Gill, V.J., *et al.* (1988) Intestinal infections in patients with the acquired immunodeficiency syndrome (AIDS). *Ann. Intern. Med.*, **108**, 328–33.
161. Smith, P.D., Macher, A.M., Bookman, M.A., *et al.* (1985) *Salmonella typhimurium* enteritis and bacteraemia in the acquired immunodeficiency syndrome. *Ann. Intern. Med.*, **102**, 207–8.
162. Spika, J.S., Waterman, S.H., Hoo, G.W.S., *et al.* (1987) Chloramphenicol-resistant *Salmonella newport* traced through hamburger to dairy farms: a major persisting source of human salmonellosis in California. *N. Engl. J. Med.*, **316**, 565–70.
163. St Louis, M.E., Morse, D.L., Potter, M.E., *et al.* (1988) The emergence of grade A eggs as a major source of *Salmonella enteritidis* infections. *JAMA*, **259**, 2103–7.
164. Strockbine, N.A., Parsonnet, J., Greene, K., *et al.* (1991) Molecular epidemiologic techniques in analysis of epidemic and endemic *Shigella dysenteriae* type 1 strain. *J. Infect. Dis.*, **163**, 406–9.
165. Tagliabue, A., Nencioni, L., Mantovani, A., *et al.* (1988) Impairment of *in vitro* natural antibacterial activity in HIV-infected patients. *J. Immunol.*, **141**, 2607–11.
166. Tauxe, R.V., Rigau-Perez, J.G., Wells, J.G., *et al.* (1985) Turtle-associated salmonellosis in Puerto Rico: hazards of the global turtle trade. *JAMA*, **254**, 237–9.
167. Taylor, D.E., Hiratsuka, K. (1990) Use of non-radioactive DNA probes for detection of *Campylobacter jejuni* and *Campylobacter coli* in stool specimens. *Mol. Cell. Probes*, **4**, 261–71.
168. Tite, J.P., Gao, X.M., Hughes, J.C., *et al.* (1990) Anti-viral immunity induced by recombinant nucleoprotein of influenza A virus. III. Delivery of recombinant nucleoprotein to the immune system using attenuated *Salmonella typhimurium* as a live carrier. *Immunology*, **70**, 540–6.
169. Tompkins, L.S., Troup, N., Labigne-Roussel, A., *et al.* (1986) Cloned, random chromosomal sequences as probes to identify *Salmonella* species. *J. Infect. Dis.*, **154**, 156–62.
170. Totten, P.A., Fennell, C.L., Tenover, F.C., *et al.* (1985) *Campylobacter cinaedi* (sp. nov.) and *Campylobacter fennelliae* (sp. nov.): two new *Campylobacter* species associated with enteric disease in homosexual men. *J. Infect. Dis.*, **151**, 131–9.
171. Tsunetsugu, Y.Y., Honda, M. (1990) Effect of cytokines on HIV release and IL-2 receptor alpha expression in monocytic cell lines. *J. Acquir. Immune. Defic. Syndr.*, **3**, 511–6.
172. Ulich, T.R., Guo, K.Z., Irwin, B., *et al.* (1990) Endotoxin-induced cytokine gene expression in vivo. II. Regulation of tumor necrosis factor and interleukin-1 alpha/beta expression and suppression. *Am. J. Pathol.*, **137**, 1173–85.
173. Valentine, J.L., Arthur, R.R., Mobley, H.L.T., *et al.* (1991) Detection of

Helicobacter pylori by using the polymerase chain reaction. *J. Clin. Microbiol.*, **29**, 689–95.

174. Venkatesan, M., Buysse, J.M., Vandendries, E., *et al.* (1988) Development and testing of invasion-associated DNA probes for the detection of *Shigella* spp. and enteroinvasive *Escherichia coli. J. Clin. Microbiol.*, **26**, 261–6.
175. Venkatesan, M.M., Buysse, J. M., Kopecko, D.J. (1989) Use of *Shigella flexneri* ipaC and ipaH gene sequences for the general identification of *Shigella* spp. and enteroinvasive *Escherichia coli. J. Clin. Microbiol.*, **27**, 2687–91.
176. Vladoianu, I.R., Chang, H.R., Pechere, J.C. (1990) Expression of host resistance to *Salmonella typhi* and *Salmonella typhimurium*: bacterial survival within macrophages of murine and human origin. *Microb. Pathog.*, **8**, 83–90.
177. Vogt, R.L., Sours, H.E., Barrett, T., *et al.* (1982) *Campylobacter* enteritis associated with contaminated water. *Ann. Intern. Med.*, **96**, 292–6.
178. Walker, R.I., Caldwell, M.B., Lee, E.C., *et al.* (1986) Pathophysiology of campylobacter enteritis. *Microbiol. Rev.*, **50**, 81–94.
179. Wetherall, B.L., Johnson, A.M. (1990) Nucleic acid probes for *Campylobacter*, in *Gene Probes for Bacteria* (eds A.J.L. Macario, E.C. de Macario), Academic Press, New York, pp. 255–93.
180. Wetherall, B.L., McDonald, P.J., Johnson, A.M. (1988) Detection of *Campylobacter pylori* DNA by hybridization with non-radioactive probes in comparison with a ^{32}P-labelled probe. *J. Med. Microbiol.*, **26**, 257–63.
181. Wheeler, A.P., Gregg, C.R. (1986) Campylobacter bacteraemia, cholecystitis, and the acquired immunodeficiency syndrome (letter). *Ann. Intern. Med.*, **105**, 804.

181a. Widjojoatmodjo, M.N., Fluit, A.C., Torensma, R., *et al.* (1991) Evaluation of the magnetic immuno PCR assay for rapid detection of *Salmonella. Eur. J. Clin. Microbiol. Infect. Dis.*, **10**, 935–8.

182. Winter, S.M., Bernard, E.M., Gold, J.W.M., *et al.* (1985) Humoral response to disseminated infection by *Mycobacterium avium – Mycobacterium intracellulare* in acquired immunodeficiency syndrome and hairy cell leukemia. *J. Infect. Dis.*, **151**, 523–7.
183. Wolfe, M.S., Armstrong, D., Louria, D.B., *et al.* (1971) Salmonellosis in patients with neoplastic disease, a review of 100 episodes at Memorial Cancer Center over a 13 year period. *Arch. Intern. Med.*, **128**, 546–54.
184. Wu, J.Y., Newton, S., Judd, A., *et al.* (1989) Expression of immunogenic epitopes of hepatitis B surface antigen with hybrid flagellin proteins by a vaccine strain of *Salmonella. Proc. Natl. Acad. Sci. U.S.A.*, **86**, 4726–30.
185. Yamamoto, T., Tamura, Y., Yokota, T. (1985) Enteroadhesion fimbriae and enterotoxin of *Escherichia coli*: genetic transfer to a streptomycin-resistant mutant of the galE oral-route live-vaccine *Salmonella typhi* Ty21a. *Infect. Immun.*, **50**, 925–8.
186. Yang, D.M., Fairweather, N., Button, L.L., *et al.* (1990) Oral *Salmonella typhimurium* (AroA). vaccine expressing a major leishmanial surface protein (gp63) preferentially induces T helper cells and protective immunity against leishmaniasis. *J. Immunol.*, **145**, 2281–5.

4

Mycobacteria

NICOLA T. RALPHS, GRAHAM J. BOULNOIS and PETER W. ANDREW

4.1 BACKGROUND

The mycobacterial diseases, tuberculosis and leprosy, are amongst the oldest infectious diseases described and are still a major cause of death and suffering worldwide. *Mycobacterium tuberculosis*, the causative agent of tuberculosis leads to 3 million deaths per year with an estimated 8 million new cases reported annually [58], and *Mycobacterium leprae*, the agent of leprosy, still physically and socially afflicts millions in the third world [72]. Approximately 50 species make up the genus *Mycobacterium* [100] and most are environmental saprophytes. However, many of the non-tuberculous, 'atypical' mycobacteria are potentially pathogenic and have been demonstrated to cause opportunistic infection (often clinically indistinguishable from tuberculosis) in immunocompromised patients [37]. In the industrialized nations of the west the virtual disappearance of leprosy and the dramatic decline of tuberculosis in this century has led to a certain degree of complacency. However, in recent years there has been a marked increase in the incidence of mycobacterial disease due in a large part to the epidemic of the acquired immunodeficiency syndrome (AIDS) [94].

Mycobacterial infections have been diagnosed by clinicians in a high percentage of AIDS sufferers. Organisms belonging to the *Mycobacterium avium-intracellulare* (MAI) complex predominate, accounting for up to 96% of all mycobacterial disease in AIDS in the USA [37]. Both *M. avium* and *M. intracellulare* have been implicated in such opportunistic

Molecular and Cell Biology of Opportunistic Infections in AIDS
Edited by Steven Myint and Alan Cann
Published in 1992 by Chapman & Hall, London ISBN 0 412 45330 4

infections, perhaps reflecting evidence that they represent different strains of the same species rather than distinct species in their own right [11]. MAI infections have been reported in 5% of AIDS patients in the UK and between 17–56% in the USA [45]. The incidence in AIDS patients declines with age and is equally distributed between the sexes [48]. Plasmid analysis of the strains suggest their origin to be environmental [69] with hospital hot water systems being one possible source via aerosolization [24, 101] and with tissue invasion probably occurring across the bronchial or intestinal mucosae [15].

M. tuberculosis is also commonly associated with AIDS; 12% of patients suffering from tuberculosis in the United States also have AIDS [12]. In the UK, there is an incidence rate of 11% in AIDS sufferers [86] and rates of between 2–20% have been reported in the USA [87]. Flora *et al.* [30] have suggested that the rate may well be even higher. In their studies, at least 9% of cases of tuberculosis in AIDS patients had not been diagnosed despite being subsequently found to be the main cause of mortality in these patients. It is thought that the likely incidence of tuberculosis in patients with HIV infection will depend on the overlap between the population infected with HIV and the population with previous tuberculous infection [29]. Therefore it is likely that in Africa cases of tuberculosis will become more and more prevalent due to the traditionally high incidence of tuberculosis in young adults, the age group most likely to be HIV-positive.

Infection with other opportunistic mycobacteria have been occasionally diagnosed in patients with HIV infection, albeit rarely. In a study of 207 patients with AIDS, Helbert *et al.* [45] identified one patient infected with *Mycobacterium ulcerans* and another with *Mycobacterium kansasii.* Other mycobacteria found to have caused disseminated disease in these patients have included *Mycobacterium chelonei*, *Mycobacterium gordonae*, *Mycobacterium xenopi* and *Mycobacterium fortuitum* [35].

Mycobacterial disease presents the clinician with a number of problems. The first is that of diagnosis; clinical features of infection may vary considerably and can include localized pulmonary disease, lymphadenitis, other localized non-pulmonary manifestations and disseminated disease. In AIDS patients the clinical features may resemble those of advanced tuberculosis with fevers, night sweats and cachexia but these are also evident in other opportunistic infections. Bacteriological diagnosis can present difficulties. At present less than 30% of mycobacterial infections can be detected by microscopy (Public Health Laboratory Leicester, personal communication). In a study of 15 AIDS patients, in whom the presence of tuberculosis was later established, none exhibited acid-fast bacilli on direct examination of broncho-alveolar lavage fluids [45]. In these instances a period of culture is needed, which can take up to 8 weeks before mycobacteria can be detected. Species identification, even of smear-positive samples, requires a similar culture time.

Increasing drug resistance is a second major problem, particularly with MAI infections. Helbert *et al.* [45] showed that 87–100% of MAI isolates studied proved resistant to capreomycin, ciprofloxacin, clofazamine, ethambutol, isoniazid, rifampicin and streptomycin. The requirement for extended, mixed chemotherapeutic regimens, often lasting many months, is also less than ideal. Lifelong treatment with isoniazid has been recommended for AIDS patients infected with *M. tuberculosis* [51].

It has been suggested that *M. tuberculosis* infection might actually hasten disease progression in HIV-infected individuals by activating CD4–positive lymphocytes and increasing virus replication [45]. It has also been proposed that exposure of HIV-infected individuals to virulent/persistent strains of mycobacteria leads first to a degree of reactivity followed by a persistent state of anergy. This could result in a long term period of immune tolerance which could interfere with the expression of cell immunity to any subsequent systemic infection caused by mycobacteria [14]. Likewise, by increasing immunosuppression in the HIV-infected patient, MAI infection could thereby help to drive the viral disease into its terminal stage. Similarly, it is likely that the AIDS-associated suppression of the cell-mediated immune response by the removal of CD4-positive lymphocytes and macrophages [28] may hasten the progression of a mycobacterial infection. Thus, AIDS patients often prove to be poorly responsive to MAI sensitins used in skin tests [15].

Attempts to improve diagnosis and treatment of mycobacterial disease have been hampered by our limited knowledge of the fundamental biology of the mycobacterial genus. There are fundamental gaps in our knowledge of mechanisms of pathogenicity or the host's response. Pathogenic mycobacteria are facultative intracellular parasites, able to survive and multiply within mononuclear phagocytes. This behaviour is obviously a critical factor in the pathogenesis of infection but many unanswered questions remain. For example, how does the microbe survive the antimicrobial activities of the normal macrophage? What is the nutritional environment within the phagosome containing the intracellular bacterium and how do mycobacteria become adapted to use the intracellular resources for their own multiplication? What is the real relationship between HIV and mycobacteria in AIDS patients?

Molecular biology potentially provides a route to solving these problems: a quick, accurate identification for diagnosis, rational design of antibiotics/drugs, and even possibly an improved, more effective and wider-ranging tuberculosis/mycobacterial vaccine. Until quite recently mycobacteria have not proved particularly amenable to genetic manipulations. Their extreme slow growth, thick cell walls and the lack of suitable cloning vectors have all inhibited progress. However, over the last few years something of a revolution has taken place in the field. Major progress has already been made and we now stand at the edge of a particularly exciting time.

4.2 DETECTION AND IDENTIFICATION OF MYCOBACTERIA

At present, microscopy of clinical samples stained for the presence of acid-fast bacilli, is still the most useful and rapid means of detection for mycobacterial infections. However, a minimum of 10^4 organisms/ml of sample is necessary for consistent detection [106]; less than 30% of mycobacterial infections are detected in the UK in this way. In the remainder, the samples need to be cultured for anything between 2–8 weeks before the presence of mycobacteria can be detected and, indeed, all samples need to be cultured for the specific biochemical tests required to effect identification. The concept that DNA probes can be used for the rapid detection and simultaneous identification of microbes is well established but their application to the clinical situation is still in the early stages of research. As we outlined earlier, the genetic manipulation of mycobacteria has proved difficult and a number of different approaches have been tried. Potential candidates for probes for mycobacteria have ranged from total genomic DNA, to random clones in plasmids or phages, and specifically designed or isolated DNA fragments [4].

Initial studies focused on using total chromosomal DNA as probes for *M. tuberculosis*-complex and *M. avium*-complex organisms [8, 83, 89]. However, the probes were not particularly sensitive; they could only match the detection limit of light microscopy. Additionally, the high degree of homology that exists between the genomes of individual members within the complexes [6] meant that intra-complex identification was not possible.

Cloned random fragments have been used to identify an MAI/*Mycobacterium paratuberculosis*-type mycobacterium in cultures grown from samples taken from Crohn's disease patients [65]. By probing Southern blots of *M. paratuberculosis*, *M. avium-intracellulare* and the Crohn's disease isolate with a cloned *Bam*H1 fragment from the same Crohn's isolate, these authors identified restriction fragment length polymorphisms (RFLP) between *M. paratuberculosis* and *M. avium-intracellulare* but none between *M. paratuberculosis* and the Crohn's disease isolate. Plasmid clones from a genomic library of *M. paratuberculosis* have been used to identify RFLPs in MAI [49]. This appears to demonstrate that highly conserved strains of MAI, which are distinct from the mycobacteria isolated from the stools of healthy subjects, infect AIDS patients [42].

We have isolated a random *M. tuberculosis* DNA restriction fragment which at low stringency (final wash in 0.1 × SSC at 42 °C for 15 min) binds to DNA from all mycobacteria but not to DNA from non-mycobacterial species. At high stringency (final wash in 0.1 × SSC at 65 °C for 10 min) this probe hybridizes only to *M. tuberculosis*- complex DNA [79].

Several different groups of workers have specifically selected DNA fragments for use as probes, often from published sequence data of mycobacterial genes. One such is from the gene coding for the 65 kDa antigen found in mycobacteria and thought to be related to the hsp60 heat-shock family [43]. Another is from the gene coding the MPB64 protein [85].

A different approach has been based on the detection of specific regions of mycobacterial rRNA [100] which makes an appealing target for detection assays as it is present in the cell in high (10^3–10^4) copy number and possesses regions characteristic for nearly every individual species. Gen-Probe Inc. (9880 Campus Point Drive, San Diego California 92121) produced three commercial products for the detection of mycobacteria based on the use of ^{125}I-labelled DNA homologous to rRNA: Rapid Diagnostic Systems for *M. tuberculosis*-complex, *M. avium*-complex and for *M. gordonnae*. They have been extensively evaluated (for example [23, 56]) and represent a considerable advance in allowing a 2-hour identification test after culture. They are, however, expensive, and the use of radionucleotides may limit their application to larger well-resourced diagnostic laboratories, and also present technical difficulties in separating bound and unbound probe. Indeed, the *M. tuberculosis* kit has been withdrawn temporarily from the market to await further evaluation.

The advent of the polymerase chain reaction (PCR) [84] which permits the amplification of specific DNA sequences and which can be used to increase a single copy of a given DNA sequence by 10^{12}-fold [62] has revolutionized detection techniques making non-radioactive detection methods feasible. Sensitivity is no longer a limitation. Indeed, the exquisite sensitivity can result in false positives due to amplification of contaminating DNA. Over the past year there has been a veritable flurry of papers reporting the application of PCR to a variety of different primer/probe systems. Specific primer/probe systems are now available for *M. tuberculosis*-complex, *M. tuberculosis*, *M. avium*-complex, *M. paratuberculosis*, *M. fortuitum* and *M. leprae* [43]. One of the first was described by Hance *et al.* [43] who amplified a 383 bp segment of the gene coding for the 65 kDa antigen which is present in all mycobacteria and then hybridized with species-specific oligonucleotide probes able to specifically identify the *M. tuberculosis*-complex, *M. avium*-complex and *M. fortuitum* related sequences. Using this system the same group were able to detect and identify *M. tuberculosis*-complex organisms directly in clinical specimens [10].

Other groups have based detection systems on the amplification of various repetitive sequences found in the mycobacterial chromosome. Eisenach *et al.* [27] amplified a 123 bp fragment known to be repeated several times in the *M. tuberculosis* chromosome, but were unable to distinguish between members of the *M. tuberculosis*-complex. DeWit *et al.* [22] amplified a different (336 bp) repetitive fragment in the *M. tuberculosis* chromosome which proved to be specific for *M. tuberculosis*. They were able to demonstrate *M. tuberculosis* DNA in all the culture-positive clinical samples they tested. Vary *et al.* [99] designed primers from the DNA sequence of IS900, the insertion sequence present in multiple copies in the genome of *M. paratuberculosis*, and were able to specifically amplify *M. paratuberculosis* DNA from the faeces of cattle with Johne's disease.

With a primer/probe system based on the amplification of a random *M. tuberculosis* DNA fragment [79], we have shown that we can detect less than 10 bacilli in sputum within 2 days of the sample being obtained (unpublished data); this is a vast improvement on the 10^4 bacilli required for microscopic detection. Shankar *et al.* [85] have reported that PCR amplification of a 240 bp region from the gene encoding the MPB64 protein is more sensitive than smear, culture and ELISA in detecting *M. tuberculosis*-complex organisms in the cerebral spinal fluid of patients with highly probable tuberculous meningitis. Several groups have tried to increase the sensitivity of single PCR-based detection systems by using a nested-primer gene amplification assay [76, 77]. In both instances an additional amplification procedure was carried out on the products of the initial amplification using a second set of primers internal to the original ones. By this method Pierre *et al.* [76] were able to detect 100% of culture-positive samples and 37% of culture-negative samples from patients with clinically diagnosed tuberculosis.

The usefulness of PCR has been expanded by the demonstration that suitable DNA for PCR can be extracted from a wide variety of clinical samples including human biopsies [103], paraffin-embedded tissues [105] and even plastic-embedded formalin-fixed tissues [41].

The recently described ligase chain reaction (LCR) [9] can allow the specific detection and identification of any mycobacterial species, because as little as one base change can be distinguished. There is, however, a danger of false negatives in such an identification system should a single base mutation occur within a particular strain. The development of PCR of RNA as opposed to DNA would also prove extremely useful to clinicians in allowing them to detect only live bacilli, since RNA is rapidly degraded in dead cells. The present PCR systems detect dead bacilli as well as live organisms, thus making the monitoring of drug therapy via PCR impractical.

4.3 EPIDEMIOLOGY

The typing of mycobacterial strains in AIDS patients is of increasing importance in allowing the identification of sources of infection. One of the earliest methods, agglutination serotyping, was reported by Griffith [40], who was able to differentiate avian tubercle bacilli from human and bovine by agglutination with absorbed antisera. It has subsequently been shown that *M. avium* and *M. intracellulare* can be divided into 23 serotypes, as reviewed by Good and Bean [36], with certain serotypes (1, 4 and 8) appearing to predominate in AIDS patients [107]. Other related approaches to typing have been explored: immunodiffusion serotyping [92] and sensitin typing [68]. However, serotyping does not differentiate between species or strains within the *M. tuberculosis*-complex and is difficult because of clumping of the bacilli [37]. There were also problems associated with the attempt of Crawford and colleagues to develop a bacteriophage typing scheme for MAI and *M. scrofulaceum* strains [21] and for *M. tuberculosis* strains [19]. Because of the limited number of phage types available, many strains were untypable, thus making the system not widely practical.

These problems have led to the investigation of genetic methods to type strains and again the approaches have been diverse. Crawford and Bates [18] first identified the presence of plasmids in MAI strains and since then much work has been done to determine whether they have any epidemiological significance. Crawford and Bates [20] detected plasmids in all the MAI strains they isolated from American AIDS patients but in only a third of MAI strains from other patients. Genetically related plasmids have been identified in environmental isolates from aerosols (75–100%), water (14–21%) and the soil (5%) [55, 69]. There appear to be at least three different groups of small (15–25 kb) plasmid types present in MAI [55] as determined by using two DNA probes (pVT2 and pLR7). MAI isolates from American AIDS patients always contain plasmids related to pLR7 and usually contain those related to pVT2. Other small plasmids, which are unrelated to pVT2 and pLR7, are only found in non-AIDS and environmental isolates. In a similar UK study [46] there was no significant difference in the rate of plasmid carriage in MAI strains isolated from patients with AIDS and from non-AIDS patients. However, they did observe a higher rate of plasmid carriage in a small group of strains from American AIDS patients. The significance of this is unclear and it is not known whether there is a real difference between AIDS-associated MAI strains from the US and from the UK. MAI plasmids have been shown indirectly to affect virulence as determined by measurements of catalase activity [75], mortality rates in beige mice and the release of oxygen metabolites by murine macrophages [31]. Jucker and Falkinham [55] raised the interesting possibility that each plasmid type might encode different virulence factors thus providing an explanation for their different distribution amongst MAI strains from AIDS and non-AIDS patients. As yet no plasmids have been identified in *M. tuberculosis*.

By far the best candidate method for epidemiological typing is restriction fragment length polymorphism which has been commonly exploited to detect genetic variability [34]. Patel *et al.* [74] used restriction endonuclease digestion patterns to distinguish two different strains of *M. tuberculosis*. Collins and DeLisle [16] used the same method to show that different strains of *M. bovis* were present in New Zealand cattle. McFadden *et al.* [66] were able to differentiate between *M. avium*, *M. intracellulare* and *M. paratuberculosis* by means of their RFLP banding patterns by probing BamH1-cut genomic DNA with a DNA probe derived from mycobacteria isolated from a Crohn's patient. The use of a range of DNA probes has allowed Hampson and colleagues [42] to demonstrate three RFLP banding patterns in MAI-complex strains. They found that 73% of the MAI-complex strains infecting AIDS patients exhibited the same pattern, whereas only 30% of non-AIDS-derived MAI were of this type.

As described previously, several groups of workers have identified DNA elements present in multiple copies per genome and which show a polymorphic banding pattern amongst the isolates investigated when used to probe Southern blots. In *M. tuberculosis*-complex several elements have been independently described [26, 47, 82, 96, 110]. Two, IS986 [47] and IS6110 [96] have been sequenced and identified as insertion-like sequences (possibly related to IS3) and appear to be virtually identical to one another. Independent isolates of *M. tuberculosis*-complex have been shown to possess a very high degree of polymorphism of restriction fragments containing IS986 DNA [47], but that the sequence was stable in strains passaged in guinea pigs over a number of generations. Using this probe system, the same group demonstrated that different *M. tuberculosis* isolates from a recent outbreak of tuberculosis showed an identical banding pattern. Thus, this element is potentially a very useful and suitable tool for the epidemiology of tuberculosis. Although no insertion sequences have yet been discovered in the MAI strains found in AIDS sufferers, repetitive sequences have been identified in *M. paratuberculosis* [17, 38], an organism very closely related to MAI. It is not unreasonable, therefore, to assume that similar elements may yet be found in MAI.

4.4 MANIPULATING MYCOBACTERIA

Molecular biology was seen to offer a fresh approach to gaining an understanding of the mechanisms of pathogenicity in mycobacteria both in AIDS and non-AIDS patients where traditional biochemical and immunological methods have had only partial success. Initially progress in the field lagged behind that for other micro-organisms because mycobacteria seemed to be refractory to many of the usual molecular genetic techniques. However, many of these problems have now been overcome and the application of recombinant DNA technology combined with the use of monoclonal antibodies to mycobacterial antigens has resulted in a plethora of studies investigating mycobacterial genes and the proteins they encode.

4.4.1 Transformation of mycobacteria

In the mid 1980s a number of groups cloned the genomes of various mycobacteria, including *M. tuberculosis*, in *E. coli* using both plasmid and phage expression vectors [13, 53, 97] and showed that at least some mycobacterial genes could be expressed in *E. coli* from their own promoters, though less efficiently than in their natural host. Many genes, though, were not expressed in *E. coli* and it was thought that this might reflect the fact that mycobacteria have a far higher G + C content (66–71%) [7] in their DNA than *E. coli* (48–52%) [73] and therefore that they must possess different transcriptional control systems. To overcome this mycobacterial genes were cloned in vectors under the control of *E. coli* expression signals [97]. However, one limitation of this method is that any product requiring a complex biosynthetic pathway might not be formed. Because of this disadvantage, attention was then directed to using *Streptomyces* species as cloning hosts on the grounds that they too possess DNA with a high G + C content (72%) and because cloning systems for this genus were already well developed [57]. However, an efficient gene transfer system for mycobacteria would be a great deal more useful; it would allow foreign genes to be transferred to and expressed in mycobacteria. Virulence genes and potential chemotherapeutic targets could be identified, characterized and tested by using non-pathogenic species such as BCG and *M. smegmatis* as cloning hosts [96]. In addition, antigens requiring complex biosynthesis could be expressed.

A variety of approaches have been tried. The use of chromosomally encoded markers as selective agents has been widely investigated [39], but did not perform consistently well. In 1964 Tokunaga and Sellers demonstrated the introduction of lytic mycobacteriophage DNA into *M. smegmatis* and attention then turned to these phages as potential candidates for the development of recombinant DNA vectors. Initially transfection efficiencies were very low (100 pfu/μg DNA) but by transfection of mycobacterial spheroplasts, the efficiency was increased to 10^4 pfu/μg DNA [52].

4.4.2 Phage vector systems

The first real success in producing a vector system capable of introducing foreign DNA into mycobacteria was achieved by Jacobs *et al.* in 1987 [52]. They constructed a recombinant shuttle plasmid (phAE1) in which an *E. coli* cosmid vector (containing an *E. coli* origin of replication, unique restriction sites and selectable markers) was inserted into a non-essential region of a mycobacteriophage DNA molecule. This plasmid was able to replicate as a plasmid in *E. coli* and as a phage in *M. smegmatis* and BCG. As such it did not require the generation of protoplasts, as phages are able to efficiently infect and replicate in both fast and slow growing mycobacteria. A problem with phAE1 was that it caused the lysis of its mycobacterial host. Intact phage particles could be isolated from plaques and then re-introduced into *M. smegmatis* or DNA isolated from phAE1 and re-introduced into *E. coli*, where they replicated as plasmids but it could not be stably maintained in the mycobacterial cell. To solve this, Snapper and colleagues [90] developed similar vectors for *M. smegmatis* and BCG from the temperate mycobacteriophage L1. They demonstrated that these L1-shuttle plasmids, like the parent phage, formed stable lysogens integrating into the mycobacterial chromosome and preventing superinfection. By cloning into these vectors, they demonstrated that heterologous genes could be expressed in mycobacteria and simultaneously identified the first selectable marker for the mycobacteria, the gene for kanamycin resistance (aminoglycoside phosphotransferase) from Tn903. Thus, foreign genes could now be cloned and manipulated in *E. coli*, introduced into mycobacterial protoplasts by transformation where they could replicate as a phage allowing the isolation of recombinant mycobacteriophage particles which could then be introduced into a variety of strains of mycobacteria by infection This work is reviewed by Jacobs *et al.* [54].

4.4.3 Plasmid vector systems

The limitations of phage-based genetic systems have, with other bacteria, been addressed by the use of plasmid-based systems. Plasmids present the possibilities of simplifying DNA manipulations and controlling copy number. However, they require an efficient means of introducing the plasmid into mycobacterial cells and the identification of a plasmid containing both a functional mycobacterial origin of replication and a selectable marker.

Zainuddin *et al.* [111] reported that an *E. coli* plasmid, pIJ666, could be introduced into *M. smegmatis* protoplasts, where it could replicate independent of any mycobacterial plasmid sequences. However, other groups have been unable to repeat this work [64] and most of the hybrid plasmids developed for mycobacterial transformation are derived from a *M. fortuitum* plasmid, pAL5000. This 4.8 kb plasmid has been sequenced [81] and a number of open reading frames determined. Snapper *et al.* [90] constructed a hybrid plasmid library in *E. coli* by randomly inserting the *E. coli* plasmid pIJ666, which contains a kanamycin resistance gene, into pAL5000. They tried unsuccessfully to transform the library into *M. smegmatis* protoplasts but succeeded in transforming it into whole *M. smegmatis* and BCG by electroporation. This resulted in the identification of the plasmid vector pYUB12. The same library was used to identify other hybrid shuttle plasmids which contained origins of replications that functioned in both *E. coli* and mycobacteria [64]. Although reasonable transformation efficiencies could be obtained using the plasmids in BCG ($>10^3$/μg DNA), the *M. smegmatis* transformation efficiencies were very low (<10/μg DNA). This was unfortunate as *M. smegmatis* would be an ideal host for molecular genetic analysis of mycobacterial genes as it is fast growing and avirulent. To rectify this problem an *M. smegmatis* mutant (mc^2155) with a high efficiency of transformation (ept mutant) was isolated which could be transformed by electroporation at efficiencies of more than 10^5/μg DNA [91].

This mutant has aided the isolation of other pAL5000 hybrid shuttle vectors. Ranes *et al.* [80] identified functional domains of pAL5000 and constructed a 'mini' *Mycobacterium–E. coli* shuttle vector (pRR3) by inserting the regions required for plasmid replication into a pUC19 backbone containing the gene for kanamycin resistance from Tn 903. Using a related shuttle plasmid (pMY10) Lazraq *et al.* [59] demonstrated for the first time that plasmids could be transferred conjugationally between Gram-negative (*E. coli*) and acid-fast (*M. smegmatis*) bacteria. They later used the same vector to transform an avirulent strain of *M. tuberculosis*, H37Ra [60] – the first time that transformation of *M. tuberculosis* had been demonstrated.

Lee *et al.* [61] have also reported the transformation of *M. tuberculosis* (in addition to *M. smegmatis* and BCG) with a series of site-specific integration-proficient plasmid vectors constructed from the mycobacteriophage L5. This type of vector system was improved upon by the very recent construction of two *E. coli*–BCG expression vectors, pMV261 (extrachromosomal) and pMV361 (integrative) by Stover *et al.* [93]. They both consist of an expression cassette (part of the regulatory region of a

BCG heat shock gene, ten unique cloning sites and a transcriptional terminator), a kanamycin resistance gene and an *E. coli* origin of replication. They differ in that pMV261 contains a mycobacterial origin of replication whereas pMV361 contains the attD and int genes of mycobacteriophage L5. They are likely to prove very useful in expressing foreign genes in mycobacteria as they have very high transformation efficiencies in BCG (>10^5/μg DNA), are reasonably small in size (4.5 kb) and possess multiple usable cloning sites. To date they have been used to express a variety of foreign antigens in BCG namely β-galactosidase, fragment C (toxC) of tetanus toxin and HIV-1 gp120 antigen [93].

A completely different approach to designing a *Mycobacterium–E. coli* shuttle vector was taken by Radford and Hodgson [78]. They constructed two vectors (pEP2 and pEP3) using the replication region from a *Corynebacterium* plasmid, which appears to be able to control plasmid replication in both Gram-negative and Gram-positive bacteria. The vectors carry resistance to either kanamycin or hygromycin, are small, contain multiple cloning sites and replicate to high copy number in various mycobacterial species and *E. coli*.

4.4.4 Other techniques

Other molecular genetic tools which have proved to be of particular value in other organisms are homologous recombination and transposon mutagenesis. Homologous recombination is the ability to replace genomic sequences with homologous DNA manipulated *in vitro*. Husson *et al.* [49] constructed a shuttle vector which could replicate autonomously at a high copy number in *E. coli* but which had to integrate into homologous DNA for survival in *M. smegmatis*. The vector contained an *E. coli* origin of replication, kanamycin and ampicillin resistance markers, *M. smegmatis* pyrF (which acts as a nutritional marker to allow both positive and negative genetic selections) and a unique cloning site for the cloning of foreign genes. Transformation of *M. smegmatis* with this vector containing the *M. leprae* 65 kDa stress protein gene resulted in the integration of the vector DNA into the genomic pyrF locus, by a double homologous recombination event, resulting in the expression of the stress protein.

Transposon mutagenesis remains a difficult procedure. Jacobs (unpublished data) achieved transposition of a mycobacterial gene library in *E. coli* with Tn5. The transposed genes were then put back into M. *smegmatis* where homologous recombination delivered the transposon to the genome. Putative mutants were then screened for alteration of phenotype. The various insertion sequence-like elements that have been described recently in *M. paratuberculosis* [17, 38] and *M. tuberculosis* [96] may have potential as mutagenic agents but this is as yet unproven. Martin *et al.* [67] isolated, cloned and sequenced a transposon-like element responsible for resistance to sulphonamides in *M. fortuitum.* They demonstrated that it could transpose to different sites in the chromosome of *M. smegmatis* but that transposition was a rare event (<0.5/μg DNA). Thus, we are some distance yet from generating libraries of transposon mutants and, for example, screening for elimination of virulence phenotypes, and hence from identifying genes for virulence determinants.

4.5 SUCCESSFUL APPLICATIONS OF MOLECULAR BIOLOGY IN MYCOBACTERIA

4.5.1 Protein antigens

One of the most successful applications of molecular biology has been in the characterization of a large number of mycobacterial antigens [109]. Mycobacterial antigens are of immense interest to researchers as they represent individual components which interact directly with the immune system. As such, a molecular biological dissection of them may lead to a greater understanding of the molecular mechanisms involved in the host's immune response to mycobacteria. This in turn could potentially lead to the identification of candidate molecules for use as subunit vaccines and as reagents for immunodiagnostic tests.

Genes coding for more than 20 of the principal protein antigens have now been cloned, sequenced and expressed, as reviewed by Young *et al.* [109]. Many have been shown to have interesting functional and regulatory properties. They can be loosely divided into two groups, the heat-shock proteins and the secreted proteins, although there is some degree of overlap between them.

4.5.2 Heat-shock proteins

Heat-shock proteins are induced in a cell or organism in response to a variety of environmental stress stimuli including heat, alcohol, oxidative radicals and toxic metal ions. They are highly conserved throughout nature and are essential for normal cell growth under all conditions. The predominant heat-shock proteins in eukaryotes belong to a family of proteins which have molecular weights of approximately 70 kDa called the hsp70 gene family. Bacteria have been shown to express a single hsp70 protein (dnaK gene product in *E. coli*) and a similar product has been identified in both *M. tuberculosis* and *M. leprae,* the 71 kDa and 70 kDa proteins respectively [32, 108]. A 65 kDa antigen has been isolated from *M. tuberculosis* and *M. leprae* [88, 108] and this is thought to belong to a second important class of stress proteins (hsp60) which includes the *E. coli* groEL gene product and proteins found in mitochondria and chloroplasts. In addition to these, an *M. leprae* 18 kDa antigen has been identified which appears to have homology with low molecular weight heat-shock proteins found in eukaryotic cells [71]. A 12 kDa antigen of *M. tuberculosis* [8] which may be homologous to the *E. coli* groES gene product which interacts with groEL has also been identified.

In *E. coli* heat-shock proteins have been shown to play a part in control of the processes of DNA replication, cell division and protein secretion. The high degree of sequence identity that exists between *E. coli* and mycobacterial heat-shock proteins [109], suggests that they might function similarly in mycobacteria. However complementation studies in the two genera have been unsuccessful, suggesting that the variable regions in the proteins are important in mediating species-specific interactions with their particular protein substrates.

4.5.3 Secreted proteins

Secreted proteins may be amongst the important mediators of interaction between the mycobacteria and their hosts and their analysis may thus provide information about the *in vivo* 'lifestyle' and behaviour of the pathogen. A variety of secreted proteins have been identified, cloned and sequenced. The secreted 23/28 kDa antigen of *M. tuberculosis*/*M. leprae* [95, 112] was subsequently identified as the enzyme superoxide dismutase. This is predominantly cytoplasmic although it is also found in large amounts in culture filtrates. Superoxide dismutase probably protects the bacterium from the toxic effects of superoxide radicals which are generated during the respiratory burst that accompanies phagocytosis of the organism by macrophages [63, 68]. Other antigens identified which, although strictly found within or associated with the cell membrane, are also found in culture filtrate are the 35 kDa [109] and 31 kDa [33] antigens of *M. leprae* and MAI respectively, the 38 kDa antigen of *M. tuberculosis*, which sequence has homology to phoS, a phosphate transport gene in *E. coli* [3], and the *M. tuberculosis* 19 kDa antigen [5] which also may play a role in nutrient transport. Major exported antigens include the BCG85 complex and the MPB64 and MPB70 antigens [109], which are synthesized with a signal peptide and are actively secreted across the membrane. Certain of the proteins encoded by the BCG85 gene complex have been shown to have fibronectin-binding activity [1], a property sometimes associated with pathogenicity as it promotes bacterial adherence to mucosal surfaces and may indirectly facilitate entry into host cells.

4.5.4 Recombinant vaccines

Very recently Stover and colleagues [93] have reported the construction and use of extrachromosomal and integrative expression vectors carrying the regulatory sequences for major BCG heat-shock proteins to allow the expression of foreign antigens (including the HIV-1 gp120 antigen) in BCG. These recombinant BCG strains were shown to be able to elicit long-lasting humoral and cellular immune responses to foreign antigens in mice. The authors foresaw a future in which these recombinant vaccine vehicles are used to allow the simultaneous expression of multiple protective antigens of different pathogens. Another approach to designing a recombinant BCG vaccine was described by Aldovini and Young [2]. They fused DNA fragments encoding HIV-1 gag, pol and env precursors to the mycobacterial hsp70 promoter and translation start-site and then inserted these into the mycobacterial autonomous replicating plasmid vector pYUB12. The recombinant plasmids were introduced into BCG by electroporation and the resulting BCG recombinant used to immunize mice. Both humoral and cell-mediated immune responses were detected. BCG offers a number of advantages over other systems as a candidate for such a recombinant vaccine vehicle. It is the most widely used vaccine in the world and has a low incidence of serious complications: it can be given at or after birth, it can be given as a single inoculum, it is a potent adjuvant, it is a live vaccine and thus continues to produce antigen by replicating *in vivo*, it is relatively heat stable and it is cheap to produce. There are, however, some quite major obstacles which will first need to be overcome. There has been considerable controversy over the protective efficacy of the BCG vaccine, with trials showing protection varying from 80% in the UK to 0% in South India [104]. The reason for failure is not entirely clear but one suggestion is that environmental mycobacteria had effectively immunized the Indian population prior to administration of the BCG vaccine. A second major problem is caused by the very success of the BCG vaccination programme itself. More than 2500 million people have been given the vaccine since 1948. Therefore there now exists a sizeable proportion of the world's population in whom subsequent immunization with a recombinant BCG vaccine would prove ineffective and, indeed, possibly dangerous (due to hypersensitivity reactions). Because of these doubts it is likely that the widespread use of a recombinant BCG vaccine, although a most promising candidate, still remains some way off.

4.6 THE FUTURE

Mycobacterial molecular biology has made great strides over the past few years. Potentially useful primer/probe systems have been developed for rapid detection and identification, and for epidemiological studies. A wide variety of *E. coli*-mycobacterial shuttle vectors and expression vectors have been created and transformation, homologous recombination and transposon mutagenesis strategies identified.

The variable effectiveness of the present BCG vaccine has stimulated the pursuit of identifying specific mycobacterial antigens which could be used in a new subunit vaccine. Such specific antigens could also form the basis of improved immunodiagnostic skin tests. The present test relies upon a delayed hypersensitivity reaction in the skin to soluble antigens known as tuberculins [25], and is unable to unequivocally differentiate between either exposure to different mycobacteria such as *M. tuberculosis* or *M. leprae* or between protective immunity and disease. The identification of such antigens will require a further understanding of the molecular basis of mycobacterial infection and disease.

Another priority will be to use a molecular biological approach to improve our understanding of how current drugs function and thus lead to the rational design of improved versions. In addition, it should be possible to identify other mycobacterial enzymes (e.g. RNA polymerase and DNA topoisomerases and gyrases) as potential targets for drug action. The genes for these enzymes could be cloned and expressed to produce sufficient quantities of the enzyme in a form useful for drug screening.

Thanks to the advances in mycobacterial molecular biology we are now in a position to understand more clearly the mycobacterial pathogens which inflict mankind in general and AIDS patients in particular, and to obtain information that might be useful in the fight against these diseases.

REFERENCES

1. Abou-Zeid, C., Ratcliff, T.L., Wiker, H.G., Harboe, M., Bennedsen, J., Rook, G.A.W. (1988) Characterisation of fibronectin-binding antigens released by *Mycobacterium tuberculosis* and *Mycobacterium bovis* BCG. *Infect. Immun.*, **56**, 3046–51.
2. Aldovini, A., Young, R.A. (1991) Humoral and cell-mediated immune responses to live recombinant BCG-HIV vaccines. *Nature*, **351**, 479–82.
3. Andersen, A.B., Hansen, E.B. (1989) Structure and mapping of antigenic domains of protein antigen b, a 38 000-molecular-weight protein of *Mycobacterium tuberculosis. Infect. Immun.*, 57, 2481–8.
4. Andrew, P.W., Boulnois, G.J. (1990) Early days in the use of DNA probes for *Mycobacterium tuberculosis* and *Mycobacterium avium* complex, in *Gene Probes for Bacteria*, (eds A.J.L. Macario and E.C. de Macario), Academic Press, San Diego, pp. 179–203.
5. Ashbridge, K.R., Booth, R.J., Watson, J.D., Lathigra, R.B. (1989) Nucleotide sequence of the 19 kDa antigen gene for Mycobacterium tuberculosis. *Nucleic Acids Res.*, **17**, 1249.

6. Baess, I. (1979) Deoxyribonucleic acid relatedness among species of slowly-growing mycobacteria. *Acta Pathol. Microbiol. Scand. (Sect. B)*, **87**, 221–6.
7. Baess, I., Mansa, B. (1978) Determination of genome size and base ratio of deoxyribonucleic acid from mycobacteria. *Acta Pathol. Microbiol. Scand. (Sect. B)*, **86**, 309–12.
8. Baird, P.N., Hall, L.M.C., Coates, A.R.M. (1988) Cloning and sequence analysis of the 10 kDa antigen gene of Mycobacterium tuberculosis. *Nucleic Acids Res.*, **16**, 9047.
9. Barany, F. (1991) Genetic disease detection and DNA amplification using cloned thermostable ligase. *Proc. Natl. Acad. Sci. U.S.A.*, **88**, 189–93.
10. Brissons-Noel, A., Gicquel, B., Lecossier, D., Levy-Freubault, V., Nassif, X., Hance, A. (1989) Rapid diagnosis of tuberculosis by amplification of mycobacterial DNA in clinical samples. *Lancet*, **i**, 1069–71.
11. Boddinghaus, B., Rogall, T., Flohr, T., Blocker, H., Bottger, E.C. (1990) Detection and identification of mycobacteria by amplification of rRNA. *J. Clin. Microbiol.*, **28**, 1751–9.
12. Chaisson, R.E., Schecter, G.F., Theuer, C.P., Rutherford, G.W., Echenberg, D.F., Hopewell, P.C. (1987) Tuberculosis in patients with the acquired immunodeficiency syndrome. *Am. Rev. Respir. Dis.*, **136**, 570–4.
13. Clark-Curtiss, J.E., Jacobs, W.R., Docherty, M.A., Ritchie, L.R., Curtis R. III. (1985) Molecular analysis of DNA and construction of genomic libraries of *Mycobacterium leprae*. *J. Bacteriol.*, **161**, 1093–102.
14. Collins, F.M. (1986) *Mycobacterium avium*-complex infections and development of the acquired immunodeficiency syndrome, casual opportunist or causal cofactor? *Int. J. Leprosy*, **54**, 458–74.
15. Collins, F.M. (1989) Mycobacterial disease, immunosuppression, and acquired immunodeficiency syndrome. *Clin. Microbiol. Rev.*, **2**, 360–77.
16. Collins, D.M., DeLisle, G.W. (1985) DNA restriction endonuclease analysis of *Mycobacterium bovis* and other members of the tuberculosis complex. *J. Clin. Microbiol.*, **21**, 562–4.
17. Collins, D.M., Gabric, D.M., DeLisle, G.W. (1989) Identification of a repetitive DNA sequence specific to *Mycobacterium paratuberculosis*. *FEMS Microbios. Letts.*, **60**, 175–8.
18. Crawford, J.T., Bates, J.H. (1979) Isolation of plasmids from mycobacteria. *Infect. Immun.*, **24**, 979–81.
19. Crawford, J.T., Bates, J.H. (1984) Phage typing of mycobacteria, in *The Mycobacteria: a Source Book, Part A*, (eds G.P. Kubica, L.G. Wayne), Marcel Dekker, New York.
20. Crawford, J.T., Bates, J.H. (1986) Analysis of plasmids in *Mycobacterium avium-intracellulare* isolates from persons with acquired immunodeficiency syndrome. *Am. Rev. Respir. Dis.*, **134**, 659–61.
21. Crawford, J.T., Fitzhugh, J.K., Bates, J.H. (1981) Phage typing of the *Mycobacterium avium-intracellulare-scrofulaceum* complex. A study of strains of diverse geographic and host origin. *Am. Rev. Respir. Dis.*, **124**, 559–67.

22. DeWit, D., Steyn, L., Shoemaker, J., Sogin, M. (1990) Direct detection of *Mycobacterium tuberculosis* in clinical specimens by DNA amplification. *J. Clin. Microbiol.*, **28**, 2437–41.
23. Drake, T.A., Hindler, J.A., Berlin, G., Bruckner, D. (1987) Rapid identification of *Mycobacterium avium* complex in culture using DNA probes. *J. Clin. Microbiol.*, **25**, 1442–5.
24. Du Moulin, G.C., Stottmeier, K.D., Pelletier, P.A., Tsang, A.Y., Hedley-Whyte, J. (1988) Concentration of *Mycobacterium avium* by hospital hot water systems. *JAMA*, **260**, 1599–601.
25. Editorial. (1984) New tuberculins. *Lancet*, **i**, 199–200.
26. Eisenach, K.D. Crawford, J.T., Bates, J.H. (1988) Repetitive DNA sequences as probes for *Mycobacterium tuberculosis. J. Clin. Microbiol.*, **26**, 2240–5.
27. Eisenach, K.D., Cave, M.D., Bates, J.H., Crawford, J.T. (1990) Polymerase chain reaction amplification of a repetitive DNA sequence specific for *Mycobacterium tuberculosis. J. Inf. Dis.*, **161**, 977–81.
28. Fauci, A.S. (1988) The human immunodeficiency virus, infectivity and mechanisms of pathogenesis. *Science*, **239**, 617–22.
29. Festenstein, F., Grange, J.M. (1991) Tuberculosis and the acquired immune deficiency syndrome. *J. Appl. Bact.*, **71**, 19–30.
30. Flora, G.S., Modilevsky, T., Antoniskis, D., Barnes, P.F. (1990) Undiagnosed tuberculosis in patients with human immunodeficiency virus infection. *Chest*, **98**, 1056–9.
31. Gangadharam, P.R.J., Perumal, V.K., Crawford, J.T., Bates, J.H. (1988) Association of plasmids and virulence of *Mycobacterium avium* complex. *Am. Rev. Respir. Dis.*, **137**, 212–4.
32. Garsia, R.J., Hellquist, L., Booth, R.J., *et al.* (1989) Homology of the 70 kilodalton antigens from *Mycobacterium leprae* and *Mycobacterium bovis3* with the *Mycobacterium tuberculosis* 71 kilodalton antigen and the conserved heat shock protein 70 of eukaryotes. *Infect. Immun.*, **57**, 204–12.
33. George, K.L., Falkinham, J.O. (1989) Identification of cytoplasmic membrane protein antigens of *Mycobacterium avium*, *Mycobacterium intracellulare* and *Mycobacterium scrofulaceum. Can. J. Microbiol.*, **35**, 529–34.
34. Gill, P., Jeffreys, A.J., Werrett, D.J. (1985) Forensic application of DNA 'fingerprint'. *Nature*, **318**, 577–9.
35. Good, R.C. (1986) Opportunistic pathogens in the genus *Mycobacterium. Ann. Rev. Microbiol.*, **39**, 347–69.
36. Good, R.C., Beam, R.E. (1984) Seroagglutination, in *The Mycobacteria: a Source Book, Part A*, (eds G.P. Kubica, L.G. Wayne), Marcel Dekker, New York, pp. 105–122.
37. Grange, J.M., Yates, M.D., Boughton, E. (1990) The avian tubercle bacillus and its relatives. *J. Appl. Bacteriol.*, **88**, 411–31.
38. Green, E.P., Tizzard, M.L.V., Moss, M.T., *et al.* (1989) Sequence and characteristics of IS900, an insertion element identified in a human Crohn's disease isolate of *Mycobacterium paratuberculosis. Nucleic Acids Res.*, **17**, 9063–72.
39. Greenberg, J., Woodley, C.L. (1984) in *The Mycobacteria: a Source Book,*

Part A, (eds G.P. Kubica, L.G. Wayne), Marcel Dekker, New York, pp. 629–39.

40. Griffith, A.S. (1925) The serological classification of mammalian and avian tubercle bacilli. *Tubercle*, 6, 417–36.
41. Grunewald, K., Feichtinger, H., Weyrer, K., Dietze, O., Lyons, J. (1990) DNA isolated from plastic-embedded tissue is suitable for PCR. *Nucleic Acids Res.*, **18**, 615.
42. Hampson, S.J., Portaels, F., Thompson, J., *et al.* (1989) DNA probes demonstrate a single highly conserved strain of *Mycobacterium avium* infecting AIDS patients. *Lancet*, **i**, 65–8.
43. Hance, A.J., Grandchamp, B., Levy-Freubault, V., *et al.* (1989) Detection and identification of mycobacteria by amplification of mycobacterial DNA. *Mol. Microbiol.*, **3**, 843–9.
44. Hartskeerl, R., deWit, M.Y., Klatzer, P. (1989) Polymerase chain reaction for the detection of *Mycobacterium leprae. J. Gen. Microbiol.*, **28**, 1751–9.
45. Helbert, M., Robinson, D., Buchanan, D., *et al.* (1990) Mycobacterial infection in patients infected with the human immunodeficiency virus. *Thorax*, **45**, 45–8.
46. Hellyer, T.J., Brown, I.N., Dale, J.W., Easmon, C.S.F. (1991) Plasmid analysis of *Mycobacterium avium-intracellulare* (MAI) isolated in the United Kingdom from patients with and without AIDS. *J. Med. Microbiol.*, **34**, 225–31.
47. Hermans, P.W.M., van Soolingen, D., Dale, J.W., *et al.* (1990) Insertion element IS986 from *Mycobacterium tuberculosis*: a useful tool for diagnosis and epidemiology of TB. *J. Clin. Microbiol.*, **28**, 2051–8.
48. Horsburg, C.R., Selik, R.M. (1989) The epidemiology of disseminated mycobacterial infection in the acquired immuno-deficiency syndrome (AIDS). *Am. Rev. Respir. Dis.*, **139**, 4–7.
49. Husson, R.N., Bradford, J.E., Young, R.A. (1990) Gene replacement and expression of foreign DNA in mycobacteria. *J. Bacteriol.*, **172**, 519–24.
50. Imaeda, T., Kaminski, Z. (1987) Identification of *Mycobacterium avium-intracellulare* complex including those isolated from AIDS patients by characterisation of their genomic DNAs. *Am. Soc. Microbiol. Abstr. Ann. Meet.*, Abstr. U-33.
51. Iseman, M.D. (1987) Is standard chemotherapy adequate in tuberculous patients infected with HIV? *Am. Rev. Respir. Dis.*, **136**, 1326.
52. Jacobs, W.R., Tuckman, M., Bloom, B.R. (1987) Introduction of foreign DNA into mycobacteria using a shuttle plasmid. *Nature*, **327**, 532–5.
53. Jacobs, W.R., Docherty, M.A., Curtis, R. III, Clark-Curtiss, J.E. (1986) Expression of *Mycobacterium leprae* genes from a *Streptococcus mutans* promoter in *Escherichia coli* K-12. *Proc. Natl. Acad. Sci. U.S.A.*, **83**, 1926–30.
54. Jacobs, W.R., Snapper, S.B., Tuckman, M., Bloom, B.R. (1989) Mycobacteriophage vector systems. *Rev. Infect. Dis.*, **11**, S404–10.
55. Jucker, M.T., Falkinham, J.O. (1990) Epidemiology of infection by nontuberculous mycobacteria. IX. Evidence for two DNA homology groups

among small plasmids in *Mycobacterium avium*, *Mycobacterium intracellulare* and *Mycobacterium scrofulaceum*. *Ann. Rev. Respir. Dis.*, **142**, 858–62.

56. Kiehn, T.E., Edwards, F.F. (1987) Rapid identification using a specific DNA probe of *Mycobacterium avium* complex from patients with acquired immunodeficiency syndrome. *J. Clin. Microbiol.*, **25**, 1551–2.
57. Kieser, T., Moss, M.T., Dale, J.W., Hopwood, D.A. (1986) Cloning and expression of *Mycobacterium bovis* BCG DNA in *Streptomyces lividans*. *J. Bacteriol.*, **168**, 72–80.
58. Kochi, A. (1991) The global tuberculosis situation and the new control strategy of the World Health Organization. *Tubercle*, **72**, 1–6.
59. Lazraq, R., Clavel-Seres, S., David, H.L., Roulland-Dussoix, D.(1990) Conjugative transfer of a shuttle plasmid from *E. coli* to *M. smegmatis*. *FEMS Microbios. Letts.*, **69**, 135–8.
60. Lazraq, R., Clavel-Seres, S., David, H.L. (1991) Transformation of distinct mycobacterial species by shuttle vectors derived from the *Mycobacterium fortuitum* pAL5000 plasmid. *Curr. Microbiol.*, **22**, 9–13.
61. Lee, M.H., Pascopella, L., Jacobs, W.R., Hatfull, G.F. (1991) Site-specific integration of mycobacteriophage L5: Integration-proficient vectors for *Mycobacterium smegmatis*, *Mycobacterium tuberculosis* and bacille Calmette-Guerin. *Proc. Natl. Acad. Sci. U.S.A.*, **88**, 3111–5.
62. Li, H., Gyllensten, U.B., Cui, X., Saiki, R.K., Erlich, H.A., Arnheim, N. (1988) Amplification and analysis of DNA sequences in single human sperm and diploid cells. *Nature*, **335**, 414–7.
63. Lowrie, D.B.. Andrew, P.W. (1988) Macrophage antimycobacterial mechanisms. *Br. Med. Bull.*, **44**, 624–34.
64. Lugosi, L., Jacobs, W.R., Bloom, B.R. (1989) Genetic transformation of BCG. *Tubercle*, **70**, 159–70.
65. McFadden, J.J., Butcher, P.D., Chiodini, R.J., Hermon-Taylor, J. (1987) Determination of genome size and DNA homology between an unclassified *Mycobacterium* species, isolated from patients with Crohn's disease and other mycobacteria. *J. Gen. Microbiol.*, **133**, 211–4.
66. McFadden, J.J., Butcher, P.D., Thompson, J., Chiodini, R.J., Hermon-Taylor, J. (1987) The use of DNA probes identifying restriction-fragment-length polymorphisms to examine the *Mycobacterium avium* complex. *Mol. Microbiol.*, **1**, 283–91.
67. Martin, C., Timm, J., Rauzier, J., Gomez-Lus, R., Davies, J. Gicquel, B. (1990) Transposition of an antibiotic resistance element in mycobacteria. *Nature*, **345**, 730–43.
68. Mayer, B.K., Falkinham, J.O. (1986) Superoxide dismutase activity of *Mycobacterium avium*, *Mycobacterium intracellulare* and *Mycobacterium scrofulaceum*. *Infect. Immun.*, **53**, 631–635.
69. Meissner, G., Anz, W. (1977) Sources of *Mycobacterium avium* complex infection resulting in human disease. *Am. Rev. Respir. Dis.*, **116**, 1057–64.
70. Meissner, P.S., Falkinham, J.O. (1986) Plasmid DNA profiles as epidemiological markers for clinical and environmental isolates of *Mycobacterium*

avium, *Mycobacterium intracellulare* and *Mycobacterium scrofulaceum*. *J. Infect. Dis.*, **153**, 325–31.

71. Nerland, A.H., Mustafa, A.S., Sweetser, D., Godal, T., Young, R.A. (1988) A protein antigen of *Mycobacterium leprae* is related to a family of small heat shock proteins. *J. Bacteriol.*, **170**, 5919–21.
72. Norland, S.K. (1985) in *Leprosy*, (ed. R.C. Hastings), Churchill Livingstone, Edinburgh, pp. 15–30.
73. Orskov, F. (1984) *Genus Escherichia*, in Bergey's Manual of Systematic Bacteriology, Vol. 1, (eds N.R. Krieg, J.G. Holt), Williams and Wilkins, London, pp. 420–3
74. Patel, R., Kach, J.T., Mounts, P. (1986) Isolation and restriction endonuclease analysis of mycobacterial DNA. *J. Gen. Microbiol.*, **132**, 541–51.
75. Pethel, M.L., Falkinham, J.O. (1989) Plasmid-influenced changes in *Mycobacterium avium* catalase activity. *Infect. Immun.*, *57*, 1714–8.
76. Pierre, C., Lecossier, D., Boussougant, Y., *et al.* (1991) Use of reamplification protocol improves sensitivity of detection of *Mycobacterium tuberculosis* in clinical samples by amplification of DNA. *J. Clin. Microbiol.*, **29**, 712–7.
77. Plikaytis, B.B., Gelber, R.H., Shinnick, T.M. (1990) Rapid and sensitive detection of *Mycobacterium leprae* using a nested-primer gene amplification assay. *J. Clin. Microbiol.*, **28**, 1913–7.
78. Radford, A.J., Hodgson, A.L. (1991) Construction and characterisation of a *Mycobacterium–Escherichia coli* shuttle vector. *Plasmid*, **25**, 49–153.
79. Ralphs, N.T., Garrett, S., Morse, R., Cookson, J.B., Andrew, P.W., Boulnois, G.J. (1991) A DNA primer/probe system for the rapid and sensitive detection of *Mycobacterium tuberculosis*-complex pathogens. *J. Appl. Bacteriol.*, 70, 221–6.
80. Ranes, M.G., Rauzier, J., Lagranderie, M., Gheorghiu, M., Gicquel, B. (1990) Functional analysis of pAL5000, a plasmid from *Mycobacterium fortuitum*: construction of a mini-Mycobacterium-*E. coli* shuttle vector. *J. Bacteriol.*, **172**, 2793–7.
81. Rauzier, J., Moniz Pereira, J., Gicquel Sanzey, B. (1988) Complete nucleotide sequence of pAL5000, a plasmid from *Mycobacterium fortuitum*. *Gene*, **71**, 315–21.
82. Reddi, P.P., Talwar, G.P., Khandekar, P.S. (1988) Repetitive DNA sequence of *Mycobacterium tuberculosis*, analysis of different hybridisation patterns with other mycobacteria. *Int. J. Leprosy*, **56**, 592–8.
83. Roberts, M.C., McMillan, C., Coyle, M.B. (1987) Whole chromosomal DNA probes for rapid identification of *Mycobacterium tuberculosis* and *Mycobacterium avium* complex. *J. Clin. Microbiol.*, **25**, 1239–43.
84. Saiki, R., Gelfand, G.H., Stoffel, S., *et al.* (1988) Primer-directed enzymatic amplification of DNA with a thermostable DNA polymerase. *Science*, **239**, 487–97.
85. Shankar, P., Manjunath, N., Mohan, K.K., *et al.* (1991) Rapid diagnosis of tuberculous meningitis by polymerase chain reaction. *Lancet*, **337**, 5–7.
86. Shaw, R.J., Rowland-Hill, C.A., Parkin, J., *et al.* (1986) Pulmonary complications of AIDS, a review of 90 English cases. *Thorax*, **41**, 730.

87. Shecter, G., Rutherford, G., Echenberg, D. (1986) Tuberculosis in AIDS patients in San Francisco, 1982–1985. *Am. Rev. Respir. Dis.*, **133** (Suppl), A184.
88. Shinnick, T.M., Vodkin, M.H., Williams, J.L. (1988) The *Mycobacterium tuberculosis* 65 kilodalton antigen is a heatshock protein which corresponds to common antigen and to the *Escherichia coli* GroEL protein. *Infect. Immun.*, **56**, 446–51.
89. Shoemaker, S.A., Fisher, J.H., Scoggin, C.H. (1985) Techniques of DNA hybridisation detect small numbers of mycobacteria with no cross-hybridisation with non-mycobacterial respiratory organisms. *Am. Rev. Respir. Dis.*, **131**, 760–3.
90. Snapper, S.B., Lugosi, L., Jekkel, A., *et al.* (1988) Lysogeny and transformation in mycobacteria: Stable expression of foreign genes. *Proc. Natl. Acad. Sci. U.S.A.*, **85**, 6987–91.
91. Snapper, S.B., Melton, R.E., Kieser, T., Jacobs, W.R. (1990) Isolation and characterisation of efficient plasmid transformation mutants of *Mycobacterium smegmatis*. *Mol. Microbiol.*, **4**, 1911–9.
92. Stanford, J.L., Grange, J.M. (1974) The meaning and structure of species as applied to mycobacteria. *Tubercle*, **55**, 143–52.
93. Stover, C.K., de la Cruz, V.F., Fuerst, *et al.* (1991) New use of BCG for recombinant vaccines. *Nature*, **351**, 456–60.
94. Styblo, K. (1990) The global aspects of tuberculosis and HIV infection. *Bull Int. Union Tuberc. Lung Dis.*, **65**, 28–32.
95. Thangaraj, H., Lamb, F., Davies, E.O., Jenner, P.J., Jeyakumar, L.H., Colston, M.J. (1990) Identification, sequencing and expression of *Mycobacterium leprae* superoxide dismutase, a major antigen. *Infect. Immun.*, **58**, 1937–42.
96. Thierry, D., Cave, M.D., Eisenach, K.D., *et al.* (1990) IS6110, an IS-like element of *Mycobacterium tuberculosis* complex. *Nucleic Acids Res.*, **18**, 188.
97. Thole, J.E.R., Dauwerse, H.G., Das, P.K., Groothuis, D.G., Schouls, L.M., van Embden, J.D.A. (1985) Cloning of *Mycobacterium bovis* BCG DNA and expression of antigens in *Escherichia coli*. *Infect. Immun.*, **50**, 800–6.
98. Tokunaga, T., Sellers, M.I. (1964) Infection of *Mycobacterium smegmatis* with D29 phage DNA. *J. Exp. Med.*, **119**, 139–49.
99. Vary, P.H., Andersen, P.R., Green, E., Hermon-Taylor, J., McFadden. (1990) Use of highly specific DNA probes and the polymerase chain reaction to detect *Mycobacterium paratuberculosis* in Johne's Disease. *J. Clin. Microbiol.*, **28**, 933–7.
100. Wayne, L.G., Kubica, G.P. (1986) Genus *Mycobacteria*, in *Bergey's Manual of Systematic Bacteriology*, Vol 2, (eds P.H.A. Sneath, N.J. Mair, M.E. Sharpe, J.G. Holt), Williams and Wilkins, London, pp. 1436–57.
101. Wendt, S.L., George, K.L., Parker, B.C., Gruft, H. Falkinham, J.O. (1980) Epidemiology of infection by nontuberculous mycobacteria. III. Isolation of potentially pathogenic mycobacteria from aerosis. *Am. Rev. Respir. Dis.*, **122**, 259–63.

102. Wilson, K.H., Blitchington, R.B., Greene, R.C. (1990) Amplification of bacterial 16S ribosomal RNA with polymerase chain reaction. *J. Clin. Microbiol.*, **28**, 1942–6.
103. Woods, S.A., Cole, S.T. (1989) A rapid method for the detection of potentially viable *Mycobacterium leprae* in human biopsies: a novel application of PCR. *FEMS Microbios. Letts.*, **65**, 305–10.
104. World Health Organization (1980) *BCG vaccination policies. Technical Report Series, no. 652, WHO*, Geneva.
105. Wright, E.S., Manos, M.M. (1990) Sample preparation from paraffin-embedded tissues, in *PCR Protocols, a Guide to Methods and Applications*, (eds M.A. Innis, D.H. Gelfand, J.J. Sninsky, T.J. White), Academic Press, London, pp. 153–8.
106. Yeager, H., Lacy, J., Smith, L.R., Lemaistre, C.A. (1964) Quantitative studies of mycobacterial populations in sputeum and saliva. *Am. Rev. Respir. Dis.*, **95**, 998–1004.
107. Young, L.S., Inderleid, C.B., Berlin, O.G., Gottlieb, M.S. (1986) Mycobacterial infections in AIDS patients with an emphasis on the *Mycobacterium avium* complex. *Rev. Infect. Dis.*, **8**, 1024–33.
108. Young, D., Lathigra, R., Hendrix, R., Sweetser, D., Young, R.A. (1988) Stress proteins are immune targets in leprosy and tuberculosis. *Proc. Natl. Acad. Sci. U.S.A.*, **85**, 4267–70.
109. Young, D., Garbe, T., Lathigra, R., Abou-Zeid, C. (1990) Protein antigens: structure, function and regulation, in *Molecular Biology of the Mycobacteria*, (ed. J.J. McFadden), Surrey University Press, Surrey.
110. Zainuddin, Z.F., Dale, J.W. (1989) Polymorphic repetitive DNA sequences in *Mycobacterium tuberculosis* detected with a gene probe from a *Mycobacterium fortuitum* plasmid. *J. Gen. Microbiol.*, **135**, 2347–55.
111. Zainuddin, Z.F., Kunze, Z.M., Dale, J.W. (1989) Transformation of *Mycobacterium smegmatis* with *Escherichia coli* plasmids carrying a selectable resistance marker. *Mol. Microbiol.*, **3**, 29–34.
112. Zhang, Y., Lathigra, R., Garbe, T., Catty, D., Young, D. (1991) Cloning and characterisation of the superoxide dismutase gene from *Mycobacterium tuberculosis*. *Mol. Microbiol.*, **5**, 381–91.

5

Entamoeba histolytica

WILLIAM M. SPICE, JORGE A. CRUZ-REYES and JOHN P. ACKERS

5.1 INTRODUCTION

Entamoeba histolytica is the cause of high levels of morbidity and mortality throughout the developing world, having been ranked as the third most important parasitic agent of disease after malaria and schistosomiasis. In 1981, it was estimated that up to 500 million people are infected annually [265]. The majority of infections are asymptomatic and apparently non-pathogenic. In most of these cases, spontaneous eradication of the parasite is thought to occur within several months [28, 153]. Some 10%, however, go on to develop invasive amoebiasis [201] in which symptoms range from mild and transient diarrhoea to fulminating colitis characterized by the passage of bloody stools. In some cases dissemination via the bloodstream may result in the production of extra-intestinal abscesses in other organs, principally the liver. Invasive amoebiasis is held to account for around 75 000 deaths per year [79].

5.1.1 Life-cycle and diagnosis

E. histolytica has a life-cycle consisting of two stages, the transmissive cyst and the vegetative trophozoite. Cysts are ingested orally from contaminated food or water, and excyst in the large bowel. Colonization of the mucosal surface by the trophozoites may subsequently lead to tissue invasion and disease; the cycle is completed by encystation and excretion of cysts. Diagnosis of amoebic infection is mainly carried out by light

Molecular and Cell Biology of Opportunistic Infections in AIDS
Edited by Steven Myint and Alan Cann
Published in 1992 by Chapman & Hall, London ISBN 0 412 45330 4

microscopic examination of faecal smears or formalin–ether concentrates of stool specimens [80]. The quadrinucleate cysts of *E. histolytica* distinguish this species from other lumen-dwelling amoebae, although there is no morphological basis for discriminating between pathogenic and non-pathogenic forms, other than by the detection of haematophagous trophozoites in some stool specimens once pathogenesis is already well under way. Both luminal and systemic amoebiasis are commonly treated with metronidazole or, in the former, with diloxanide furoate [187, 201].

5.2 PATHOGENICITY OF AMOEBIC INFECTION

An issue of fundamental importance in the study of amoebiasis is whether amoebae isolated from symptomatic and asymptomatic infections represent genotypically distinct groups, or two potentially interconvertible phenotypes of the same organism [1, 143, 222]. While recent investigations of the molecular biology of this organism (reviewed in section 5.2) strongly suggest separate genetic identities for these two groups, reports that at least some non-pathogenic isolates are capable of converting to the pathogenic form *in vitro* merit further detailed examination [145, 148].

The eventual resolution of this question will not only increase our basic understanding of amoebiasis, but will also be of practical benefit to the clinical management of infected individuals. Thus, if it were shown that carriers of non-pathogenic amoebae have no future risk of disease, treatment in the majority (90%) of cases would be rendered unnecessary, and would permit substantial reductions in drug expenditure [8, 87]. Conversely, if it were shown that a harmless commensal could become invasive as a result of unknown changes in the luminal environment, pre-emptive treatment of asymptomatic cystpassers might be considered more appropriate [143, 222].

5.2.1 Historical perspectives

Controversy has surrounded the question of pathogenicity in *E. histolytica* since the early 20th century. The trophozoite was recognized in two forms, a small, luminal amoeba termed 'minuta' which gave rise to cysts [61] and a larger, haematophagous form associated with tissue pathology. In the 'unicist' hypothesis advanced by Dobell [56] and labelled 'Promethean' by Elsdon-Dew [62], all infections with *E. histolytica* were thought to result in damage to the gut. To account for asymptomatic cases, the rate of host repair was assumed to be sufficient to suppress any outward manifestations of disease. The advent of *in vitro* culture [22] and the observation that the amoebae could be grown in a medium containing bacteria and starch, contributed towards Dobell's later recognition of the commensal nature of some infections [57], although the original idea continued to be enthusiastically endorsed by some workers until as recently as 1970 [43, 44, 47, 65].

The opposing view, expressed by Kuenen and Swellengrebel [108] and by Mathis and Mercier [133], was that the 'minuta' amoeba represented a commensal stage which could, under appropriate conditions, produce the pathogenic tissue form. Only the former was considered to be capable of giving rise to cysts, while the latter was placed outside the normal cycle of transmission and infection.

In 1925, Brumpt, recognizing the wide geographical variations in the proportion of infected individuals who would subsequently develop invasive disease, proposed the existence of two morphologically indistinguishable species which he termed *E. dysenteriae* and *E. dispar*, to represent pathogenic and non-pathogenic *E. histolytica* respectively [30]. This became known as the 'dualistic' hypothesis, but failed to gain wide acceptance owing to the lack of any practical basis for differentiation between the two amoebae.

More recently, the 'pluralist' hypothesis propounded by Hoare [82], Neal [155], Elsdon-Dew [62, 63] and others has attributed regional variations in the levels of disease to the existence of a range of strains all belonging to the single species *E. histolytica*, but varying in pathogenicity from the asymptomatic commensal to the highly virulent. This approach bears similarities to the original concept of Kuenen and Swellengrebel [108].

5.2.2 Recent developments

Little further progress was made in this debate until Martinez-Palomo *et al.* reported in 1973 that only pathogenic amoebae were agglutinable with low concentrations of concanavalin A [130], thereby providing the first concrete evidence in support of Brumpt's notion of two species. Subsequently, Sargeaunt and Williams [227, 228] showed that amoebae isolated from symptomatic and asymptomatic infections produced specific electrophoretic banding patterns using two isoenzymes, hexokinase and phosphoglucomutase. This observation was demonstrated with 6000 isolates collected worldwide [224], and the patterns were found to be stable over long periods in culture [229].

Over the last 10 years a range of other cellular, biochemical and immunological properties have been used to differentiate pathogenic from non-pathogenic *E. histolytica*, although until relatively recently, few have shed any light on the basis of pathogenicity in this parasite. The growing consensus that *E. histolytica* comprises two genetically discrete groups of organisms was challenged in the late 1980s by Mirelman *et al.* [145, 148], with a report that cloned isolates defined as non-pathogenic by isozyme analysis could undergo an apparent conversion to the pathogenic phenotype. One of the distinguishing features of pathogenic amoebae is their ability to grow in the absence of bacteria, and it was while attempting to derive an axenic, non-pathogenic cell line that the switch in phenotype was observed. The procedure was based on the antibiotic-mediated removal from xenic culture of associated faecal flora, combined with the addition of gamma-irradiated bacteria as a substitute food source. The change was accompanied by the acquisition of pathogen-related activities such as cytotoxicity against Chinese hamster ovary (CHO) cell monolayers and the production of hepatic lesions in hamsters. Similar 'conversions' have now been reported from other laboratories [11] (Diamond, personal communication), although the event has not yet been demonstrated *in vivo* [205]. Such efforts are hampered by the lack of any suitable animal model for amoebic colitis.

The renewed uncertainty surrounding the relationship between pathogenic and nonpathogenic *E. histolytica* has generated a substantial amount of research effort over the past two years aimed at resolving this question through the application of recombinant DNA techniques. As a result, not only is it becoming increasingly clear that the two forms of infection should indeed be attributed to separate organisms, but an overall picture of the molecular basis of pathogenicity has also begun to emerge. These developments will be reviewed in the following sections.

5.3 GENETIC ORGANIZATION

E. histolytica has traditionally been regarded as a primitive, anaerobic organism, lacking mitochondria and many other organelles normally associated with higher eukaryotic cells [122, 131]. There is no extra-nuclear DNA, and the absence of evidence for sex in *E. histolytica* has precluded conventional genetic analysis, although recent *in vitro* studies have suggested that genetic exchange may occur [21, 225, 238].

5.3.1 Nuclear division and organization

Nuclear division, which is apparently not always accompanied by cell division [142], occurs by a pleuromitotic process [161], in which the nuclear membrane remains intact and microtubules form 'V'-shaped arrays rather than a conventional spindle. Four stages in nuclear division have been recognized: interphase, early anaphase, late anaphase and telophase, but a metaphase is apparently absent [161]. A nuclear bud appears during prophase, which enlarges and finally separates to produce the daughter nucleus after approximately 120 minutes.

Electron microscopic studies of the nuclear distribution of nucleic acids in *E. histolytica* have been carried out on autoradiographed sections of pulse-labelled amoebae [4, 5, 7], and on sections stained with phosphotungstic acid [142]. A region of DNA containing peripheral chromatin, originally reported by Sharma *et al.* [234], has been identified as a site of extensive RNA synthesis. This is thought to represent a functional equivalent of the nucleolus [3, 142], a similar arrangement having been observed in *Amoeba proteus* [67]. A centrally located endosome, containing tightly packed DNA, may be involved in DNA condensation prior to division [3, 164]. Meza *et al.* [142] have speculated that the endosome could be brought into association with the peripheral chromatin during replication and transcription. RNA, possibly heteronuclear, and DNA are also distributed throughout the nucleus, the latter mainly during interphase.

5.3.2 Chromosomes

In common with many protozoa, the absence of a nuclear condensation stage during division has hindered analysis of the amoeba karyotype. In electron microscopic studies using colchicine-synchronized cultures, Solis *et al.* [239] have reported four or five electron dense bodies which could correspond to chromosomes, while in 1988 Orozco *et al.* [161], using the same system, observed 12–16 chromosome-like structures associated with the microtubular arrays.

The pulsed field gel electrophoresis (PFGE) technique of Schwartz and Cantor [232], by which large (100–2000 kb) DNA molecules can be resolved on agarose gels, has enabled studies on chromosome sizes and numbers in many protozoan parasites to be conducted [64, 90, 102, 220, 263], although the chromosomes of *E. histolytica* have not, for the most part, proved amenable to this procedure [142]. However, densitometric analysis of PFGE gels did reveal up to nine bands over a size range of 280–2000 kb, which were interpreted as comprising about 16 large and 31 small chromosomes [261]. These results, though preliminary, provide an important basis for further characterization of the *E. histolytica* karyotype.

5.3.3 Organization of chromatin

Organization of the peripheral chromatin has been shown to correspond broadly with that found in other cells, although certain differences have been observed. In higher eukaryotes, extended DNA fibres have a regular beaded appearance arising from the presence of nucleosomes, which are 10 nm diameter particles consisting of histones associated with DNA sequences of between 160 and 250 bp [106, 252]. Limited digestion with micrococcal nuclease produces a ladder of DNA fragments on agarose gels corresponding to multimers of the nucleosomal unit.

Electron microscopy of phosphotungstic acid stained chromatin spreads from *E. histolytica* revealed nucleosomes spaced at often irregular intervals [257]. Mild digestion yielded a smear on ethidium bromide stained agarose gels, and radioactive DNA probes for both actin and ribosomal genes also produced a smear on Southern blots. Only a non-transcribed probe from the extra-chromosomally situated ribosomal episome (section 5.2.7) gave a clear ladder, leading the authors to conclude that little of the DNA in amoeba chromatin is nucleosome protected.

The nucleosomal proteins extracted from *E. histolytica* chromatin differed substantially from typical histones in molecular weight. In addition, synthetic oligonucleotides based on consensus sequences for the H3 and H4 histones failed either to detect histone-encoding clones in a genomic library or to permit amplification by PCR of homologous sequences from total DNA, suggesting that histone genes in this parasite have diverged considerably from the highly conserved sequences of other eukaryotes [257].

5.3.4 Ribosomal RNA

Ribosomal RNA species in *E. histolytica* are known to be highly labile, due to the tight association of RNAse with ribonucleoprotein complexes [14, 183]. Using diethylpyrocarbonate to inhibit RNAse activity during the isolation procedure, Heebner and Albach [81] described rRNA species from the axenic strain HK-9 with sedimentation coefficients of 4S, 17S and 25S in ratios of 1 : 2 : 3. The 25S rRNA dissociates readily at 55 °C into products of 5.8S, 16S and 17S, indicating that the molecules are non-covalently bound *in vivo* [15, 23]. An additional 5S rRNA has been detected by denaturing polyacrylamide gel electrophoresis [6]. Ribosomes are commonly organized into helical arrays in the cytoplasm of both cysts and trophozoites, as observed by electron microscopy [60, 211].

5.3.5 Characterization of DNA

The characterization of nucleic acids in *E. histolytica* was facilitated by the development in 1968 of axenic culture methods [53], whereby amoebal DNA could be studied in isolation from the DNA of original bacterial associates. Caution should be exercised, however, in the extrapolation of information concerning gene expression under axenic conditions to amoebae living in the xenic environment of the gut lumen where pathogenesis is initiated, especially in the light of the known influence of certain bacteria in modifying amoebal virulence [144, 147, 179, 271].

DNA content, %GC and ploidy

Most studies of the DNA content of *E. histolytica* trophozoites have used the diphenylamine method of Burton [33], and reported values are subject to considerable variation. For example, DNA content for strain HK-9 ranged from 0.45–1.5 pg/cell [75, 113], while another widely used strain, NIH:200, had 0.4 pg/cell [207]. In contrast, the DNA content of strain HM2 was reported to fluctuate between 3.0–8.9 ± 0.48 pg/cell [113]. While such relatively high values require confirmation, it has been suggested that changes in ploidy [113], possibly through the production of multinucleate trophozoites, as noted both in our own laboratory and elsewhere [113, 142], could be contributory factors. Pathogenic *E. histolytica* isolated from lesions in experimental animals has been reported to contain four times more DNA/cell than the same strain grown in culture [124].

While interstrain variations in the GC content of DNA are also evident from a number of studies, more consistency has been found in comparisons of the same strain by different workers. Based on thermal denaturation measurements, values quoted for HK-9 range from 22.4–24% [74, 113, 207], compared with 25.4–27.3% for NIH:200 [75, 207] and 27% for strains HM2 and HM3 [113]. A surprisingly high figure of 38% was calculated for the widely used HMl strain by Perez-Mutul [168], although this was obtained by buoyant density measurement which appears to produce somewhat higher values than thermal denaturation [207].

On the basis of GC measurements, Gelderman *et al.* [75] estimated the size of the HK-9 haploid genome to be $10^{7.5}$ (3.16×10^7) base pairs, a figure which has been supported by Valdes *et al.* [261] whose calculations were founded on the numbers and sizes of chromosomes in strain HM-1 detected by PGF electrophoresis. Taking this value to be equivalent to 0.033 pg of DNA per haploid genome, Byers [34] used Gelderman's figure of 0.45 pg of DNA per HK-9 trophozoite to propose a ploidy level of 14n.

Genetic complexity and repetitive DNA

Little information is available concerning genetic complexity and gene number in *E. histolytica*. From the T_m data of Gelderman [75], Byers [34] has inferred that approximately 50% of the genome may be composed of repetitive sequences. Assuming a DNA content per cell of 0.4 pg, which is equivalent to approximately 4×10^8 base pairs, and an average gene size of 2000 bp, Torres-Guerrero *et al.* [257] have calculated a maximum number of 200 000 genes. However, the proposed ploidy level of 14n would reduce this figure to about 14 000 genes per haploid genome. Furthermore, if only 50% of the genome is composed of single copy sequences, this figure would be reduced to 7000, while if *E. histolytica* is to any extent typical of higher eukaryotes, in which less than 20% of single copy sequences are transcribed into mRNA [253, 257], the total number of genes could be as few as 1500.

Repetitive DNA in *E. histolytica* was first visualized by Cruz-Hernandez *et al.* [48], who reported the presence of bands in total DNA which had been digested with the restriction enzymes *Eco*RI and *Hin*DIII and electrophoresed on agarose gels. Subsequently, Bhattacharya *et al.* [17], using the same enzymes, conducted a comprehensive study of repetitive DNA from five *E. histolytica* strains and other *Entamoeba* species and detected characteristic banding patterns in every case. Southern blots hybridized with ribosomal probes from *Plasmodium berghei* produced signals corresponding to the prominent bands observed on gels, from which it was concluded that the ribosomal DNA accounts for a large proportion of repetitive DNA and at least 10% of total DNA in *E. histolytica*.

DNA purification

It has been noted by several authors that the DNA of *E. histolytica*, like ribosomal RNA, is difficult to purify intact and in sufficient quantities for useful analysis [3, 48, 59, 142, 168]. The trophozoites are slow growing and osmotically fragile [53], while co-purification of carbohydrate and potent nucleases during conventional phenol extraction presents further problems. As a result, many workers have adopted the preparation of nuclei as a preliminary step in DNA isolation [59, 84, 250].

5.3.6 Gene sequence analysis

Perhaps partly because of these technical factors, the first report of gene cloning in this parasite did not appear until 1987, when Edman *et al.* [59] and Huber *et al.*[84] independently described the sequence of actin. This is one of the most abundantly transcribed genes in *E. histolytica*, comprising 1% of clones in a representative cDNA library [84], while actin itself accounts for 10% of total cell protein [69]. Analysis of cDNA and genomic sub-clones revealed a number of interesting features, including the absence of introns, and an unusual codon bias involving a large number (83%) of silent 3rd position base changes to A or T. This has given rise to a discrepancy between amino acid and nucleotide sequence homologies with human cytoplasmic actin (89% compared with 69% respectively, by Edman *et al.* [59]). The length of the 3′ non-transcribed region was found to be variable in a number of clones and both typical and atypical polyadenylation signals were reported. The 5′ non-transcribed region was comparatively short (11 nucleotides) and was not variable in length [59]. These observations, taken together with the appearance of multiple bands on Southern blots of strain HM-1, suggest the existence of a multi-gene family of between four and seven members [59, 84].

Analysis by Huber *et al.* [85] of apoferredoxin, another abundantly transcribed *E. histolytica* gene, disclosed similarities with the actin genes, including the absence of intervening sequences, a short 5′ non-transcribed sequence and a strong 3rd position codon bias for A/T, characteristics also shared by a gene encoding a 125 kDa surface antigen [58] (section 3.2.1).

5.3.7 Characterization of ribosomal genes

Cloning of the highly abundant DNA species previously reported to contain ribosomal genes [17] led to the discovery by Huber *et al.* [86] and Bhattacharya *et al.* [18] that these sequences are carried on a 24.6 kb extra chromosomal DNA molecule present in about 200 copies per genome. A similar arrangement has been reported in the free living soil amoeba *Naegleria gruberi* [39] and for several other schizopyrenoid amoebae [40]. The episomal genes were mapped to two large inverted repeat regions of at least 5.2 kb, each of which represents a single ribosomal cistron. Inverted repeats have also been detected on linear ribosomal elements in other organisms including *Tetrahymena* and *Physarum polycephalum* [95, 264]. The palindromic arrangement of these genes is thought to guard against recombinational events [163]. Intervening regions of repeated, non-coding DNA were also detected in the *E. histolytica* episome both upstream and downstream of the coding regions, possibly serving a regulatory function during transcription [76, 206]. Evidence for the existence of at least one chromosomal copy of the ribosomal genes has been found by Bhattacharya *et al.* [18] and by Valdes *et al.* [261] from Southern blots of low concentration agarose gels and PFGE gels respectively.

An AT-rich origin of replication was identified in the episome downstream of the ribosomal inverted repeats which was able to confer stable ARS (autonomous replication sequence) activity on an ARS-deficient yeast vector, YIp5 [77]. The isolation of an amoebal ARS is an important step towards the development of a genetic transformation system for *E. histolytica*.

5.3.8 Molecular differentiation of pathogenic and non-pathogenic strains

Ribosomal sequences

The presence of homologous ribosomal elements in non-pathogenic *E. histolytica* was described by Garfinkel *et al.* [72]. A genomic clone, H6.6, spanning the entire length of the ribosomal cistron, produced distinct banding patterns when used to probe *Eco*RI-digested DNA from pathogenic and non-pathogenic isolates. Two were specific for the former and one for the latter, suggesting that pathogenic *E. histolytica* may consist of two or more genotypic subsets. Interestingly, when radioactively labelled H6.6 was used to probe Southern blots of the two strains reported previously by Mirelman *et al.* [145, 148] as having undergone phenotypic conversion, the banding patterns obtained were identical to those recognized for clinical pathogenic isolates.

Furthermore P145, another episomal clone derived from non-coding DNA downstream of the inverted repeats, was found to hybridize exclusively to DNA of pathogenic isolates under high stringency conditions after a short exposure. This was the first example of a DNA sequence specific to pathogenic *E. histolytica*, and was reported together with an analogous probe, B133, isolated from a genomic library of a non-pathogen and with exclusive specificity for this form [72].

A faint hybridization signal (approximately seven-fold weaker) was, however, obtained on long exposure when P145 was hybridized to non-pathogenic DNA, and *vice versa*. Although the 43.6% sequence homology between the two probes might have been sufficient to account for this observation, Garfinkel *et al.* [72] instead proposed that each form of *E. histolytica* harbours one or a few master copies of the corresponding sequences. The subsequent report by Mirelman *et al.* [146] that the pathogen-specific element P145 could be amplified by PCR from the DNA of the cloned, non-pathogenic strain SAW1734R, lent support to this contention, leading the authors to propose selective gene amplification as a possible mechanism for previously observed phenotypic conversions. Further to this result, sequence comparisons between P145 and the PCR product obtained from the non-pathogen would now prove useful in confirming the precise identity of the latter.

Genotypic divergence

In 1989, Tannich *et al.* [250] described the sequence of a clone isolated from an HM1 cDNA expression library using an antiserum pre-enriched for pathogen-specific antibodies. The clone, cEH-Pl, was then used to screen a non-pathogen cDNA library under low stringency conditions, from which a homologous clone, cEH-NPl, was purified. Sequencing of the two clones revealed a 12% divergence in the inferred amino acid sequence, and Southern blot hybridizations with a range of pathogenic and non-pathogenic isolates of *E. histolytica* produced banding patterns specific for each form.

A number of other recent reports concerning gene sequences from both pathogenic and non-pathogenic *E. histolytica* [58, 251] (sections 5.2.1 and 5.3.2) have supported the earlier proposal of Sargeaunt and co-workers that the two forms are genetically distinct. In addition, recent results from our laboratory have disclosed the existence of a pathogen-specific RFLP lying within the ribosomal gene cistron (45a); further work is in progress to identify the precise nature of this difference. Details of divergent sequences, where they may relate to the molecular basis of pathogenicity, will be discussed in the next section.

The growing evidence for genotypic divergence is such that it is becoming increasingly difficult to envisage a model of genetic organization that could also accommodate phenotypic conversion. An approach towards resolving this apparent conflict would be use PCR to establish whether gene sequences specific for each form could be amplified in their respective counterparts. The successful application of PCR by Tannich *et al.* [248] to amplify specific fragments of the gene described previously [250] from both forms of the organism suggests that it should also be possible to design primers capable of detecting any variant sequences present in low copy number.

5.4 MOLECULAR PATHOGENESIS

Pathogenesis in amoebiasis is produced by complex interactions involving the invasive processes of the parasite [186], the virulence of the infecting isolate [271], the physiological [54, 152] and immune status of the host [258], and bacterial associates in the gut lumen [177, 178, 180, 181]. While aspects of all of these determinants have been studied at the cellular level, molecular analysis has so far been largely confined to the mechanisms of host tissue invasion.

As its name implies, *E. histolytica* has the potential to express a wide range of destructive properties, enabling it to penetrate deep into the intestinal submucosa and to form necrotic lesions in solid organs [37, 243, 244]. The following stages in parasite invasion can be defined:

1. Binding of trophozoites to carbohydrate moieties on host epithelial cells (or secretions) via cell surface adhesins and receptors;
2. Detachment and cytolysis of target cells;
3. Phagocytosis [159, 208] and intracellular degradation of host cellular

components, including neutrophils [51, 68, 219] and erythrocytes [191, 259];
4. Translocation through the tissue by pseudopodial action [129, 235].

Investigations of the molecular basis of invasion have concentrated on the first two stages, adhesion and tissue destruction, and are consequently of greater interest in the present context. This section will also describe a number of recently isolated genes encoding novel cell surface molecules with potential involvement in pathogenesis, making particular reference, where possible, to comparisons between pathogenic and non-pathogenic *E. histolytica*.

5.4.1 Adhesion to host tissues

Several lectin-like activities have been detected in *E. histolytica*, indicating the presence of receptors with specificities for a number of different carbohydrate groups. Kobiler and Mirelman identified a 220 kDa membrane-bound lectin, with high affinity for *N*-acetyl glucosamine-containing molecules such as chitotriose and hyaluronic acid [105, 140]. Antibodies raised against the lectin partially inhibited binding of trophozoites to erythrocytes and mammalian cells and prevented ingestion of red blood cells [210]. Similar inhibitory effects were also reported using antibodies to an immuno-dominant 112 kDa surface adhesion [12]. Trophozoite adherence to target cells also depends on the maintenance of microfilament function, as demonstrated by the inhibition of amoebic adherence to CHO cells at low temperatures [141,190] and by cytochalasins B and D [191].

The Gal/GalNAc lectin

The most intensively studied receptor of *E. histolytica* is the 170 kDa galactose/*N*-acetyl galactosamine-inhibitable lectin first described by Ravdin and Guerrant [191] and proposed as a potential immunoprotective reagent due to its high immunogenicity [189] and its central role in pathogenesis [173–175, 186, 197]. Ravdin and Guerrant [191] demonstrated the specificity of this lectin for galactose and *N*-acetyl-galactosamine (GalNAc) residues by the ability of these sugars to inhibit binding of trophozoites to CHO cells *in vitro*; cytotoxicity was also blocked, although phagocytosis was reported to be unaffected [191, 193]. Similar results were obtained using fixed rat and human colonic mucosa [194]. In addition, galactose inhibited adherence to bacteria and erythrocytes [191] and to rat and human colonic mucins [35, 38].

Purified lectin was found to prevent amoebic adherence to CHO cells in a galactose inhibitable and competitive manner [175]. The nature of the carbohydrate receptors recognized by the lectin was confirmed with a range of CHO glycosylation-deficient mutants; binding to a cell line lacking O- and *N*-linked Gal and GalNAc residues was virtually abolished, while enhancement was observed in mutants with a higher terminal galactose content [200]. The 170 kDa lectin is linked by disulphide bonds to a 35 kDa subunit which is non-immunogenic and does not participate in sugar binding [170], although a possible identity with the fibronectin receptor reported by Talamas-Rohana and Meza [247] has been suggested [125]. Polyclonal and monoclonal antibodies raised against the 170 kDa lectin were effective in preventing adherence to CHO cells [170, 198], although in a later study, Petri *et al.* [176] reported that two out of a panel of six monoclonal antibodies (Mabs) increased binding; such activation could be due to changes in the conformation or state of aggregation of the molecule. This result was reflected in the finding that while individual sera from 25 of 44 patients with amoebic liver abscess inhibited trophozoite adherence to CHO cells, a further 16 sera increased binding.

The lectin epitopes recognized by the two enhancing Mabs were present on both pathogenic and non-pathogenic amoebae, whereas the remaining four were only found in pathogenic isolates; Mabs to three of these were inhibitory, while one had no effect [171]. Based on these antigenic differences, the authors were able to classify 50 isolates of *E. histolytica* as pathogenic or non-pathogenic by radioimmunoassay. The presence of a functional lectin in the non-pathogenic amoebae suggests that the molecule is important in colonization of the intestine in addition to its role in pathogenesis.

Beside its potential application in clinical diagnosis, immunization with the Gal/GalNAc lectin has also been shown to confer immunity in the gerbil liver abscess model. In a series of trials, Petri and Ravdin [174] reported complete protection in 67% of immunized animals. However, an effective vaccine would presumably exclude either of the antigenic determinants shown by Petri *et al.* [171] to enhance trophozoite adherence, thus ruling out the use of whole lectin as an immunogen.

Gal/GalNAc lectin genes

Gene sequences for the Gal/GalNAc lectin from pathogenic *E. histolytica* derived independently by Tannich *et al.* [249] and Mann *et al.* [125] are 12.5% divergent, suggesting the possible existence of two or more genes. This proposal is supported by variations observed in *N*-terminal amino acid sequences [126] and by the appearance of multiple bands on Southern blots of genomic DNA probed with a full length cDNA clone [249]. A single 4 kb message was detected by Northern blotting.

Both of the reported genes lack intervening sequences, and indicate a protein of 1270–1276 amino acids with a molecular weight of approximately 143 kDa. Glycosylation may account for the higher value of 170 kDa obtained from preparations of the purified lectin, since amoebae pre-incubated with the glycosylation inhibitor tunicamycin lost their ability to adhere to CHO cells [126]. The derived protein sequence predicts a short, conserved C-terminal cytoplasmic tail of approximately 40 residues. This region contains putative phosphorylation sites and may participate in signal transduction [125]. A short, adjacent hydrophobic domain is proposed as a transmembrane region. This is followed by a large, extracellular portion of more than 1200 residues, and is variously reported to contain nine [249] and sixteen [126] glycosylation sites. The initial two-thirds is cysteine-rich, possibly conferring resistance to proteolysis [126]. A short, hydrophobic signal peptide is located at the *N*-terminal. Significant homologies with other known genes were not found, and no sequences for the corresponding non-pathogenic lectin gene have yet been reported. Future definition of sequences encoding pathogen-specific epitopes can be expected to facilitate the development of recombinant vaccines.

5.4.2 Other surface proteins

In addition to the elucidation of adherence mechanisms in *E. histolytica*, interest in the identification of other cell surface properties with possible involvement in pathogenicity has led to a number of recent reports describing membrane proteins of as yet undisclosed function [20, 58, 240, 242, 254–256]. Those for which gene sequences have already been characterized are discussed below.

125 kDa immunodominant antigen

Edman *et al.* reported a highly immunoreactive 125 kDa antigen from lysates of both pathogenic and non-pathogenic *E. histolytica* [58]. A monoclonal antibody prepared against the partially purified antigen induced cap formation with live trophozoites, indicating a surface location for the molecule. A monospecific antibody eluted from phage lysates of a cDNA clone encoding part of the sequence for this antigen gave the same result.

Analysis of two genomic clones for this gene, named M17, revealed an open reading frame of 3345 bp, which was in broad agreement with the mRNA size of approximately 3000 bp obtained by Northern blotting. The 5′ flanking region of the gene was found to contain only 17 nucleotides, in common with the short non-transcribed 5′ sequences previously observed in both actin and ferredoxin genes [59, 84, 85]. Additional similarities were found in the presence of an unusual putative promoter sequence of TATTTAAA, in contrast to the more common TATAAATA motif, and a 5′-ATTCA-3′ sequence present at the transcriptional start site of both the M17 and actin genes. Southern blotting indicated that the gene was present as a single copy.

The inferred amino acid sequence of M17 revealed that part of it was almost identical to the clone reported previously by Tannich *et al.* [250]. Although both were initially isolated from the pathogenic strain HM-1, a 1% divergence in amino acid sequence was detected. Comparison with a homologous gene from a non-pathogen revealed a 12.1% divergence at the amino acid level, in good agreement with the finding of Tannich *et al.*, and a 10.3% difference in nucleotide sequence. The non-pathogens used in the two studies differed in amino acids by 1.3%.

The sequence reported by Edman *et al.* is rich in asparagine (8.2%), threonine (8.1%), serine (7.6%) and tyrosine (6.3%) residues, and contains 17 *N*-linked glycosylation sites. A hydrophobic, 35 residue *N*-terminal region may be membrane-located, while the polar C-terminal portion is possibly exposed. A stretch of 31 amino acids in this region showed close homologies with the beta chains of human fibronectin receptor and mouse integrin, although the functional significance of this association remains to be determined. No other sequence similarities were found.

Serine-rich protein

A gene encoding a serine-rich protein with an inferred molecular weight of 25 kDa was isolated by Stanley *et al.* [240] by differential screening of an HM-1 cDNA library with cDNA probes derived from HM-1 and the closely related Laredo amoeba. Many of the 52 serine residues, out of a total of 233, are present in tandemly-repeated motifs of 8 and 12 amino acids; the repeats occur between residues 82 and 194, and are preceded by a highly charged region. A hydrophobic domain is situated at each end of the protein, possibly indicating membrane insertion sites.

No homologies were found with other sequences at either the nucleotide or amino acid levels, although the protein bears a strong structural resemblance to the malarial circumsporozoite protein [101], with respect to the presence of multiple dodecapeptide repeats preceded by a charged region, and terminal hydrophobic domains.

An antiserum raised against the recombinant protein identified bands of 46 and 52 kDa on Western blots of *E. histolytica* membrane preparations, but the molecule was apparently absent from the soluble fraction. Post-translational modification was suggested to account for the approximately two-fold discrepancy between predicted and actual sizes of the protein. The antiserum reduced binding to CHO cells by 70%, implying a possible role in adhesion.

Cysteine-rich protein

Torian *et al.* have described a 29 kDa surface protein from axenic trophozoites of *E. histolytica* with an unusually high cysteine content [254]. A 700 bp cDNA clone encoding the molecule was selected from an expression library of strain H-302 : NIH using a rabbit antiserum pre-enriched for surface antigens. The clone hybridized with three other axenic strains but gave no signal with several other protozoan species. Non-pathogenic *E. histolytica* were not included.

Polyclonal and monoclonal antibodies to the fusion protein detected a 29 kDa band on Western blots of axenic amoeba whole lysates, suggesting that a length of sequence encoding some 40 *N*-terminal amino acids was missing from the original clone. Western blots of pathogenic isolates grown in polyxenic culture also produced a 29 kDa band with some, but not all of the Mabs tested, suggesting variation in the expression of certain epitopes, while two non-pathogens were weakly immunoreactive.

Immunoprecipitation of ^{125}I-surface-labelled amoebae using a mouse Mab gave a 29 kDa band on autoradiographs; localization of the antigen on the cell surface was also indicated by uniform staining of fixed trophozoites using indirect immunofluorescence. Analysis of the deduced amino acid sequence of the antigen showed that cysteine accounted for 7% of the 227 residues present, and that almost half of these occurred within the first 18 *N*-terminal residues, including two doublets and a triplet. No functional significance has yet been assigned to this configuration, although Torian *et al.* [254] report that a similar motif is present in a host cell binding protein of a porcine coronavirus and in a wool keratin protein. A six amino acid hydrophilic sequence was identified as a potential immunogenic epitope. It is not yet known whether the adjacent *N*-terminal portion of the sequence presumed missing from the cDNA clone contains a similarly high proportion of cysteine.

5.4.3 Histolysis

Target cell killing by *E. histolytica* trophozoites, observed *in vitro*, takes place within minutes of initial contact [192], and incubation of amoebae with a mammalian cell monolayer at a cell ratio of 1 : 1 can result in complete destruction in approximately 60 minutes [160]. The sequence of events accompanying epithelial cell detachment and lysis was described by Martinez-Palomo *et al.* [129] using time-lapse microcinematography and electron microscopy. Amoebae adhered strongly to an MDCK cell monolayer, shown by distortion of microvilli; this was followed by the opening of intercellular junctions, surface blebbing, disruption of the plasma membrane, cell detachment and phagocytosis. In addition, portions of host cell cytoplasm were directly pinched off and engulfed. Initial attachment was marked by a lowering of transepithelial resistance, as measured in an Ussing chamber [129].

This, and other studies on the physical interactions between amoebae and target tissues, has underlined the importance of amoebic motility and microfilament function to attachment [142], phagocytosis [159] and target cell displacement by pseudopodial action [235]. The structural approach has been complemented by a large number of investigations over the past 14 years aimed at establishing the biochemical basis of histolysis. While yielding considerable insights, such studies have needed to rely extensively on somewhat artificial *in vitro* systems.

Tissue dissolution by *E. histolytica* results from a combination of cytolethal, contact dependent mechanisms, and non-cytotoxic proteolytic enzymes which act on components of the extracellular matrix; the latter are secreted, with the exception of collagenase (section 5.3.2). There is so far little evidence to support the existence of secreted cytotoxins as previously suggested [26, 128, 169], although non-enzyme based enterotoxic activities capable of inducing fluid secretion in *in vivo* gut models have been reported [36, 136, 137].

5.4.4 Cytotoxic mechanisms

Calcium influx

The role of calcium in cytolysis was studied by Ravdin *et al.* [199], who reported that pre-treatment of CHO cells and human neutrophils with the slow Na–Ca ion channel inhibitor, verapamil, reduced amoebic killing but was ineffective when used to pre-treat trophozoites. Tetrodotoxin, a fast sodium channel inhibitor had no effect on either cell type, implying that the process was calcium dependent. This was later supported in experiments using the fluorescent calcium probe Fura-2 [195]. Thirty to three hundred seconds after initial contact with the trophozoite, a substantial and irreversible rise in target cell calcium concentration was followed by an increase in non-specific membrane permeability and cell death. No change in overall or local calcium levels were detected in the amoebae. The cytotoxic effect of calcium influx is widely known [195], and it is thus also likely to contribute to cell killing by trophozoites. Calcium influx was inhibited by galactose, suggesting a mediating role for the Gal/GalNAc lectin.

Phospholipase A

A phospholipase A enzyme, initially identified as a haemolytic activity in whole amoeba lysates [114, 215], is calcium dependent and membrane bound [112]. Lopez-Revilla and Said-Fernandez [114] correlated the level of activity with virulence. A second phospholipase A enzyme was detected in the soluble fraction and is calcium independent [112]. Killing of target CHO cells was inhibited with the chelating agents EDTA and EGTA, and with specific phospholipase A antagonists such as Rosenthal's inhibitor and phosphatidylcholine [196]. Destruction of the target cell membrane by the generation of lysophospholipids and free fatty acids was proposed as the mode of action of this enzyme [196, 216].

Protein kinase C

Protein kinase C is a ubiquitous enzyme with a central role in intracellular signal transduction [73, 156]. Weikel *et al.* [269] investigated the possible involvement of an amoebal protein kinase C in the regulation of cytolysis by observing the effects of phorbol 12-myristate 13-acetate (PMA), a specific activator of this enzyme, on the rate of CHO cell destruction by the pathogenic strain HM-1. Exposure of amoebae to 10^{-6}–10^{-7}M PMA by either brief pre-incubation (5 min), or direct addition to the assay system, resulted in a 100% enhancement of cell killing. Cytotoxicity was significantly reduced by sphingosine, a specific inhibitor of protein kinase C. PMA did not enhance cytolytic activity in two avirulent clones of HM-1. While these results indicate the involvement of a protein kinase C in the regulation of cytolysis by *E. histolytica*, the mechanisms involved in this process have yet to be defined.

Amoebapore

A membrane pore forming protein of 13–15 kDa, named 'amoebapore', was described by Lynch *et al.* [123] and Young *et al.* [273]. The molecule has been detected in highly dense, sub-cellular particles in axenic trophozoites [213], and exists as a dimer in two isoforms [212]. Using synthetic unilamellar lipid vesicles, Rosenberg *et al.* [212] demonstrated spontaneous insertion by amoebapore into synthetic unilamellar lipid vesicles, producing large pores which were non-selective for ions and were resistant to changes in transmembrane potential. Although studies on cytotoxicity have not been conducted to date, a reported correlation between amoebapore levels and strain virulence [100], together with the involvement of similar molecules in pathogenesis by other parasites [158, 272] and in cell killing by T-lymphocytes [96] suggests a likely role in cytotoxicity for amoebapore.

Non-cytotoxic activities

Enzymic degradation of extracellular matrix components, basement membrane proteins and connective tissue has long been considered to play a central role in tissue invasion by *E. histolytica* [16, 89, 97, 99, 119, 138, 150, 154]. Most, if not all, of these activities reported from *in vitro* studies are mediated by collagenase and cysteine proteinases, and levels of expression have been widely correlated with virulence [13, 70, 71, 97, 98, 117, 118, 150].

Collagenolysis

In 1982, Munoz *et al.* [150] described the destruction of human collagen films by HM-1 trophozoites, showing a higher specificity for type I collagen over type III. The EDTA-inhibitable, contact dependent activity was located in the membrane fraction of freeze-thawed lysates and was not secreted. In contrast, a second collagenolytic activity was detected in conditioned medium by Gadasi and Kessler [70] using the same strain, and could be eluted from plasma membranes with 0.5 M NaCl, suggesting that the enzyme was not integrally bound.

Cysteine proteinases

Cysteine proteinases constitute a large proportion of the overall proteolytic activity found in whole cell lysates of axenic *E. histolytica* [13, 167, 231]. The partial purification of an acid proteinase has been reported [138, 230], but little is known of its substrate specificity, sub-cellular location or possible involvement in histolysis. The more widely reported neutral cysteine proteinase secreted by trophozoites [26, 66, 99], is serum inhibitable [26, 119] and has been reported to induce fluid secretion in ligated intestinal loops *in vivo* [66, 119, 260].

It is at present unclear whether the various molecular weights reported for this activity, including values of 16 kDa [118], 26–29 kDa [115, 231] and 56 kDa [99, 203], should be attributed to different molecules with similar properties, multiple forms of a single enzyme [115] or purification artefacts [99].

The enzyme(s) described by Keene *et al.* [99] and Schulte and Scholze [231] degrade(s) fibronectin, laminin and human collagen types I, IV and V, resembling cathepsin B in its ability to cleave the synthetic substrate Z-Arg-Arg-NHMec [118], (although Luaces and Barrett [115] also noted several dissimilarities with this enzyme). Destruction of the cell anchoring proteins fibronectin and laminin is likely to account for the non-lethal rounding-up and detachment of epithelial cells from *in vitro* monolayers frequently noted by Lushbaugh *et al.* [119–121] and others [24, 25, 135, 136, 190].

Keene *et al.* [99] proposed that the collagenolytic activity corresponded to that reported by Gadasi and Kessler [70], and that the enzyme may act in concert with the membrane-associated metallocollagenase of Munoz *et al.* [150]. The completely purified cysteine proteinase degraded human glomerular basement membrane [115]. Ortiz-Ortiz *et al.* [162], and later Reed *et al.* [202, 204], reported that a cysteine proteinase activates the alternative and classical pathways of complement by cleavage of the C3 component. Other possible functions such as plasminogen activation and elevation of vascular permeability were suggested by Keene *et al.* [99]. The sensitivity of this enzyme to serum, and its functional dependence on a reducing environment, imply that a close physical association between the amoeba and its target tissue *in vivo* is probably required for effective digestion [99].

Secretion of a cysteine proteinase activity by non-pathogenic isolates of *E. histolytica* was described by Reed *et al.* [203], and was found to be significantly lower than for pathogens; similar results were obtained by de Meester *et al.* [49] using an iodinated diazopeptidyl inhibitor of cysteine proteinases to label enzymes from whole lysates of the two forms. Reed *et al.* [203] identified the 56 kDa molecule as being pathogen-specific.

Cysteine proteinase genes

An antiserum raised against the 27 kDa cysteine proteinase purified by Schulte and Scholze [231] was used by Tannich *et al.* [251] to isolate a cDNA clone from an HM-1 expression library. Nucleotide and inferred amino acid sequences of this clone, and a homologous clone derived by hybridization to a non-pathogenic cDNA library, were found to diverge by 16% and 17% respectively. Sequence alignment with a papain gene revealed that all residues known to participate in the presumed active site are present in each form.

Quantitative analysis of mRNA using a 28-mer oligonucleotide synthesized from a region conserved in both sequences, showed a level of expression 10- to 100-fold higher in three pathogenic isolates compared with two non-pathogens. This finding is in broad agreement with the previous comparative estimates of proteinase activity made by Reed *et al.* [203] and de Meester *et al.* [49], and suggests that quantitative rather than qualitative differences may account for the effects observed. Citing the intracellular sorting system of yeast, Tannich *et al.* [251] hypothesized that the over-production of a primarily intracellular proteinase by pathogenic amoebae could activate a secretory pathway, thereby adventiously conferring the ability to degrade extracellular tissue components.

Southern blotting of digested genomic DNA with the pathogenic cDNA clone gave a multiple banding pattern, indicating the possible existence of several gene copies. Sequence analysis of the clone revealed a putative preproenzyme containing a hydrophobic leader sequence and a region equivalent to 80 amino acids which could be cleaved off to produce the active enzyme. The predicted molecular weight of 24.1 kDa for the mature protein was in sufficient agreement with that of the purified enzyme [251].

Using PCR based on consensus sequences for eukaryote cysteine proteinases, Eakin *et al.* [188] obtained a partial gene sequence from axenic amoebae which shared only 43% homology with the corresponding portion of the sequence reported by Tannich *et al.* This gene may encode the acid proteinase described earlier [138, 230], or another neutral cysteine proteinase.

5.5 AMOEBIASIS IN AIDS

5.5.1 Sexually-transmitted amoebiasis

E. histolytica is a common sexually transmitted parasite among western male homosexuals. Surveys conducted in Berlin [270], London [8], New York [182], San Francisco [127] and Seattle [185] have recorded prevalence rates of 16.3%, 20%, 20%, 28.6% and 30%, respectively.

Sexually acquired amoebiasis is generally assumed to be asymptomatic [226, 267]. Most isolated parasites produce non-pathogenic isoenzyme patterns [8, 132, 184], and infected individuals are reported to be seronegative for anti-amoebal antibodies [139, 270]. However, several authors have argued that non-pathogenic *E. histolytica* may cause morbidity in homosexuals, questioning the presumption of Allason-Jones *et al.* [9] that treatment is not required in such cases [107, 166, 172, 241]. McMillan *et al.* [139] found that, while the incidence of diarrhoea in a group of 35 infected homosexuals was not significantly higher than in a non-infected control group, the histology of the rectal mucosa was abnormal in 63% of the former, and enteric symptoms persisted for a longer period.

Furthermore, positive serology, which is indicative of invasive disease, has been reported in up to 20% of Japanese male homosexuals [245, 246], and pathogenic amoebae have been isolated from four symptomatic and seropositive subjects [157]. Evidence of sexually acquired, pathogenic *E. histolytica* in homosexuals has also been recorded in San Francisco [165], Italy [2] and the UK [32].

It would appear, therefore, that sexually transmitted *E. histolytica* is associated largely, though not exclusively, with the non-pathogenic form. While the reasons for this are at present unclear, it is possible that hitherto unknown differences in transmission patterns between the two types of amoebae may be involved.

5.5.2 *E. histolytica* in AIDS

In view of the incidence of both non-pathogenic and pathogenic *E. histolytica* in male homosexuals, it is perhaps surprising that virtually no cases of invasive amoebiasis have been reported in AIDS patients, despite evidence that immunity to invasive amoebiasis is T-cell mediated [217], and previous observations that the severity of amoebiasis is increased in immunosuppressed individuals [93, 110].

In several recent studies, sexually acquired *E. histolytica* isolates cultured from HIV infected patients were exclusively non-pathogenic by isoenzyme analysis, and in no case was any correlation found with enteric symptoms [9, 31, 205]. The absence of amoebic disease in homosexual AIDS patients has been borne out in other surveys of parasite-related morbidity [109, 237].

While this phenomenon may be related in part to the small numbers of pathogenic strains endemic in the west, studies undertaken in Africa [41, 42, 233], where invasive amoebiasis is endemic, have likewise failed to detect amoebic disease in AIDS patients, with the exception of one report [262]. These observations prompted Lucas [116] to describe amoebiasis as a 'missing disease' in AIDS, and led Curry *et al.* [46] to suggest the existence of a negative association.

5.5.3 Interactions with HIV

Induction of HIV replication

A mitogenic activity for human lymphocytes, present in pathogenic *E. histolytica* extracts, was described by Diamantstein *et al.* [52] and Kettis and Sundquist [103]. The level of this activity was subsequently correlated with virulence *in vitro* and was shown to be associated with the Gal/GalNAc inhibitable lectin [218]. It is not known whether this property is also expressed by non-pathogens. Conclusive evidence that the Gal/GalNAc lectin is responsible for observed T-cell proliferation is not yet available, although similar effects produced by other lectins such as Con A and phytohaemagglutinin [111, 149] indicate a likely involvement.

Based on observations that T-lymphocyte proliferation is an essential precursor of HIV replication [274], and that the onset of AIDS can be correlated with infection by intestinal parasites and other sexually transmitted diseases [268], Croxson *et al.* [45] investigated the possibility that the reported mitogenic activity of *E. histolytica* could also induce replication of HIV *in vitro*.

Using crude sonicates of the pathogenic strain HK-9, together with appropriate controls, these workers were able to detect only a low grade lymphoblastic response in long-term cultures of peripheral blood mononuclear cells (PBMC) derived from 15 HIV-infected individuals. However, virus replication was demonstrated in five of these cases, indicating the potential of *E. histolytica* as an agent of HIV induction. The relatively low level of mitogenic activity observed in HK-9 sonicates, compared with the earlier finding of Salata and Ravdin [218], could in part be explained by the differences in virulence of the strains used, since HK-9 is known to be less virulent than HM-1 [71, 151]. Given that most, if not all amoebic infection in AIDS is apparently caused by non-pathogenic isolates, it would now be of considerable interest to establish whether this form is also capable of inducing HIV replication.

Detection of HIV in trophozoites

Reports during the 1970s of the occurrence of viruses in *E. histolytica* [19, 55, 83, 134] has led to speculation that amoebae could act as vectors for HIV. To test this hypothesis, Brown *et al.* [29] incubated HIV-infected HUT-78 cells with HM-1 trophozoites, which were found to lack CD4-like receptors. Following phagocytosis of the infected cells, HIV was detected immunocytochemically in amoebae for up to 48 hours, but in 15 separate experiments, co-culture of HIV-harbouring trophozoites with uninfected PBMC at a 1 : 1 ratio failed to result in transmission. In a further experiment, amoebae isolated from the stools of seven AIDS patients and one with AIDS related complex were tested for HIV p18 and p25 core antigens; 2% of trophozoites from one case were positive for HIV, but, as before, it was not possible to demonstrate transmission to uninfected PBMC. These results suggest that the uptake of HIV by *E. histolytica* is a non-specific event, and that, once internalized, the virus is not transmissible.

5.6 MOLECULAR DIAGNOSIS

5.6.1 Standard procedures

The laboratory diagnosis of amoebiasis depends on the microscopic identification of trophozoites or cysts in stools, and on the detection of anti-amoebal antibodies in blood. A comprehensive account of diagnostic procedures is given by Walsh [265]. The limitations of these techniques are numerous, and to some extent account for the lack of reliable epidemiological data from several regions of the world. For example, stool examination depends extensively on the detection of cysts rather than trophozoites, which tend to be present only during periods of acute dysentery. The numbers of cysts shed may vary widely, requiring at least three separate samples to achieve 80–90% sensitivity [91], which is both time consuming and often impracticable.

Secondly, lymphocytes with ingested red blood cells may readily be mistaken for haematophagous trophozoites, especially in old or fixed samples where the important criterion of amoebic motility cannot be applied [104]. Thirdly, *E. histolytica* cysts can only be distinguished from those of some other commensal amoebae by the number of nuclei present, requiring the skills of highly trained and experienced staff who may not always be available [214, 236]. Fourthly, since the cysts of pathogenic and non-pathogenic strains are morphologically identical, differentiation between the two forms requires prior culture of amoebae over several days in order to obtain sufficient numbers for isoenzyme analysis [223].

5.6.2 Immunological methods

Immunological detection of *E. histolytica* antigen in stools by ELISA has been widely reported [10, 50, 78, 209], although in every case, the immunological reagents employed are raised against trophozoites rather than cysts, since the latter cannot be produced *in vitro*. Given that trophozoites are absent from the stools of the majority of infected individuals, such assays are of limited use in diagnosis, as shown in the relatively low sensitivities reported. In a recent study, Jain *et al.* [88] found that, while 100% of faecal specimens containing trophozoites were positive by ELISA, only 40% of cyst-passers were correctly diagnosed. A higher detection rate of 80% was reported by Kabil *et al.* [92] for asymptomatic carriers, but this test also produced false positives in 30% of samples containing unrelated parasites.

Diagnosis based on positive serology is hampered by the inability to distinguish between past and present infections. Furthermore, detection of antibodies in cases of amoebic colitis, which comprise the majority of pathological conditions, is only 50–80% sensitive, compared with 85–90% in patients with liver abscess [266].

5.6.3 Molecular diagnosis

The application of molecular biology techniques to the diagnosis of amoebic infection provides an opportunity to improve on the sensitivity, specificity and rapidity of conventional procedures. Two recent reports have exploited the abundance of non-transcribed sequences derived from the high copy number extrachromosomal episome of *E. histolytica* (section 5.2.7). Samuelson *et al.* [221] independently isolated the 145 base pair tandemly repeated episomal element reported elsewhere by Garfinkel *et al.* [72] as being specific for pathogenic amoebae. When used as a radiolabelled probe under low stringency conditions, however, trophozoites and cysts of both pathogenic and non-pathogenic isolates were detected in unfractionated stool samples spotted onto nylon membranes. A sensitivity of 1.0 and a specificity of 0.93 were recorded in a blind trial with 123 clinical samples, 25 of which were positive for *E. histolytica* as previously judged by microscopy. In separate experiments, hybridization signals were obtained after overnight autoradiography with as few as eight cultured trophozoites per spot; the level of detection in faecal material was estimated to be approximately ten-fold lower.

Bracha *et al.* [27] utilized the same probe, P145, together with B133, the corresponding non-pathogenic element, previously reported to discriminate between the two forms at high stringency [72], to classify 81 clinical isolates obtained from several regions of the world. Specific hybridization signals were obtained with as few as 200 cultured amoebae spotted onto nylon membranes. However, detection of either form in stools was not attempted.

Tannich and Burchard [248] also developed a test to distinguish pathogenic from non-pathogenic trophozoites, based on the gene previously described by Tannich *et al.* [250] and Edman *et al.* [58] as encoding a 125 kDa surface antigen of *E. histolytica* (sections 5.2.8 and 5.3.2). A 482 base pair fragment of the gene containing identical 3′ and 5′ ends was amplified from each form of the parasite by PCR. Specific internal sequence differences were then detected by digestion with selected restriction enzymes, followed by agarose gel electrophoresis. PCR products could be obtained directly from cell lysates of as few as ten cultured trophozoites, although it is unclear whether yields from this number of amoebae were sufficient to perform restriction analysis. The application of this test in direct faecal analysis has yet to be demonstrated.

An ideal molecular diagnostic method for amoebiasis needs to be sufficiently sensitive to detect *E. histolytica* cysts in stools and also specific enough to differentiate between pathogenic and non-pathogenic amoebae. The test described by Samuelson *et al.* meets the first criterion, so eliminating the need for prior culture of isolates. From the results of Bracha *et al.* it is reasonable to expect that the same probe applied at high stringency would identify only pathogenic cysts, thus fulfilling the second criterion, although whether this would be accompanied by a critical loss of sensitivity remains to be determined. Reliance on the preparation of radioactive probes has obvious limitations for use in the field, where non-isotopic labelling methods would be more practicable.

The PCR-based test of Tannich and Burchard, while dispensing with DNA hybridization, nevertheless depends on the availability of equipment to perform PCR and gel electrophoresis, and has yet to be demonstrated with cyst template DNA. However, provided that the technical difficulties associated with performing PCR in the presence of faecal contaminants can be overcome, this approach could prove suitable for use in relatively specialized centres.

5.7 CONCLUSION

Studies on the molecular biology of *E. histolytica*, though only recently initiated, have already yielded some important insights into the genetic organization and pathogenic mechanisms of this parasite. Perhaps the most significant development has been the demonstration of genomic differences between pathogenic and non-pathogenic forms, providing definitive support for Brumpt's original notion of two species. In the light of this evidence, chemotherapy for non-pathogenic *E. histolytica* can be assumed to have little advantage, although further investigation is needed into reports of mucosal damage associated with such infections in homosexual men.

The application of molecular biological methods in amoebiasis research will also improve our understanding of many other aspects of genetic organization, including the *E. histolytica* karyotype, the control of replication and gene expression, ploidy, the differentiation of trophozoites into cysts, and genetic exchange. The development of a transfection system will provide a particularly powerful tool for addressing these and other questions.

The study of pathogenic mechanisms in *E. histolytica* at the cellular level is being increasingly complemented by molecular analysis, with particular reference to parasite–target cell binding, cytolysis and tissue dissolution. This approach should also prove valuable in determining the molecular basis of amoeba–bacterium relations, virulence, interactions with the host immune system and the control and co-ordination of pathogenic processes.

Contrary to some expectations, there is no evidence that amoebiasis is an opportunistic infection in AIDS. However, the reported mitogenicity of parasite antigens for T-lymphocytes, and the potentiation of HIV replication require closer evaluation.

DNA probes have been reported which are capable of detecting *E. histolytica* cysts and trophozoites in stools. This technique offers several advantages over conventional microscopy, particularly with regard to speed and sample throughput. Discrimination between pathogenic and non-pathogenic forms has also been demonstrated for small numbers of cultured trophozoites; the eventual development of a pathogen-specific

DNA test for use on faecal specimens will be of considerable benefit in both epidemiological and clinical applications.

ACKNOWLEDGEMENTS

Some of the authors' work was supported by grants from the Wellcome Trust and Universidad Nacional Autonoma de Mexico.

REFERENCES

1. (1985). Is that amoeba harmful or not? [editorial]. *Lancet*, i, 732–4.
2. Aceti, A., Pennica, A., Ippolito, G., *et al.* (1987) Antiamebic antibodies in homosexual men [letter]. *N. Engl. J. Med.*, **316**, 692–2.
3. Albach, R.A. (1989) Nucleic acids of *Entamoeba histolytica*. *J. Protozool.*, **36**, 197–205.
4. Albach, R.A., Booden, T., Boonlayangoor, P., Downing, S. (1977) Entamoeba histolytica: autoradiographic analysis of nuclear sites of RNA synthesis. *Exp. Parasitol.*, **42**, 248–59.
5. Albach, R.A., Booden, T., Boonlayangoor, P., Downing S. (1980) Concepts of function of peripheral non-chromatin and endosome in Entamoeba histolytica. *Arch. Invest. Med. (Mex.)*, **11**, 63–74.
6. Albach, R.A., Prachayasittikul, V., Heebner, G.M. (1984) Isolation and characterization of RNA of *Entamoeba histolytica*. *Mol. Biochem. Parasitol.*, **12**, 261–72.
7. Albach, R.A., Shaffer, J.G., Watson, R.H. (1966) Autoradiographic studies of H^3 thymidine uptake in *Entamoeba histolytica* in CLG medium. *J. Protozool.*, **13**, 349–55.
8. Allason-Jones, E., Mindel, A., Sargeaunt, P., Williams, P. (1986) Entamoeba histolytica as a commensal intestinal parasite in homosexual men. *N. Engl. J. Med.*, **315**, 353–6.
9. Allason-Jones, E., Mindel, A., Sargeaunt, P.G., Katz, D. (1988) Outcome of untreated infection with *Entamoeba histolytica* in homosexual men with and without HIV antibody. *Br. Med. J.*, **297**, 654–7.
10. Anand, P., Malaviya, B., Das, P., Mateen, M.A., Habibullah, C.M., Das, S.R. (1985) Multilayer-enzyme linked immunosorbent assay (ML-ELISA) for detection of *Entamoeba histolytica* trophozoite copro-antigen. *Immunol. Invest.*, **14**, 443–53.
11. Andrews, B.J., Mentzoni, L., Bjorvatn, B. (1990) Zymodeme conversion of isolates of *Entamoeba histolytica*. *Trans. R. Soc. Trop. Med. Hyg.*, **84**, 63–5.
12. Arroyo, R., Orozco, E. (1987) Localization and identification of an *Entamoeba histolytica* adhesin. *Mol. Biochem. Parasitol.*, **23**, 151–8.
13. Avila, E.E., Sanchez Garza, M., Calderon, J. (1985) *Entamoeba histolytica* and *E. invadens*, sulfhydryl-dependent proteolytic activity. *J. Protozool.*, **32**, 163–6.
14. Barker, D.C. (1976) Differentiation of *Entamoeba* patterns of nucleic acids

and ribosomes during encystment and excystation, in *Biochemistry of Parasites and Host–parasite Relationships*, Elsevier, Amsterdam, p. 253.
15. Barker, D.C., Swales, L.S. (1972) Characteristics of ribosomes during differentiation from trophozoite to cyst in axenic *Entamoeba* spp. *Cell Differ.*, **1**, 297–306.
16. Becker, I., Perez-Tamayo, R., Montfort, I., Alvizouri, A.M., Perez-Montfort, R. (1988) *Entamoeba histolytica*: role of amebic proteinases and polymorphonuclear leukocytes in acute experimental amebiasis in the rat. *Exp. Parasitol.*, **67**, 268–80.
17. Bhattacharya, S., Bhattacharya, A., Diamond L.S. (1988) Comparison of repeated DNA from strains of *Entamoeba histolytica* and other *Entamoeba*. *Mol. Biochem. Parasitol.*, **27**, 257–62.
18. Bhattacharya, S., Bhattacharya, A., Diamond, L.S., Soldo, A.T. (1989) Circular DNA of *Entamoeba histolytica* encodes ribosomal RNA. *J. Protozool.*, **36**, 455–8.
19. Bird, R.G., McCaul, T.F. (1976) The rhabdoviruses of *Entamoeba histolytica* and *Entamoeba invadens*. *Ann. Trop. Med. Parasitol.*, **70**, 81–93.
20. Blakely, P., Sargeaunt, P.G., Reed, S.L. (1990) An immunogenic 30-kDa surface antigen of pathogenic clinical isolates of *Entamoeba histolytica*. *J. Infect. Dis.*, **162**, 949–54.
21. Blanc, D., Nicholls, R., Sargeaunt, P.G. (1989) Experimental production of new zymodemes of *Entamoeba histolytica* supports the hypothesis of genetic exchange. *Trans. R. Soc. Trop. Med. Hyg.*, **83**, 787–90.
22. Boeck, W.C., Drbohlav, J. (1925) The cultivation of *Entamoeba histolytica*. *Amer. J. Hyg.*, **5**, 371–407.
23. Boonlayangoor, P., Otten, M., Booden, T., Albach, R.A. (1978) Isolation and sedimentation analysis of RNA from *Entamoeba histolytica*. *Arch. Invest. Med. Mex.*, **9**, 121–8.
24. Bos, H.J. (1979) *Entamoeba histolytica*: cytopathogenicity of intact amebae and cellfree extracts; isolation and characterization of an intracellular toxin. *Exp. Parasitol.*, **47**, 369–77.
25. Bos, H.J., Leijendekker, W.J., van den Eijk, A.A. (1980) *Entamoeba histolytica*: cytopathogenicity, including serum effects on contact-dependent and toxin-induced lysis of hamster kidney cell monolayers. *Exp. Parasitol.*, **50**, 342–57.
26. Bos, H.J., van den Eijk, A.A. (1980) Serum-inhibited toxicity of *Entamoeba histolytica*. *Arch. Invest. Med. (Mex.)*, **11**, 135–41.
27. Bracha, R., Diamond, L.S., Ackers, J.P., Burchard, G.D., Mirelman, D. (1990) Differentiation of clinical isolates of *Entamoeba histolytica* by using specific DNA probes. *J. Clin. Microbiol.*, **28**, 680–4.
28. Bray, R.S., Harris, W.G. (1977) The epidemiology of infection with *Entamoeba histolytica* in the Gambia, West Africa. *Trans. R. Soc. Trop. Med. Hyg.*, **71**, 401–7.
29. Brown, M., Reed, S.L., Levy, J.A., Busch, M., McKerrow, J.H. (1991) Detection of HIV-1 in *Entamoeba histolytica* without evidence of transmission to human cells. *Aids*, **5**, 93–6.

30. Brumpt, E. (1925) Etude sommaire del 'Entamoeba dispar' n.sp. Amibe a kystes quadrinuclees, parasite de l'homme. *Bull. Acad. Med. (Paris)*, **94**, 943–52.
31. Burchard, G.D., Hufert, F.T., Mirelman, D. (1991) Characterization of 20 *Entamoeba histolytica* strains isolated from patients with HIV infection. *Infection*, **19**, 164–9.
32. Burnham, W.R., Reeve, R.S., Finch, R.G. (1980) *Entamoeba histolytica* infection in male homosexuals. *Gut*, **21**, 1097–9.
33. Burton, K. (1956) A study of the conditions and mechanism of the diphenylamine reaction for the colorimetiric estimation of deoxyribonucleic acid. *Biochem. J.*, **62**, 315–23.
34. Byers, T. (1986) Molecular biology of DNA in *Acanthamoeba*, *Amoeba*, *Entamoeba* and *Naegleria*. *Int. Rev. Cytol.*, **99**, 311–41.
35. Chadee, K., Johnson, M.L., Orozco, E., Petri, W.A.J., Ravdin, J.I. (1988) Binding and internalization of rat colonic mucins by the galactose/*N*-acetyl-*D*-galactosamine adherence lectin of *Entamoeba histolytica*. *J. Infect. Dis.*, **158**, 398–406.
36. Chadee, K., Keller, K., Forstner, J., Innes, D.J., Ravdin, J.I. (1991) Mucin and nonmucin secretagogue activity of *Entamoeba histolytica* and cholera toxin in rat colon. *Gastroenterology*, **100**, 986–97.
37. Chadee, K., Meerovitch, E. (1985) The pathology of experimentally induced cecal amebiasis in gerbils (*Meriones unguiculatus*). Liver changes and amebic liver abscess formation. *Am. J. Pathol.*, **119**, 485–94.
38. Chadee, K., Petri, W.A. Jr, Innes, D.J., Ravdin, J.I. (1987) Rat and human colonic mucins bind to and inhibit adherence lectin of *Entamoeba histolytica*. *J. Clin. Invest.*, **80**, 1245–54.
39. Clark, C.G., Cross, G.A. (1987) rRNA genes of *Naegleria gruberi* are carried exclusively on a 14-kilobase-pair plasmid. *Mol. Cell Biol.*, 7, 3027–31.
40. Clark, C.G., Cross, G.A. (1988) Circular ribosomal RNA genes are a general feature of schizopyrenid amoebae. *J. Protozool.*, **35**, 326–9.
41. Colebunders, R., Lusakumuni, K., Nelson, A.M., *et al.* (1988) Persistent diarrhoea in Zairian AIDS patients: an endoscopic and histological study. *Gut*, **29**, 1687–91.
42. Conlon, C.P., Pinching, A.J., Perera, C.U., Moody, A., Luo, N.P., Lucas, S.B. (1990) HIV related enteropathy in Zambia: a clinical, microbiological, and histological study. *Am. J. Trop. Med. Hyg.*, **42**, 83–8.
43. Craig, C.F. (1944) *The Ecology, Diagnosis and Treatment of Amebiasis*, Williams and Wilkins, Baltimore.
44. Craig, C.F., Faust, E.C. (1951) Clinical Parasitology, 5th edn, Henry Kimpton, London.
45. Croxson, S., Mildvan, D., Mathews, H., Poiesz, B.J. (1988) *Entamoeba histolytica* antigen-specific induction of human immunodeficiency virus replication. *J. Clin. Microbiol.*, **26**, 1304–8.
45a. Cruz-Reyes, J.A., Spice, W.M., Rehman, T. *et al.* (1992) Ribosomal DNA sequences in the differentiation of pathogenic and non-pathogenic isolates of *Entamoeba histolytica*. *Parasitology*, **104**, 239–46.

46. Curry, A., Turner, A.J., Lucas, S. (1991) Opportunistic protozoan infections in human immunodeficiency virus disease, review highlighting diagnostic and therapeutic aspects. *J. Clin. Pathol.*, **44**, 182–93.
47. D'Antoni, J.S. (1952) Concepts and misconceptions in amebiasis. *Amer. J. Trop. Med. Hyg.*, **1**, 146–54.
48. de la Cruz-Hernandez, F., Ocadiz, J.R., Orozco, E. (1986) Analysis of the organization of the DNA of various strains and clones of *Entamoeba histolytica. Arch. Invest. Med. (Mex.)*, **17 Suppl 1**, 107–12.
49. de Meester, F., Shaw, E., Scholze, H., Stolarsky, T., Mirelman, D. (1990) Specific labelling of cysteine proteinases in pathogenic and nonpathogenic *Entamoeba histolytica. Infect. Immun.*, **58**, 1396–401.
50. del Moro, R., Oliva, A., Herion, P., Capin, R., Ortiz Ortiz, L. (1987) Diagnosis of *Entamoeba histolytica* in feces by ELISA. *J. Clin. Lab. Anal.*, **1**, 322–5.
51. Densen, P., Mandell, G.L. (1978) Gonococcal interactions with polymorphonuclear neutrophils. Importance of the phagosome for bactericidal activity. *J. Clin. Invest.*, **62**, 1161–71.
52. Diamantstein, T., Klos, M., Gold, D., Hahn, H. (1981) Interaction between *Entamoeba histolytica* and the immune system. I. Mitogenicity of *Entamoeba histolytica* extracts for human peripheral T lymphocytes. *J. Immunol.*, **126**, 2084–6.
53. Diamond, L.S. (1968) Techniques of axenic cultivation of *Entameba histolytica Schaudinn*, 1903 and *E. histolytica*-like amebae. *J. Parasitol.*, **54**, 1047–56.
54. Diamond, L.S. (1982) Amebiasis, Nutritional implications. *Rev. Inf. Dis.*, **4**, 843.
55. Diamond, L.S., Mattern, C.F., Bartgis, I.L. (1972) Viruses of *Entamoeba histolytica*. I. Identification of transmissible virus-like agents. *J. Virol.*, **9**, 326–41.
56. Dobell, C. (1919) *The Amoebae Living in Man*, Medical Research Committee, London.
57. Dobell, C. (1931) Researches on the intestinal protozoa of monkeys and man. IV. An experimental study of the *histolytica*-like species of *Entamoeba* living naturally in macaques. *Parasitol.*, **23**, 1–72.
58. Edman, U., Meraz, M.A., Rausser, S., Agabian, N., Meza, I. (1990) Characterization of an immuno-dominant variable surface antigen from pathogenic and nonpathogenic *Entamoeba histolytica. J. Exp. Med.*, **172**, 879–88.
59. Edman, U., Meza, I., Agabian, N. (1987) Genomic and cDNA actin sequences from a virulent strain of *Entamoeba histolytica. Proc. Natl. Acad. Sci. U.S.A.*, **84**, 3024–8.
60. el Hashimi, W., Pittman, F. (1970) Ultrastructure of *Entamoeba histolytica* trophozoites obtained from the colon and from *in vitro* cultures. *Am. J. Trop. Med. Hyg.*, **19**, 215–26.
61. Elmassian, M. (1909) Sur une nouvelle espece amibienne chez l'homme, *Entamoeba minuta* n.sp. Morphologie-Evolution-Pathogenie. *Zent. Bakt. (Orig.)*, **52**, 335.

62. Elsdon-Dew, R. (1968) The epidemiology of amoebiasis. *Adv. Parasitol.*, **6**, 1–62.
63. Elsdon-Dew, R. (1971) Amebiasis as a world problem. *Bull. N.Y. Acad. Med.*, **47**, 438–47.
64. Fan, J.B., Korman, S.H., Cantor, C.R., Smith, C.L. (1991) *Giardia lamblia*: haploid genome size determined by pulsed field gel electrophoresis is less than 12 Mb. *Nucleic Acids Res.*, **19**, 1905–8.
65. Faust, E.C., Russell, P.F., Jung, R.C. (1970) Clinical Parasitology, 8th edn, Lea and Febiger, Philadelphia.
66. Feingold, C., Bracha, R., Wexler, A., Mirelman, D. (1985) Isolation, purification, and partial characterization of an enterotoxin from extracts of *Entamoeba histolytica* trophozoites. *Infect. Immun.*, **48**, 211–8.
67. Flickinger, C.J. (1974) The fine structure of our species of amoeba. *J. Protozool.*, **21**, 59.
68. Frost, P., Smith, J., Frost, H. (1978) The radiolabelling of lymphocytes and tumour cells with 111 Indium (39991). *Proc. Soc. Exp. Biol. Med.*, **157**, 61–5.
69. Gadasi, H. (1982) Isolated *Entamoeba histolytica* actin does not inhibit DNAse-I activity. *Biochem. Biophys. Res. Commun.*, **104**, 158–64.
70. Gadasi, H., Kessler, E. (1983) Correlation of virulence and collagenolytic activity in *Entamoeba histolytica. Infect. Immun.*, **39**, 528–31.
71. Gadasi, H., Kobiler, D. (1983) *Entamoeba histolytica*: correlation between virulence and content of proteolytic enzymes. *Exp. Parasitol.*, **55**, 105–10.
72. Garfinkel, L.I., Giladi, M., Huber, M., *et al.* (1989) DNA probes specific for *Entamoeba histolytica* possessing pathogenic and nonpathogenic zymodemes. *Infect. Immun.*, **S7**, 926–31.
73. Garrison, J.C. (1985) Possible roles of protein kinase C in cell function. *Ann. Rep. Med. Chem.*, **20**, 227–36.
74. Gelderman, A.H., Bartgis, I.L., Keister, D.B., Diamond, L.S. (1971) A comparison of genome sizes and thermal-denaturation-derived base composition of DNAs from several members of *Entamoeba (histolytica group). J. Parasitol.*, **57**, 912–16.
75. Gelderman, A.H., Keister, D.B., Bartgis, I.L., Diamond, L.S. (1971) Characterization of the deoxyribonucleic acid of representative strains of *Entamoeba histolytica*, *E. histolytica*-like amebae, and *E. moshkovskii*. *J. Parasitol.*, **57**, 906–11.
76. Grimaldi, G., Di Nocera, P.P. (1988) Multiple repeated units in *Drosophila melanogaster* ribosomal DNA spacer stimulate rRNA precursor transcription. *Proc. Natl. Acad. Sci. U.S.A.*, **85**, 5502–6.
77. Grodberg, J., Salazar, N., Oren, R., Mirelman, D. (1990) Autonomous replication sequences in an extrachromosomal element of a pathogenic *Entamoeba histolytica. Nucleic Acids Res.*, **18**, 5515–9.
78. Grundy, M.S. (1982) Preliminary observations using a multi-layer ELISA method for the detection of *Entamoeba histolytica* trophozoite antigens in stool samples. *Trans. R. Soc. Trop. Med. Hyg.*, **76**, 396–400.
79. Guerrant, R.L. (1986) Amebiasis: introduction, current status, and research questions. *Rev. Infect. Dis.*, **8**, 218–27.

80. Healy, G.R. (1988) Diagnostic techniques for stool samples, in *Amebiasis: Human Infection by Entamoeba histolytica*, (ed. J.I. Ravdin), John Wiley, New York, pp. 635–49.
81. Heebner, G.M., Albach, R.A. (1982) High salt SDS-DEP technic for isolation of 'intact' rRNA from *Entamoeba histolytica. Arch. Invest. Med. (Mex.)*, **13**, 23–8.
82. Hoare, C.A. (1952) The commensal phase of *Entamoeba histolytica. Exp. Parasitol.*, **1**, 411–27.
83. Hruska, J.F., Mattem, C.F., Diamond, L.S., Keister, D.B. (1973) Viruses of *Entamoeba histolytica.* 3. Properties of the polyhedral virus of the HB-301 strain. *J. Virol.*, **11**, 129–36.
84. Huber, M., Garfinkel, L., Gider, C., Mirelman, D., Revel, M., Rozenblatt, S. (1987) *Entamoeba histolytica*: cloning and characterization of actin cDNA. *Mol. Biochem. Parasitol.*, 24, 227–35.
85. Huber, M., Garfinkel, L., Gider, C., Mirelman, D., Revel, M., Rozenblatt, S. (1988) Nucleotide sequence analysis of an *Entamoeba histolytica* ferredoxin gene. *Mol. Biochem. Parasitol.*, **31**, 27–33.
86. Huber, M., Koller, B., Gitler, C., *et al.* (1989) *Entamoeba histolytica* ribosomal RNA genes are carried on palindromic circular DNA molecules. *Mol. Biochem. Parasitol.*, **32**, 285–96.
87. Jackson, T.F. (1987) *Entamoeba histolytica* cyst passers – to treat or not to treat? [editorial]. *S. Fr. Med. J.*, **72**, 657–8.
88. Jain, U., Patel, M.T., Desai, P.K., Kaliraj, P. (1990) Development of simple and stable ELISA for detection of coproantigen in intestinal amoebiasis. *Indian J. Exp. Biol.*, **28**, 671–5.
89. Jarumilinta, R., Maegraith, B.G. (1969) Enzymes of *Entamoeba histolytica. Bull. World Health Organ.*, **41**, 269–73.
90. Johnson, P.J., Borst, P. (1986) Mapping of VSG genes on large expression-site chromosomes of *Trypanosoma brucei* separated by pulsed-field gradient electrophoresis. *Gene*, **43**, 213–20.
91. Juniper, K. (1979) Amoebiasis. *Clin. Gastroenterol.*, 7, 3.
92. Kabil, S.M., Hassan, M.M., Atta, M., el Ghysha, O. (1990) ELISA in detection of *Entamoeba histolytica* antigens in stools. *J. Egypt. Soc. Parasitol.*, **20**, 673–6.
93. Kanani, S.R., Knight, R. (1969) Amoebic dysentery precipitated by corticosteroids. *Br. Med. J.*, **3**, 114.
94. Kapler, G.M., Coburn, C.M., Beverley, S.M. (1990) Stable transfection of the human parasite *Leishmania major* delineates a 30-kilobase region sufficient for extrachromosomal replication and expression. *Mol. Cell Biol.*, **10**, 1084–94.
95. Karrer, K.M., Gall, J.G. (1976) The macronuclear ribosomal DNA of *Tetrahymena pyriformis* is a palindrome. *J. Mol. Biol.*, **104**, 421–53.
96. Kawasaki, A., Shinkai, Y., Kuwana, Y., *et al.* (1990) Perforin, a pore-forming protein detectable by monoclonal antibodies, is a functional marker for killer cells. *Int. Immunol.*, **2**, 677–84.
97. Keene, W.E., Hidalgo, M.E., Orozco, E., McKerrow, J.H. (1990) *Entamoeba*

histolytica: correlation of the cytopathic effect of virulent trophozoites with secretion of a cysteine proteinase. *Exp. Parasitol.*, **71**, 199–206.

98. Keene, W.E., McKerrow, J.H. (1985) Isolation and characterization of the neutral thiol proteinase of virulent *Entamoeba histolytica. Fed. Proc.*, **44**, 1335.
99. Keene, W.E., Petitt, M.G., Allen, S., McKerrow, J.H. (1986) The major neutral proteinase of *Entamoeba histolytica. J. Exp. Med.*, **163**, 536–49.
100. Keller, F., Walter, C., Lohden, U., Hanke, W., Bakker-Grunwald, T., Trissl, D. (1988) Pathogenic and non-pathogenic *Entamoeba*: pore formation and hemolytic activity. *J. Protozool.*, **3S**, 359–65.
101. Kemp, D.J., Coppel, R.L., Anders, R.F. (1987) Repetitive proteins and genes of malaria. *Ann. Rev. Microbiol.*, **41**, 181–208.
102. Kemp, D.J., Corcoran, L.M., Coppel, R.L., *et al.* (1985) Size variation in chromosomes from independent cultured isolates of *Plasmodium falciparum. Nature*, **315**, 347–50.
103. Kettis, A.A., Sundqvist, K.G. (1982) Activation of lymphocytes from healthy donors and patients with amebiasis by extracts of *Entamoeba histolytica. Arch. Invest. Med. (Mex.)*, **13**, 245–8.
104. Knight, R. (1974). Amoebiasis. *Trop. Doct.*, **4**, 6–11.
105. Kobiler, D., Mirelman, D., Mattern, C.F. (1981) Lectin and toxin-like activities of *Entamoeba histolytica*, comparison of properties. *Am. J. Trop. Med. Hyg.*, **30**, 955–9.
106. Kornberg, R.D. (1977) Structure of chromatin. *Ann. Rev. Biochem.*, **46**, 931–54.
107. Krogstad, D.J. (1986) Isoenzyme patterns and pathogenicity in amebic infection [editorial]. *N. Engl. J. Med.*, **315**, 390–1.
108. Kuenen, W.A., Swellengrebel, N.H. (1913) Die *Entamobendes* Menschen und ihre praktische Bedeutung. *Centralblatt Bakteriol.*, **71**, 378.
109. Laughon, B.E., Druckman, D.A., Vernon, A., *et al.* (1988) Prevalence of enteric pathogens in homosexual men with and without acquired immunodeficiency syndrome. *Gastroenterology*, **94**, 984–93.
110. Lewis, E.A., Antia, A.U. (1969) Amoebic colitis: review of 295 cases. *Trans. R. Soc. Trop. Med. Hyg.*, **63**, 633–8.
111. Lipsky, P.E., Ellner, J.J., Rosenthal, A.S. (1976) Phytohemagglutinin-induced proliferation of guinea-pig thymus-derived lymphocytes. I. Accessory cell dependence. *J. Immunol.*, **116**, 868–75.
112. Long-Krug, S.A, Fischer, K.J, Hysmith, R.M., Ravdin, J.I. (1985) Phospholipase A enzymes of *Entamoeba histolytica*: description and subcellular localization. *J. lnfect. Dis.*, **152**, 536–41.
113. Lopez-Revilla, R., Gomez, R. (1978) *Entamoeba histolytica, E. invadens*, and *E. moshkovskii*: fluctuations of the DNA content of axenic trophozoites. *Exp. Parasitol.*, **44**, 243–8.
114. Lopez-Revilla, R., Said-Femandez, S. (1980) Cytopathogenicity of *Entamoeba histolytica*: hemolytic activity of trophozoite homogenates. *Am. J. Trop. Med. Hyg.*, **29**, 209–12.
115. Luaces, A.L., Barrett, A.J. (1988) Affinity purification and biochemical

characterization of histolysin, the major cysteine proteinase of *Entamoeba histolytica*. *Biochem. J.*, **250**, 903–9.

116. Lucas, S.B. (1990) Missing infections in AIDS. *Trans. R. Soc. Trop. Med. Hyg.*, **84 Suppl 1**, 34–8.
117. Lushbaugh, W.B., Hofbauer, A.F., Kairalla, A.A., Cantey, J.R., Pittman, F.E. (1984) Relationship of cytotoxins of axenically cultivated *Entamoeba histolytica* to virulence. *Gastroenterology*, **86**, 1488–95.
118. Lushbaugh, W.B., Hofbauer, A.F., Pittman, F.E. (1985) *Entamoeba histolytica*: purification of cathepsin B. *Exp. Parasitol.*, **59**, 328–36.
119. Lushbaugh, W.B, Kairalla, A.B., Cantey, J.R., Hofbauer, A.F., Pittman F.E. (1979) Isolation of a cytotoxin-enterotoxin from *Entamoeba histolytica*. *J. Infect. Dis.*, **139**, 9–17.
120. Lushbaugh, W.B., Kairalla, A.B., Hofbauer, A.F., Amaud, P., Cantey, J.R., Pittman, F.E. (1981) Inhibition of *Entamoeba histolytica* cytotoxin by alpha 1 antiprotease and alpha 2 macroglobulin. *Am. J. Trop. Med. Hyg.*, **30**, 575–85.
121. Lushbaugh, W.B., Kairalla, A.B., Loadholt, C.B., Pittman, F.E. (1978) Effect of hamster liver passage on the virulence of axenically cultivated *Entamoeba histolytica*. *Am. J. Trop. Med. Hyg.*, **27**, 248–54.
122. Lushbaugh, W.B., Miller, J.H. (1988) The morphology of *Entamoeba histolytica*, in *Amebiasis: Human Infection* by Entamoeba Histolytica, (ed. J.I. Ravdin), John Wiley, New York, pp. 41–68.
123. Lynch, E.C., Rosenberg, I., Gitler, C. (1982) An ion-channel forming protein produced by *Entamoeba histolytica*. *EMBO J.*, **1**, 801–4.
124. Mackenstedt, U., Schmidt, M., Raether, W., Mehlhorn, H., Uphoff, M. (1990) Increase of DNA content in tissue stages of *Entamoeba histolytica* strain SFL 3. *Parasitol. Res.*, **76**, 373–8.
125. Mann, B.J., Petri, W.A. Jr. (1991) Cell surface proteins of *Entamoeba histolytica*. *Parasitol. Today*, **7**, 173–6.
126. Mann, B.J., Torian, B.E., Vedvick, T.S., Petri, W.A. Jr. (1991) Sequence of a cysteine-rich galactose-specific lectin of *Entamoeba histolytica*. *Proc. Natl. Acad. Sci. U.S.A.*, **88**, 3248–52.
127. Markell, E.K., Havens, R.F., Kuritsubo, R.A., Wingerd, J. (1984) Intestinal protozoa in homosexual men of the San Francisco Bay area: prevalence and correlates of infection. *Am. J. Trop. Med. Hyg.*, **33**, 239–45.
128. Martinez-Palomo, A. (1982) *The Biology of* Entamoeba histolytica, Research Studies Press, Chichester, pp. 1–161.
129. Martinez-Palomo, A., Gonzalez Robles, A., Chavez, B., Orozco, E., Fernandez Castelo, S., Cervantes, A. (1985) Structural bases of the cytolytic mechanisms of *Entamoeba histolytica*. *J. Protozool.*, **32**, 166–75.
130. Martinez-Palomo, A., Gonzalez-Robles, A., de la Torre, M. (1973) Selective agglutination of pathogenic strains of *Entamoeba histolytica* induced by Con. A. *Nature New. Biol.*, **245**, 186–7.
131. Martinez-Palomo, A. (1986) Biology of *Entamoeba histolytica*, in *Amebiasis*, (ed. A. Martinez-Palomo), Elsevier, Amsterdam, pp. 11–44.

132. Mathews, H.M., Moss, D.M., Healy, G.R., Mildvan, D. (1986) Isoenzyme analysis of *Entamoeba histolytica* isolated from homosexual men. *J. Infect. Dis.*, **153**, 793–5.
133. Mathis, C., Mercer, L. (1916) La division simple chez *Entamoeba dysenteriae. Comptes Rendus des Seances de la Societe be Biologie (Paris)*, **79**, 982–4.
134. Mattern, C.F., Diamond, L.S., Daniel, W.A. (1972) Viruses of *Entamoeba histolytica*. II. Morphogenesis of the polyhedral particle. *J. Virol.*, **9**, 342–58.
135. Mattern, C.F., Keister, D.B., Natovitz, P.C. (1980) *Entamoeba histolytica* 'toxin': fetuin neutralizable and lectin-like. *Am. J. Trop Med. Hyg.*, **29**, 26–30.
136. McGowan, K., Deneke, C.F., Thorne, G.M., Gorbach, S.L. (1982) *Entamoeba histolytica* cytotoxin: purification, characterization, strain virulence, and protease activity. *J. lnfect. Dis.*, **146**, 616–25.
137. McGowan, K., Kane, A., Asarkof, N., *et al.* (1983) *Entamoeba histolytica* causes intestinal secretion: role of serotonin. *Science*, **221**, 762–4.
138. McLaughlin, J., Faubert, G. (1977) Partial purification and some properties of a neutral sulfhydryl and an acid proteinase from *Entamoeba histolytica. Can. J. Microbiol.*, **23**, 420–5.
139. McMillan, A., McNeillage, G.J. (1984) Comparison of the sensitivity of microscopy and culture in the laboratory diagnosis of intestinal protozoal infection. *J. Clin. Pathol.*, 37, 809–11.
140. Meza, I., Cazares, F., Rosales-Encina, J.L., Talamas-Rohana, P., Rojkind, M. (1987) Use of antibodies to characterize a 220-kilodalton surface protein from *Entamoeba histolytica. J. Infect. Dis.*, **156**, 798–805.
141. Meza, I., Sabanero, M., Cazares, F., Bryan, J. (1983) Isolation and characterization of actin from *Entamoeba histolytica. J. Biol. Chem.*, **258**, 3936–41.
142. Meza, I., Torres-Guerrero, H.K., Meraz, M.A. (1990) Molecular organizations of *Entamoeba histolytica*, in *Amebiasis, Infection and Disease by* Entamoeba histolytica), (ed. R.R. Kretschmer), CRC Press, Boca Raton, pp. 19–42.
143. Mirelman, D. (1987) Effect of cultural conditions and bacterial associates on the zymodemes of *Entamoeba histolytica. Parasitol. Today*, **3**, 37–40.
144. Mirelman, D. (1987) Ameba–bacterium relationship in amebiasis. *Microbiol. Rev.*, **51**, 272–84.
145. Mirelman, D., Bracha, R., Chayen, A., Aust-Kettis, A., Diamond, L.S. (1986) *Entamoeba histolytica*: effect of growth conditions and bacterial associates on isoenzyme pattems and virulence. *Exp. Parasitol.*, **62**, 142–8.
146. Mirelman, D., Bracha, R., Rozenblatt, S., Garfinkel, L.I. (1990) Repetitive DNA elements characteristic of pathogenic *Entamoeba histolytica* strains can also be detected after polymerase chain reaction in a cloned nonpathogenic strain. *Infect. Immun.*, 58, 1660–3.
147. Mirelman, D., Bracha, R., Sargeaunt, P.G. (1984) *Entamoeba histolytica*: virulence enhancement of isoenzyme-stable parasites. *Exp. Parasitol.*, 57, 172–7.

148. Mirelman, D., Bracha, R., Wexler, A., Chayen, A. (1986) Changes in isoenzyme patterns of a cloned culture of nonpathogenic *Entamoeba histolytica* during axenization. *Infect. Immun.*, **54**, 827–32.
149. Mizerski, J., Chartrand, S., Gajlpeczalska, K. (1981) Con. A induced suppressor cells in man. I. Induction and characterization of suppressor cells. *Immunobiology*, **158**, 270–81.
150. Munoz, M.L., Calderon, J., Rojkind, M. (1982) The collagenase of *Entamoeba histolytica. J. Exp. Med.*, **155**, 42–51.
151. Munoz, M.L., Rojkind, M., Calderon, J., Tanimoto, M., Arias-Negrete, S., Martinez Palomo, A. (1984) *Entamoeba histolytica*: collagenolytic activity and virulence. *J. Protozool.*, **31**, 468–70.
152. Murray, M.J., Murray, A., Murray, C.J. (1980) The salutary effect of milk on amoebiasis and its reversal by iron. *Br. Med. J.*, **280**, 1351–2.
153. Nanda, R., Baveja, U., Anand, B.S. (1984) *Entamoeba histolytica* cyst passers: clinical features and outcome in untreated subjects. *Lancet*, **ii**, 301–3.
154. Neal, R.A. (1960) Enzymic proteolysis by *Entamoeba histolytica*: biochemical characteristics and relationships with invasiveness. *Parasitology*, **50**, 531–50.
155. Neal, R.A. (1966). Experimental studies on *Entamoeba* with reference to speciation. *Adv. Parasitol.*, **4**, 1–51.
156. Nishizuka, Y. (1986) Studies and perspectives of protein kinase C. *Science*, **233**, 305–12.
157. Nozaki, T., Motta, S.R., Takeuchi, T., Kobayashi, S., Sargeaunt, P.G. (1989) Pathogenic zymodemes of *Entamoeba histolytica* in Japanese male homosexual population. *Trans. R. Soc. Trop. Med. Hyg.*, **83**, 525.
158. Ojcius, D.M., Ding-E Young, J. (1990) A role for pore-forming proteins in pathogenesis by parasites. *Parasitol. Today*, **6**, 163–5.
159. Orozco, E., Guameros, G., Martinez-Palomo, A., Sanchez, T. (1983) *Entamoeba histolytica.* Phagocytosis as a virulence factor. *J. Exp. Med.*, **158**, 1511–21.
160. Orozco, E., Martinez-Palomo, A., Lopez-Revilla, R. (1978) Un modelo in vitro para el estudio cuantitativo de la virulencia de *Entamoeba histolytica. Arch. Invest. Med. (Mex.)*, **9** (**Suppl. 1**), 257–60.
161. Orozco, E., Solis, F.J., Dominguez, J., Chavez, B., Hernandez F. (1988) *Entamoeba histolytica*: cell cycle and nuclear division. *Exp. Parasitol.*, **67**, 85–95.
162. Ortiz-Ortiz, L., Capin, R., Capin, N.R., Sepulveda, B., Zamacona G. (1978) Activation of the alternative pathway of complement by *Entamoeba histolytica. Clin. Exp. Immunol.*, **34**, 10–18.
163. Palmer, J.D., Thompson, W.F. (1982) Chloroplast DNA rearrangements are more frequent when a large inverted repeat sequence is lost. *Cell*, **29**, 537–50.
164. Pan, C.T., Geiman, Q.M. (1955) Comparative studies of intestinal amoebae. 1. Distribution and cyclic changes of the nucleic acids of *Entamoeba histolytica* and *Entamoeba coli. Amer. J. Hyg.*, **62**, 66–79.

165. Pearce, R.B. (1983) Intestinal protozoal infections in AIDS [letter]. *Lancet*, **ii**, 51.
166. Pearce, R.B., Abrams, D.I. (1987) *Entamoeba histolytica* in homosexual men [letter]. *N. Engl. J. Med.*, **316**, 690–1.
167. Perez-Montfort, R., Ostoa-Saloma, P., Velazquez-Medina, L., Montfort, I., Becker, I. (1987) Catalytic classes of proteinases of *Entamoeba histolytica. Mol. Biochem. Parasitol.*, **26**, 87–97.
168. Perez-Mutul, J., Cazares, F., Meza, I. (1986) Isolation and initial characterization of DNA from *Entamoeba histolytica* HM1: IMSS. *Arch. Invest. Med. Mex.*, **17 Suppl. 1**, 113–9.
169. Perez-Tamayo, R. (1986) Pathology of amebiasis, in *Amebiasis*, (ed. A. Marinez-Palomo), Elsevier, Amsterdam, pp. 45–94.
170. Petri, W.A, Jr., Chapman, M.D., Snodgrass, T., Mann, B.J., Broman, J., Ravdin, J.I. (1989) Subunit structure of the galactose and *N*-acetyl-*D*-galactosamine-inhibitable adherence lectin of *Entamoeba histolytica. J. Biol. Chem.*, **264**, 3007–12.
171. Petri, W.A. Jr., Jackson, T.F., Gathiram, V., *et al.* (1990) Pathogenic and nonpathogenic strains of *Entamoeba histolytica* can be differentiated by monoclonal antibodies to the galactose-specific adherence lectin. *Infect. Immun.*, **S8**, 1802–6.
172. Petri, W.A. Jr., Ravdin, J.I. (1986) Treatment of homosexual men infected with *Entamoeba histolytica* [letter]. *N. Engl. J. Med.*, **315**, 393.
173. Petri, W.A. Jr., Ravdin, J.I. (1987) Cytopathogenicity of *Entamoeba histolytica*: the role of amebic adherence and contact-dependent cytolysis in pathogenesis. *Eur. J. Epidemiol.*, **3**, 123–36.
174. Petri, W.A. Jr., Ravdin, J.I. (1991) Protection of gerbils from amebic liver abscess by immunization with the galactose-specific adherence lectin of *Entamoeba histolytica. Infect. Immun.*, **59**, 97–101.
175. Petri, W.A. Jr., Smith, R.D., Schlesinger, P.H, Murphy, C.F. Ravdin, J.I. (1987) Isolation of the galactose-binding lectin that mediates the *in vitro* adherence of *Entamoeba histolytica. J. Clin. Invest.*, **80**, 1238–44.
176. Petri, W.A. Jr., Snodgrass, T.L., Jackson, T.F., *et al.* (1990) Monoclonal antibodies directed against the galactose-binding lectin of *Entamoeba histolytica* enhance adherence. *J. Immunol.*, **144**, 4803–9.
177. Phillips, B.P., Bartgis, I.L. (1954) Effects of growth *in vitro* with selected microbial associates and of encystation and excystation, on the virulence of *Entamoeba histolytica* for the guinea pig. *Amer. J. Trop. Med. Hyg.*, **3**, 621–7.
178. Phillips, B.P., Diamond, L.S., Bartgis, I.L., Stuppler, S.A. (1981) Results of intracaecal innoculation of germ-free and conventional guinea-pigs and germ-free rats with axenically cultivated *Entamoeba histolytica. J. Protozool.*, **31**, 571–3.
179. Phillips, B.P., Gorstein, F. (1966) Effects of different species of bacteria on the pathology of enteric amebiasis in monocontaminated guinea pigs. *Am. J. Trop. Med. Hyg.*, **15**, 863–8.
180. Phillips, B.P., Wolfe, P.A., Bartgis, I.L. (1958) Studies on the ameba–bacteria

relationship II. Some concepts on the etiology of the disease. *Amer. J. Trop. Med. Hyg.*, 7, 392–9.

181. Phillips, B.P., Wolfe, P.A., Rees, C.W., Gordon, H.A., Wright, W.H., Reyniers, J.A. (1955) Studies on the ameba–bacteria relationship in amebiasis, comparative results of the intracaecal innoculation of germ-free, monocontaminated and conventional guinea-pigs with *Entamoeba histolytica. Amer. J. Trop. Med. Hyg.*, **4**, 675–92.
182. Phillips, S.C., Mildvan, D., William, D.C. (1981) Sexual transmision of enteric protozoa and helminths in a venereal disease clinic population. *N. Engl. J. Med.*, **305**, 603–6.
183. Prachayasittikul, V., Albach, R.A. (1982) RNA depolymerase in *Entamoeba histolytica*, soluble vs ribosomal. *Arch. Invest. Med. (Mex.)*, **13**, 29–35.
184. Proctor, E.M., Wong, Q., Yang, J., Keystone, J.S. (1987) The electrophoretic isoenzyme patterns of strains of *Entamoeba histolytica* isolated in two major cities in Canada. *Am. J. Trop. Med. Hyg.*, **37**, 296–301.
185. Quinn, T.C, Stamm, W.E., Goodell, S.E., *et al.* (1983) The polymicrobial origin of intestinal infections in homosexual men. *N. Engl. J. Med.*, **309**, 576–82.
186. Ravdin, J.I. (1986) Pathogenesis of disease caused by *Entamoeba histolytica*: studies of adherence, secreted toxins, and contact-dependent cytolysis. *Rev. Infect. Dis.*, **8**, 247–60.
187. Ravdin, J.I. (1988) Intestinal disease caused by *Entamoeba histolytica*, in *Amebiasis: Human infection with* Entamoeba histolytica, (ed. J.I. Ravdin), John Wiley, New York, pp. 495–510.
188. Ravdin, J.I. (1989). Amebiasis now. *Am. J. Trop. Med. Hyg.*, **41**, 40–8.
189. Ravdin, J.I. (1990) *Entamoeba histolytica*: pathogenic mechanisms, human immune response, and vaccine development. *Clin. Res.*, **38**, 215–25.
190. Ravdin, J.I., Croft, B.Y., Guerrant, R.L. (1980) Cytopathogenic mechanisms of *Entamoeba histolytica. J. Exp. Med.*, **152**, 377–90.
191. Ravdin, J.I., Guerrant, R.L. (1981) Role of adherence in cytopathogenic mechanisms of *Entamoeba histolytica.* Study with mammalian tissue culture cells and human erythrocytes. *J. Clin. Invest.*, **68**, 1305–13.
192. Ravdin, J.I., Guerrant, R.L. (1982) A review of the parasite cellular mechanisms involved in the pathogenesis of amebiasis. *Rev. Infect. Dis.*, **4**, 1185–207.
193. Ravdin, J.I., Guerrant, R.L. (1982) Separation of adherence, cytolytic, and phagocytic events in the cytopathogenic mechanisms of *Entamoeba histolytica. Arch. Invest. Med. (Mex.)*, **13**, 123–8.
194. Ravdin, J.I., John, J.E., Johnston, L.I., Innes, D.J., Guerrant, R.L. (1985) Adherence of *Entamoeba histolytica* trophozoites to rat and human colonic mucosa. *Infect. Immun.*, **48**, 292–7.
195. Ravdin, J.I., Moreau, F., Sullivan, J.A., Petri, W.A. Jr., Mandell, G.L. (1988) Relationship of free intracellular calcium to the cytolytic activity of *Entamoeba histolytica. Infect. Immun.*, **56**, 1505–12.
196. Ravdin, J.I., Murphy, C.F., Guerrant, R.L., Long Krug, S.A. (1985) Effect of antagonists of calcium and phospholipase A on the cytopathogenicity of *Entamoeba histolytica. J. Infect. Dis.*, **S2**, 542–549.

197. Ravdin, J.I., Murphy, C.F., Salata, R.A., Guerrant, R.L., Hewlett E.L. (1985) *N*-acetyl-*D*-galactosamine-inhibitable adherence lectin of *Entamoeba histolytica*. I. Partial purification and relation to amoebic virulence *in vitro*. *J. Infect. Dis.*, **151**, 804–15.
198. Ravdin, J.I., Petri, W.A., Murphy, C.F., Smith, R.D. (1986) Production of mouse monoclonal antibodies which inhibit *in vitro* adherence of *Entamoeba histolytica* trophozoites. *Infect. Immun.*, **53**, 1–5.
199. Ravdin, J.I., Sperelakis, N., Guerrant, R.L. (1982) Effect of ion channel inhibitors on the cytopathogenicity of *Entamoeba histolytica*. *J. Infect. Dis.*, **146**, 335–40.
200. Ravdin, J.I., Stanley, P., Murphy, C.F., Petri, W.A. Jr. (1989) Characterization of cell surface carbohydrate receptors for *Entamoeba histolytica* adherence lectin. *Infect. Immun.*, **57**, 2179–86.
201. Reed, S.L., Braude, A.I. (1988) Extraintestinal disease: clinical syndromes, diagnostic profile and therapy, in, *Amebiasis: Human Infection by* Entamoeba histolytica, (ed. J.I. Ravdin), John Wiley, New York, pp. 511–32.
202. Reed, S.L., Curd, J.G., Gigli, I., Gillin, F.D., Braude, A.I. (1986) Activation of complement by pathogenic and nonpathogenic *Entamoeba histolytica*. *J. Immunol.*, **136**, 2265–70.
203. Reed, S.L., Keene, W.E., McKerrow, J.H. (1989) Thiol proteinase expression and pathogenicity of *Entamoeba histolytica*. *J. Clin. Microbiol.*, **27**, 2772–7.
204. Reed, S.L., Keene, W.E., McKerrow, J.H., Gigli, I. (1989) Cleavage of C3 by a neutral cysteine proteinase of *Entamoeba histolytica*. *J. Immunol.*, **143**, 189–95.
205. Reed, S.L., Wessel, D.W., Davis, C.E. (1991) *Entamoeba histolytica* infection and AIDS. *Am. J. Med.*, **90**, 269–71.
206. Reeder, R.H. (1984). Enhancers and ribosomal gene spacers. *Cell*, **38**, 349–51.
207. Reeves, R.E., Lushbaugh, T.S., Montalvo, F.E. (1971) Characterization of deoxyribonucleic acid of *Entamoeba histolytica* by cesium chloride density centrifugation. *J. Parasitol.*, **57**, 939–44.
208. Rodriguez, M.A., Orozco, E. (1986) Isolation and characterization of phagocytosis and virulence-deficient mutants of *Entamoeba histolytica*. *J. Infect. Dis.*, **154**, 27–32.
209. Root, D.M., Cole, F.X., Williamson, J.A. (1978). The development and standaridization of an ELISA method for the detection of *Entamoeba histolytica* antigens in fecal samples. *Arch. Invest. Med. (Mex.)*, **9**, 203–10.
210. Rosales-Encina, J.L., Meza, I., Lopez De Leon, A., Talamas-Rohana, P., Rojkind, M. (1987) Isolation of a 220-kilodalton protein with lectin properties from a virulent strain of *Entamoeba histolytica*. *J. Infect. Dis.*, **156**, 790–7.
211. Rosenbaum, R.M., Wittner, M. (1970) Ultrastructure of bacterized and axenic trophozoites of *Entamoeba histolytica* with particular reference to helical bodies. *J. Cell Biol.*, **45**, 367–82.
212. Rosenberg, I., Bach, D., Loew, L.M., Gitler, C. (1989) Isolation, characterization and partial purification of a transferable membrane channel

(amoebapore) produced by *Entamoeba histolytica. Mol. Biochem. Parasitol.*, **33**, 237–47.

213. Rosenberg, I., Gitler, C. (1985) Subcellular fractionation of amoebapore and plasma membrane components of *Entamoeba histolytica* using self-generating Percoll gradients. *Mol. Biochem. Parasitol.*, **14**, 231–48.
214. Ruebush, T.K. II, Juranek, D.D., Brodsky, R.E. (1978) Diagnoses of intestinal parasites by state and territorial public health laboratories. *J. Inf. Dis.*, **138**, 114–7.
215. Said-Femandez, S., Lopez-Revilla, R. (1982) Subcellular distribution and stability of the major hemolytic activity of *Entamoeba histolytica* trophozoites. *Z. Parasitenkd.*, **67**, 249–54.
216. Said-Femandez, S., Lopez-Revilla, R. (1988) Free fatty acids released from phospholipids are the major heat-stable hemolytic factor of *Entamoeba histolytica* trophozoites. *Infect. Immun.*, **56**, 874–9.
217. Salata, R.A., Martinez Palomo, A., Murray, H.W., *et al.* (1986) Patients treated for amebic liver abscess develop cell-mediated immune responses effective *in vitro* against *Entamoeba histolytica. J. Immunol.*, **136**, 2633–9.
218. Salata, R.A., Ravdin, J.I. (1986) Review of the human immune mechanisms directed against *Entamoeba histolytica. Rev. Infect. Dis.*, **8**, 261–72.
219. Salata, R.A., Schacter, B.Z., Ellner, J.J. (1983) Recruitment of OKM1 staining lymphocytes with selective binding to K-562 tumour targets by interferon. *Clin. Exp. Immunol.*, **52**, 185–90.
220. Samaras, N., Spithill, T.W. (1987) Molecular karyotype of five species of *Leishmania* and analysis of gene locations and chromosomal rearrangements. *Mol. Biochem. Parasitol.*, **25**, 279–91.
221. Samuelson, J., Acuna Soto, R., Reed, S., Biagi, F., Wirth, D. (1989) DNA hybridization probe for clinical diagnosis of *Entamoeba histolytica. J. Clin. Microbiol.*, **27**, 671–6.
222. Sargeaunt, P.G. (1987) The reliability of *Entamoeba histolytica* zymodemes in clinical diagnosis. *Parasitol. Today*, **3**, 40–3.
223. Sargeaunt, P.G. (1988) Zymodemes of *Entamoeba histolytica*, in *Amebiasis: Human Infection with* Entamoeba histolytica, (ed. J.I. Ravdin), John Wiley, New York, pp. 370–87.
224. Sargeaunt, P.G., Jackson, T,F., Wiffen, S., *et al.* (1987) The reliability of *Entamoeba histolytica* zymodemes in clinical laboratory diagnosis. *Arch. Invest. Med. (Mex.)*, **18**, 69–75.
225. Sargeaunt, P.G., Jackson, T.F., Wiffen, S.R., Bhojnani, R. (1988) Biological evidence of genetic exchange in *Entamoeba histolytica. Trans. R. Soc. Trop. Med. Hyg.*, **82**, 862–7.
226. Sargeaunt, P.G., Oates, J.K., MacLennan, I., Oriel, J.D., Goldmeier, D. (1983) *Entamoeba histolytica* in male homosexuals. *Br. J. Vener. Dis.*, **59**, 193–5.
227. Sargeaunt, P.G., Williams, J.E. (1979) Electrophoretic isoenzyme patterns of the pathogenic and non-pathogenic intestinal amoebae of man. *Trans. Roy. Soc. Trop. Med. Hyg.*, **73**, 225–7.
228. Sargeaunt, P.G., Williams, J.E., Grene, J.D. (1978) The differentiation of

invasive and non-invasive *Entamoeba histolytica* by isoenzyme electrophoresis. *Trans. R. Soc. Trop. Med. Hyg.*, **72**, 519–21.

229. Sargeaunt, P.G., Williams, J.E., Neal, R.A. (1980) A comparative study of *Entamoeba histolytica* (NIH, 200, HK9, etc.), '*E. histolytica*-like' and other morphologically identical amoebae using isoenzyme electrophoresis. *Trans. R. Soc. Trop. Med. Hyg.*, **74**, 469–74.
230. Scholze, H., Werries, E. (1984) A weakly acidic protease has a powerful proteolytic activity in *Entamoeba histolytica. Mol. Biochem. Parasitol.*, **11**, 293–300.
231. Schulte, W., Scholze, H. (1989). Action of the major protease from *Entamoeba histolytica* on proteins of the extracellular matrix. *J. Protozool.*, **36**, 538–43.
232. Schwartz, D.C., Cantor, C.R. (1984) Separation of yeast chromosome-sized DNAs by pulsed field gradient gel electrophoresis. *Cell*, **37**, 67–75.
233. Sewankambo, N., Mugerwa, R.D., Goodgame,R., *et al.* (1987) Enteropathic AIDS in Uganda. An endoscopic, histological and microbiological study. *Aids*, **1**, 9–13.
234. Sharma, N.N., Albach, R.A., Shaffer, J.G. (1969) Autoradiographic studies of uridine-S-H3 uptake in *Entamoeba histolytica* in the CLG medium. *N. Engl. J. Med.*, **281**, 1004–11.
235. Smith, J.M., Meerovitch, E. (1982) *Entamoeba histolytica*: the dynamics of penetrating a meniscus. *Can. J. Zool.*, **60**, 2124–8.
236. Smith, J.W. (1978) Identification of fecal parasites in the special parasitology survey of the College of American Pathologists. *Am. J. Clin. Pathol.*, **72** (**Suppl. 2**), 371.
237. Smith, P.D., Lane, H.C., Gill, V.J., *et al.* (1988) Intestinal infections in patients with the acquired immunodeficiency syndrome (AIDS). Etiology and response to therapy. *Ann. Intern. Med.*, **108**, 328–33.
238. Solis, F., Orozco, E. (1991) *Entamoeba histolytica*: generation and characterization of hybrid clones. *Exp. Parasitol.*, **72**, 236–42.
239. Solis, F.J., Chavez, B., Orozco, E. (1986) Phases of nuclear division in *Entamoeba histolytica. Arch. Invest. Med. (Mex.)*, **17 Suppl 1**, 95–9.
240. Stanley, S.L. Jr., Becker, A., Kunz-Jenkins, C., Foster, L., Li, E. (1990) Cloning and expression of a membrane antigen of *Entamoeba histolytica* possessing multiple tandem repeats. *Proc. Natl. Acad. Sci. U.S.A.*, **87**, 4976–80.
241. Sullam, P.M., Slutkin, G., Gottlieb, A.B., Mills, J. (1986) Paromomycin therapy of endemic amebiasis in homosexual men. *Sex. Transm. Dis.*, **13**, 151–5.
242. Tachibana, H., Kobayashi, S., Kato, Y., Nagakura, K., Kaneda, Y., Takeuchi, T. (1990) Identification of a pathogenic isolate-specific 30 000-Mr antigen of *Entamoeba histolytica* by using a monoclonal antibody. *Infect. Immun.*, **58**, 955–60.
243. Takeuchi, A., Phillips B.P. (1975) Electron microscope studies of experimental *Entamoeba histolytica* infection in the guinea pig. I. Penetration of the intestinal epithelium by trophozoites. *Am. J. Trop. Med. Hyg.*, **24**, 34–48.

244. Takeuchi, A., Phillips, B.P. (1976) Electron microscopic studies of experimental *Entamoeba histolytica* infection in the guinea pig. II. Early cellular and vascular changes accompanying invasion of the lamina propria. *Virchows Arch. [B] Cell. Pathol.*, **20**, 1–13.
245. Takeuchi, T., Miyahira Y., Kobayashi, S., Nozaki, T., Motta, S.R., Matsuda, J. (1990) High seropositivity for *Entamoeba histolytica* infection in Japanese homosexual men: further evidence for the occurrence of pathogenic strains. *Trans. R. Soc. Trop. Med. Hyg.*, **84**, 250–1.
246. Takeuchi, T., Okuzawa, E., Nozaki, T., *et al.* (1989) High seropositivity of Japanese homosexual men for amebic infection [letter]. *J. Infect. Dis.*, **S9**, 808.
247. Talamas-Rohana, P., Meza, I. (1988) Interaction between pathogenic amebas and fibronectin, substrate degradation and changes in cytoskeleton organization. *J. Cell Biol.*, **106**, 1787–94.
248. Tannich, E., Burchard, G.D. (1991) Differentiation of pathogenic from nonpathogenic *Entamoeba histolytica* by restriction fragment analysis of a single gene amplified *in vitro*. *J. Clin. Microbiol.*, **29**, 250–5.
249. Tannich, E., Ebert, F., Horstmann, R.D. (1991) Primary structure of the 170-kDa surface lectin of pathogenic *Entamoeba histolytica*. *Proc. Natl. Acad. Sci. U.S.A.*, **88**, 1849–53.
250. Tannich, E., Horstmann, R.D., Knobloch, J., Arnold, H.H. (1989) Genomic DNA differences between pathogenic and nonpathogenic *Entamoeba histolytica*. *Proc. Natl. Acad. Sci. U.S.A.*, **86**, 5118–22.
251. Tannich, E., Scholze, H., Nickel, R., Horstmann, R.D. (1991) Homologous cysteine proteinases of pathogenic and nonpathogenic *Entamoeba histolytica*. Differences in structure and expression. *J. Biol. Chem.*, **266**, 4798–803.
252. Thoma, F., Koller, T. (1981) Unravelled nucleosomes, nucleosomal beads and higher order structures of chromatin, influence of non-histone components and H1. *J. Mol. Biol.*, **149**, 709–33.
253. Thomas, C.A. Jr. (1971) The genetic organization of chromosomes. *Ann. Rev. Genet.*, **5**, 237–56.
254. Torian, B.E., Flores, B.M., Stroeher, V.L., Hagen, F.S., Stamm, W.E. (1990) cDNA sequence analysis of a 29-kDa cysteine-rich surface antigen of pathogenic *Entamoeba histolytica*. *Proc. Natl Acad. Sci. U.S.A.*, **87**, 6358–62.
255. Torian, B.E., Lukehart, S.A., Stamm, W.E. (1987) Use of monoclonal antibodies to identify, characterize, and purify a 96 000-dalton surface antigen of pathogenic *Entamoeba histolytica*. *J. Infect. Dis.*, **156**, 334–43.
256. Torian, B.E., Reed, S.L., Flores, B.M., *et al.* (1990) The 96-kilodalton antigen as an integral membrane protein in pathogenic *Entamoeba histolytica*: potential differences in pathogenic and nonpathogenic isolates. *Infect. Immun.*, **58**, 753–60.
257. Torres-Guerrero, H., Peattie, D.A., Meza, I. (1991) Chromatin organization in *Entamoeba histolytica*. *Mol. Biochem. Parasitol.*, **45**, 121–30.
258. Trissl, D. (1982) Immunology of *Entamoeba histolytica* in human and animal hosts. *Rev. Infect. Dis.*, **4**, 1154–84.

259. Trissl, D., Martinez-Palomo, A., de la Tolte, M., de la Hoz, R., Suarez, E. (1978) Surface properties of *Entamoeba*: increased rates of human erythrocyte phagocytosis in pathogenic strains. *J. Exp. Med.*, **148**, 1137–43.
260. Udezulu, I.A., Leitch, G.J., Bailey, G.B. (1982) Use of indomethacin to demonstrate enterotoxic activity in extracts of *Entamoeba histolytica* trophozoites. *Infect. Immun.*, **36**, 795–801.
261. Valdes, J., de la Cruz-Hernandez, F., Ocadiz, R., Orozco, E. (1990) Molecular karyotype of *Entamoeba histolytica* and *Entamoeba invadens*. *Trans. R. Soc. Trop. Med. Hyg.*, **84**, 537–41.
262. Van de Perre, P., Rouvroy, D., Lepage, P., *et al.* (1984) Acquired immunodeficiency syndrome in Rwanda. *Lancet*, **ii**, 62–5.
263. Van der Ploeg, L.H., Schwartz, D.C., Cantor, C.R., Borst, P. (1984) Antigenic variation in *Trypanosoma brucei* analyzed by electrophoretic separation of chromosome-sized DNA molecules. *Cell*, **37**, 77–84.
264. Vogt, V.M., Braun, R. (1976) Structure of ribosomal DNA in *Physarum polycephalum*. *J. Mol. Biol.*, **106**, 567–87.
265. Walsh, J.A. (1986) Problems in recognition and diagnosis of amebiasis: estimation of the global magnitude of morbidity and mortality. *Rev. Infect. Dis.*, **8**, 228–38.
266. Walsh, J.A. (1988) Prevalence of *Entamoeba histolytica* infection, in *Amebiasis: Human Infection by Entamoeba histolytica*, (ed. J.I. Ravdin), John Wiley, New York, pp. 93–105.
267. Weber, J. (1985) Sexually acquired parasitic infections in homosexual men. *Parasitol. Today*, **1**, 93–5.
268. Weber, J.N., Wadsworth, J., Rogers, L.A., *et al.* (1986) Three-year prospective study of HTLV-III/LAV infection in homosexual men. *Lancet*, **i**, 1179–82.
269. Weikel, C.S., Murphy, C.F., Orozco, E., Ravdin, J.I. (1988) Phorbol esters specifically enhance the cytolytic activity of *Entamoeba histolytica*. *Infect. Immun.*, **56**, 1485–91.
270. Weinke, T., Friedrich Janicke, B., Hopp, P., Janitschke, K. (1990) Prevalence and clinical importance of *Entamoeba histolytica* in two high-risk groups: travellers returning from the tropics and male homosexuals. *J. Infect. Dis.*, **161**, 1029–31.
271. Wittner, M., Rosenbaum, R.M. (1970) Role of bacteria in modifying virulence of *Entamoeba histolytica*. Studies of amebae from axenic cultures. *Am. J. Trop. Med. Hyg.*, **19**, 755–61.
272. Young, J.D., Lowrey, D.M. (1989) Biochemical and functional characterization of a membrane-associated pore-forming protein from the pathogenic ameboflagellate *Naegleria fowleri*. *J. Biol. Chem.*, **264**, 1077–83.
273. Young, J.D., Young, T.M., Lu, L.P., Unkeless, J.C., Cohn, Z.A. (1982) Characterization of a membrane pore-forming protein from *Entamoeba histolytica*. *J. Exp. Med.*, **156**, 1677–90.
274. Zagury, D., Bernard, J., Leonard, R., *et al.* (1986) Long-term cultures of HTLV-infected T-cells: a model of cytopathology of T-cell depletion in AIDS. *Science*, **231**, 850–3.

6

Cryptosporidium and *Isospora*

ALAIN BONNIN, EDUARDO DEI-CAS and PATRICK CAMERLYNCK

6.1 INTRODUCTION

The coccidian *Cryptosporidium* spp., *Isospora belli*, *Sarcocystis suihominis* and *S. bovihominis* are intestinal parasites of humans. *Coccidia* are obligate parasitic intracellular protozoa, with a typical apical complex seen with the electron microscope. There are species with a single host (monoxenous) and species with a definitive (or final) host and an intermediate host (heteroxenous). The parasite specificity is higher to the final host, in which the sexual multiplication and the formation of the oocyst containing the infective stages (sporozoites) take place. Asexual endogenous or tissue stages develop in the intermediate host.

Human coccidia and related human sporozoa are shown in Table 6.1. Only *Toxoplasma*, *Cryptosporidium* and *I. belli* infections are associated with the acquired immune deficiency syndrome (AIDS). The last two protozoans cause self-limiting diarrhoeal illness in immunocompetent hosts, whereas they are the most commonly identified agents of chronic diarrhoea, clinically indistinguishable, in AIDS patients [66]. Cryptosporidiosis and isosporiasis are considered as opportunistic infections in the CDC definition for AIDS [13].

Only recently have *Cryptosporidium* spp. been recognized as important intestinal pathogenic protozoa in patients suffering from AIDS [66]. Cryptosporidiosis can be transmitted directly from person to person or animal to man, and indirectly by fomites, water or food. However, as *Cryptosporidium* spp. were reported in many homeotherm and poikilotherm host species, there is also a large zoonotic potential reservoir of parasite [11]. In contrast, *I. belli* only infects humans and no other reservoir hosts are recognized [28].

Molecular and Cell Biology of Opportunistic Infections in AIDS
Edited by Steven Myint and Alan Cann
Published in 1992 by Chapman & Hall, London ISBN 0 412 45330 4

Table 6.1 Human coccidia and other human sporozoan parasites

Species	*Vertebrate reservoirs*	*Life cycle*	*Experimental hosts*	*Associated with AIDS*
Cryptosporidium spp.	Mammals, birds, fishes, reptiles	M	Mainly rodents	Yes
Isospora belli	Human	M	–	Yes
I. natalensis	Human?	M?	–	?
Sarcocystis bovihominis	Human	H (ox)	–	No
S. suihominis	Human	H (pig)	–	No
Toxoplasma *gondii*	Cat, other mammals, birds	H (homeotherms)	Rodents	Yes
Plasmodium spp.[a]	Human	H (anopheles)	Primates	No
Babesia spp.	Cattle, dogs, rodents	H (ticks)	Mainly rodents	No

M, monoxenous; H, heteroxenous; invertebrate or intermediate hosts are in parentheses.
[a] *Plasmodium falciparum*, *P. vivax*, *P. ovale*, *P. malariae*.

6.2 *CRYPTOSPORIDIUM*

6.2.1 Introduction

Organisms of the genus *Cryptosporidium* are widespread coccidian protozoans that infect the digestive and respiratory tract of many domestic and wild animals, including man. Since Tyzzer's original description in 1907, *Cryptosporidium* spp. have been considered to be rare commensal organisms. Increased interest by veterinary workers began when *Cryptosporidium* spp. were discovered to be the cause of turkey and bovine enteritis, and in the ensuing years, *Cryptosporidium* has appeared as a significant pathogen in young livestock animals.

The first cases of human cryptosporidiosis were reported in 1976, and most subsequent reports concerned immunocompromised patients. However, human cryptosporidiosis was still considered as a rare disease. This concept was transformed with the emergence of the acquired immune deficiency syndrome. Indeed, *C. parvum* appeared as one of the major pathogens responsible for opportunistic infections in AIDS patients. Since then, new techniques for rapid and reliable identification of the parasite in stool samples have been developed and have led the way to widespread investigation of the epidemiology of cryptosporidiosis. These studies demonstrated that cryptosporidiosis is a worldwide cause of diarrhoea in immunocompetent persons, especially in young children and in developing countries.

The extreme resistance of the oocysts accounts for the widespread dissemination of the parasite and for large waterborne outbreaks. The severity of the disease in AIDS patients and the usual resistance of *C. parvum* to all chemotherapeutic compounds evaluated so far emphasize the need for original approaches in the treatment and prevention of cryptosporidiosis.

6.2.2 Molecular aspects of taxonomy

The assignation of the genus *Cryptosporidium* to the phylum *Apicomplexa* is determined by the ultrastructure of the invasive stages, the zoites, that possess an apical complex of highly differentiated organelles, the rhoptries and the micronemes [20], structures probably involved in host cell invasion and characteristic of the phylum [25]. Thus, on the basis of its life cycle and ultrastructural features, *Cryptosporidium* spp. are closely related to other apicomplexan genera. Analysis performed at the molecular level has not yet confirmed this taxonomic position, since comparison of RNA sequences from the semi-conserved regions of the small ribosomal subunit did not demonstrate any especially close relationship between *Cryptosporidium* and the other *Apicomplexa* [33].

Cryptosporidium was initially assumed to be host specific, and 19 distinct species had been named by 1984 [37]. Yet clinical observations and cross-transmission studies have demonstrated that some isolates were not host specific, suggesting the possibility of a single species genus [73]. However, in 1984 Levine [37] nominated four species of *Cryptosporidium*, infecting mammals (*C. muris*), birds (*C. meleagridis*), fish (*C. nasorum*) and reptiles (*C. crotali*). Later studies by Upton and Current [80] indicated that, on the basis of morphological and biological features, Tyzzer's original distinction of two mammalian species, *C. muris* and *C. parvum* was valid. Similarly, in birds, Current *et al.* [21] distinguished *C. baileyi* from *C. meleagridis*. Although it was recently demonstrated that *C. baileyi* may produce illness in immunocompromised patients [24], cases of clinical cryptosporidiosis in man and other mammals are now considered to be due to *C. parvum* [17]. Thus, most of the data summarized in this chapter have been obtained with isolates identified either as *C. parvum*, or at least collected from the stools of mammals suffering from diarrhoea.

Molecular techniques have offered more precise methods of classifying *Cryptosporidium* isolates. Field inversion gel electrophoresis has shown that the migration pattern of sporozoite chromosomal DNA from one isolate of *C. baileyi* differed distinctly from the pattern obtained with five *C. parvum* isolates [45]. Indeed, immunofluorescence reactivity of an oocyst wall specific monoclonal antibody also distinguished the avian isolate of the parasite [45]. Similarly, Tilley *et al.* [71] have recently used autoradiography to demonstrate that, although *C. baileyi*, *C. muris* and *C. parvum* share common proteins, several molecules associated with the oocyst wall are species specific. Thus, molecular structural analysis seems to confirm the species distinction established on morphological and biological grounds.

Another question regarding the taxonomy of *Cryptosporidium* is whether there is intraspecific variation. Mead *et al.* [45] failed to demonstrate any differences between five isolates of *C. parvum* compared by field inversion gel electrophoretic analysis of chromosomal DNA and by their reactivity with anti oocyst wall monoclonal antibodies. However, analysis of sporozoite peptides by two-dimensional gel electrophoresis has provided evidence for a discrete biochemical variation among isolates [47]. The biological significance of this variation, especially with respect to differences in strain virulence or host adaptation, requires further investigation.

6.2.3 Life cycle

The life cycle of *Cryptosporidium* is monoxenous. Like other *Coccidia*, it consists of merogony, gametogony and sporogony (Figure 6.1). Contamination occurs through ingestion of thick-wall sporulated oocysts that release infective sporozoites in the gastrointestinal tract of the host. Transmission electron microscopy has revealed that the dissolution of a suture in the oocyst wall occurs during excystation and permits the release of sporozoites [56]. Unlike other *Coccidia*, *Cryptosporidium* may excyst *in vitro* without exposure of the oocysts to reducing conditions, trypsin or bile salts [26, 55, 59], suggesting that the excystation stimuli of *Cryptosporidium* are different from those used by other *Coccidia*.

Free sporozoites exhibit gliding and flexing movements [20]. They probe the microvillous region of the host cell and gain access to the parasitophorous vacuole, an intracellular compartment surrounded by a membrane derived from the host cell plasmalemma [40, 44]. Although this process of invasion is highly conserved among the sporozoans [25], *Cryptosporidium* is unique within the phylum in the location of the parasitophorous vacuole in the microvillous region at the surface of the host cell [20, 40, 44].

Following internalization, the parasite rounds up and undergoes dedifferentiation until all cytoplasmic organelles are resorbed. Redifferentiation of the trophozoite is characterized by the generation of rough

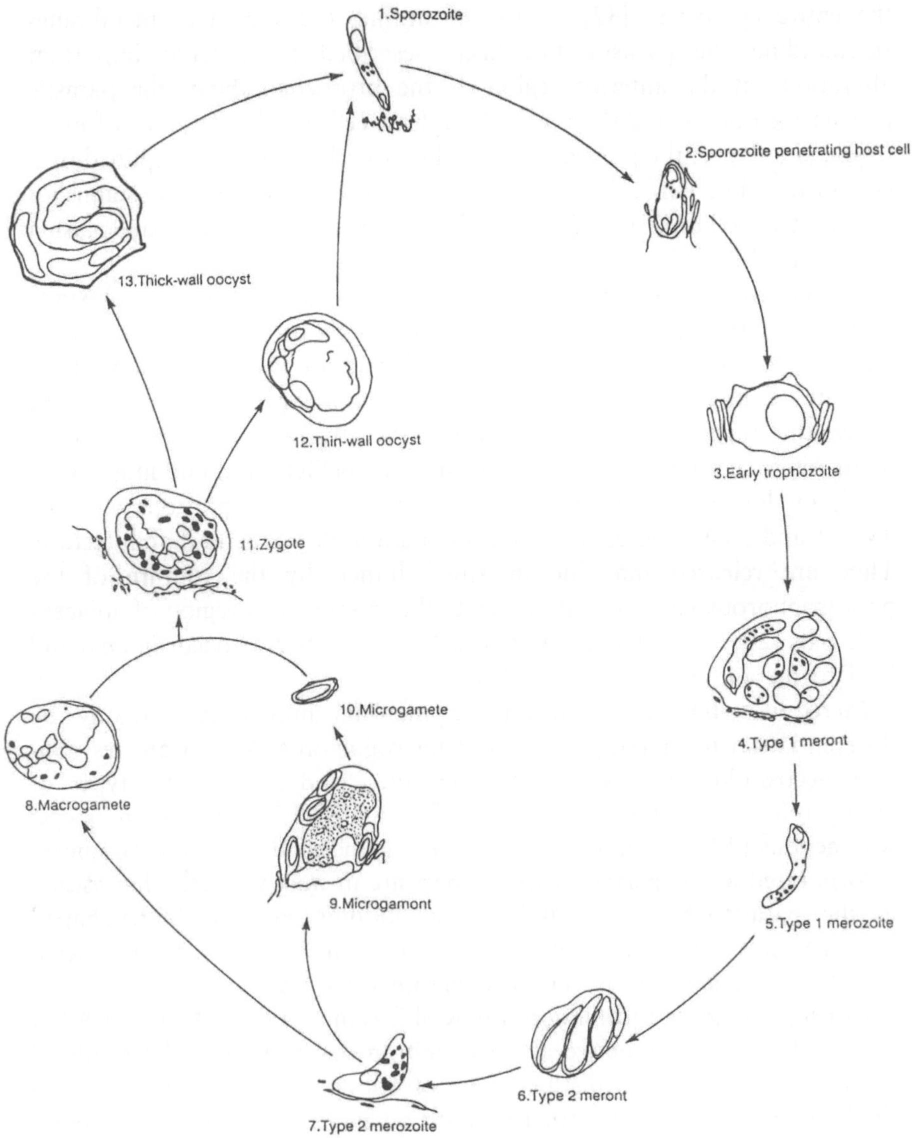

Figure 6.1 Representation of the life-cycle of *Cryptosporidium parvum* (diagrams are adapted from electron micrographs). (1) to (7), merogony; (8) to (10), gametogony; (11) to (13), sporogony. The development of the parasite takes place in the brush border region of epithelial cells, in the intestinal tract of the host. Merozoites are released in the intestinal lumen and invade adjacent microvilli. Thick wall oocysts are excreted in the environment with the stools, whereas thin wall oocysts probably excyst *in situ*. Recycling of type 1 merozoites has been described.

endoplasmic reticulum which develops adjacent to the nucleus until it fills the entire cytoplasm [82]. In the meantime, the system of membranes surrounding the parasite has been described to undergo important alterations at the anterior region of the protozoan. First, the parasite plasma membrane and the internalized host cell membrane join to form a fusion site. Later, the parasite part of this complex becomes amplified by a system of folds, while the host cell membrane disintegrates to establish a direct contact between the host cell cytoplasm and the parasite plasma membrane [20, 44].

During the process of schizogony, the nucleus of the trophozoite undergoes either three divisions (type 1 meronts) or two divisions (type 2 meronts) to give rise to small nuclei which migrate to the periphery of the parasite (Figure 6.1). Subsequently, the parasite pellicle invaginates to cover the nuclei, and individualize eight (type 1) or four (type 2) merozoites, surrounded by a three-membrane pellicle, and budding from a residuum located at the basilar region of the parasitophorous vacuole. Type 1 and type 2 merozoites have a common ultrastructural architecture. They are released into the intestinal lumen by the rupture of the parasitophorous vacuole and penetrate the microvillous region of adjacent host cells [20, 82]. A recycling of type 1 merozoites may occur *in vivo* and lead to an autoinfective cycle [20].

Merozoites released from type 2 meronts initiate gametogony by differentiation in macrogamonts and microgamonts. Macrogametes contain electron-lucent polysaccharide granules, and two distinct types of electron-dense wall-forming bodies (WF1 and WF2) involved in oocyst wall genesis [20]. The nucleus of the microgamonts divides into 16 nuclei, incorporated at the periphery of the parasite in microgametes that escape in the parasitophorous vacuole by a budding process. Bullet-shaped microgametes are released in the intestinal lumen and fertilize macrogametes by penetration of microgamete into macrogamete [72].

During oocyst wall formation, the wall-forming bodies (WF1 and WF2) release their contents between the membranes surrounding the fertilized macrogamete [20]. Two distinct types of oocysts have been described on the basis of the wall ultrastructure. Both types are sporulated and contain four naked sporozoites when mature. Double-wall ('thick-wall') oocysts are discharged in the faeces and result in dissemination of the parasite in the environment and host contamination. Thin-wall oocysts that have a single layered wall excyst in the same host and may be responsible for autoinfection.

Complete development of the parasite *in vitro* has been reported in several cell lines such as human foetal lung, primary chicken kidney and porcine kidney cells [18], and in human colonic cancer cells [23]. Asexual development of *C. parvum* has also been achieved in a cell line derived from a carcinoma of the small intestine in rats [9]. Recycling of type 1 merozoites seems to occur *in vitro* [9, 18]. *C. parvum* was also reported to undergo complete development from sporozoites to oocysts in chickens embryos [19].

6.2.4 Molecular structure; interaction with the infected host

Antibodies to *Cryptosporidium* have been detected in the serum [12, 74, 77, 78], faeces [31] and colostrum [70] of humans and animals infected with *Cryptosporidium*, but the respective role of immunoglobulins and cell mediated immunity in clearing the parasite remains controversial. Clinical and experimental data suggest that anti-cryptosporidial antibodies may have a role in host defences. Chronic cryptosporidiosis has been reported in patients with agammaglobulinaemia [64] or IgA deficiency [32]. A rise in serum antibodies occurs in the course of resolving cryptosporidiosis in immunocompetent animals and humans [74, 12], and the relationship between oocyst clearing and appearance of specific faecal IgA [31] emphasizes the likely role of mucosal immunity. Indeed, duodenal perfusion of immune bovine colostrum has been effective in the treatment of cryptosporidiosis in immunodeficient patients [79]. Sporozoite infectivity is reduced by exposure to immune bovine serum, immune bovine colostrum or anti-sporozoite monoclonal antibodies [5, 27, 53, 58, 59]. Thus, in the past few years, much attention has focused on the molecular structure of the parasite and on the characterization of the molecules and epitopes recognized by humoral and mucosal antibodies.

SDS-PAGE analysis of cryptosporidial oocyst extract on Coomassie blue or silver stained gels demonstrated a complex molecular structure with about 50 different molecules ranging from 14 kDa to more than 330 kDa [35, 38, 39, 71]. Twenty-one bands ranging in size from 14 to 200 kDa were detected in purified oocyst wall preparations [39], and oocyst surface iodination showed that molecules at 15.5, 32, 42, 47.5, 79 and 96 kDa were consistently present in *C. parvum* isolates [39]. Metabolic radiolabelling of sporozoites also resulted in a complex protein pattern including several high molecular weight molecules [58]. Indeed, silver stained gels of sporozoite extract separated by two-dimensional gel electrophoresis showed over 300 distinct polypeptide molecules and therefore confirmed the complex molecular structure of the invasive stage [47].

Oocyst and sporozoite antigens which induce serum antibodies have been identified in mice [38], calves [35, 46, 58], horses [46], rabbits [35], goats [39], lambs [31] and humans [39, 46, 77]. Among these antigens (Table 6.2), a 20–23 kDa sporozoite surface molecule sharing epitopes with higher molecular weight proteins [41] was recognized by most

Table 6.2 Main antigens of *Cryptosporidium* characterized by immunoblotting or immunoprecipitation

Molecular weight (kDa)	*Immunolocalization*	*Parasitic extract*	*Antibody*	*Reference*
23		Oocyst (calf)	Human sera	77
20	Sporozoite (surface)	Sporozoite (calf)	Human, calf, horse sera	46
15.5, 23, 32		Oocyst (human)	Human, goat sera	39
15, 23, 180		Oocyst (calf)	Lamb sera	31
15, 23, 116, 180		Oocyst (calf)	Lamb faecal IgA	31
9–10		Oocyst/sporozoite (goat)	Bovine colostral IgA	70
14.5–16.5		Oocyst/sporozoite (goat)	Bovine colostral IgA, IgG_1, IgG_2, IgM	70
20	Sporozoite (surface) Merozoite	Sporozoite (calf)	Mab C6B6	46
20	Sporozoite (surface) Merozoite	Sporozoite (calf)	Mabs C6B6, C8C5	3
25–200	Sporozoite (polar) Merozoite	Sporozoite (calf)	Mab C4A1	3
28, 55, 98	Sporozoite–merozoite (surface)	Sporozoite (calf)	Mab 17–41[b]	5, 58
ND	Sporozoite–merozoite (surface)	Sporozoite (calf)	Mab 18–44[b]	5, 58
40	Oocyst wall	Oocyst (calf)	Mab 18–280–2	1
63–210	Sporozoite–merozoite (apical region)	Oocyst (calf)	Mab HAD	7
130, 210[a]	Sporozoite–merozoite (micronemes)	Oocyst (calf)	Mab TOU	7
40–250[a]	Oocyst wall, sexual stages	Oocyst (calf)	Mab OW–IGO	6

ND, molecular weight not determined; Mab, monoclonal antibody.
[a] carbohydrate epitope.
[b] neutralizing antibody.

human, calf and horse immune sera [46, 77]. This antigen was localized by immunoelectron microscopy in the pellicle, at the apical end of sporozoites and merozoites [41]. Tilley *et al.* [70] recently identified both this highly immunogenic 20–23 kDa molecule and a less immunogenic 18–20 kDa glycosylated surface molecule of the sporozoite.

Identification of antibodies which invoke mucosal antibody response was achieved using faecal extracts and immune bovine colostrum. Studies in experimentally infected lambs demonstrated that molecules of 15, 23 and 180 kDa were consistently recognized by faecal gut-associated IgA and serum antibodies [31]. Immunoblots of *C. parvum* oocyst or sporozoite extract with hyperimmune bovine colostrum have demonstrated that a great number of parasite antigens react with colostral immunoglobulins [70]. Among these molecules, a highly reactive 9–10 kDa antigen was shown to react with IgA antibodies only, whereas a 14.5–16.5 kDa antigen reacted with IgA, IgG_1, IgG_2 and IgM colostral antibodies.

Only a few cryptosporidial antigens have been characterized and immunolocalized with monoclonal antibodies. Anusz *et al.* [1] identified a 40 kDa molecule localized at the exterior surface of the oocyst wall by indirect immunofluorescence. Monoclonal antibodies which bind to the 20–23 kDa sporozoite surface molecule characterized with immune sera also recognized merozoites recovered from the ileum of infected mice [3, 46]. Riggs *et al.* [58] and Bjorneby *et al.* [5] characterized two surface epitopes shared by sporozoites and merozoites, and demonstrated that monoclonal antibodies recognizing these epitopes neutralize sporozoite and merozoite infectivity. Arrowood *et al.* [3] identified a polar antigen of the sporozoite that showed a complex blot pattern. Similarly, two antigens located at the anterior pole of sporozoites and merozoites which showed a complex immunoblot pattern with high molecular weight molecules were characterized with monoclonal antibodies. One of these antibodies, which recognized a carbohydrate epitope, reacted on immunoelectron microscopy with the micronemes of the zoites [7]. A monoclonal antibody recognizing the oocyst wall was shown to react with a carbohydrate epitope of two major molecules at 250 and 40 kDa. Immunolocalization of the corresponding antigen in early and developing macrogametes suggested that this glycoprotein is released in the parasitophorous vacuole from electron-lucent vesicles of the macrogametes, before being condensed in the oocyst wall [6]. These findings confirm the results from Luft *et al.* [38] who suggested that carbohydrates form a significant proportion of the epitopes that bind to antibodies in the immune response to *C. parvum.*

Currently available data thus suggest that the protection against *Cryptosporidium* requires the presence of antibodies. However, AIDS patients may experience chronic cryptosporidiosis despite a specific immunoglobulin response [78]. Recent experiments also lend support to the idea that antibodies alone are insufficient to control infection [68]. Consequently, it appeared important to investigate the role of cell mediated immunity in the resistance to *Cryptosporidium*. Infection of adult nude mice resulted in persistent cryptosporidiosis [76]. Moreover, passive transfer of lymphoid cells from spleen and mesenteric lymph nodes of immune mice cured infected mice [76]; experimentally infected $CD4^+$ lymphocyte-depleted mice developed severe cryptosporidiosis, suggesting that the CD 4^+ T-lymphocyte subset is involved in this cell mediated protection [76]. However, these observations do not exclude the possibility that the effector mechanisms of this T-cell dependent immunity are antibodies. Characterization of the molecules and epitopes that induce cell mediated immunity will probably be an important stage in understanding the control mechanisms operative in cryptosporidiosis. *In vitro* experiments have already been performed, and have demonstrated that spleen lymphocytes from mice that had previously received multiple oral exposure to *C. parvum* could undergo blastogenesis when exposed to oocyst extract [84].

6.2.5 Pathogenesis

Immunodeficient patients infected with *C. parvum* usually experience severe and chronic life threatening watery diarrhoea, resulting in major electrolytic imbalance, weight loss and malabsorbtion [48, 64]. In addition to the jejunum and the ileum, which are predominantly involved, infection of the whole intestinal tract may occur. Dissemination to the oesophagus, stomach, liver, biliary tract, pancreatic ducts and bronchial tree and pulmonary parenchyma have also been reported. Observation at post-mortem of *Cryptosporidium* in the lumen of submucosal colonic blood vessels in the course of disseminated cryptosporidiosis suggests that such dissemination may occur through haematogenous spread [30].

The precise pathophysiological mechanism of *Cryptosporidium*-induced clinical disorders remains unknown. The histological lesions described in intestinal cryptosporidiosis (moderate villous atrophy, increased crypt size and minimal inflammatory infiltrate of the lamina propria) [48, 64], are non-specific. The clinical features of the diarrhoea, which resembles a cholera-like syndrome, suggests an enterotoxin mediated activity. Recent clinical and experimental data favour this hypothesis. A dramatic improvement of the diarrhoea was reported in a few patients treated with sandstatin, a somatostatin analogue that reduces intestinal secretion and enhances water and electrolyte reabsorbtion [15]. Indeed, it has been demonstrated that cell-free medium or supernatant from homogenates of cultured cryptosporidia can stimulate chlorine secretion in the ileal mucosa of rabbits without causing histological alterations [29].

6.2.6 Molecular approaches to diagnosis and treatment

Diagnosis

Laboratory diagnosis is usually made by recognition of the characteristic oocysts in the faeces. Direct examination of stools, even after concentration procedures, is not reliable in routine practice, and specific staining techniques are needed. Similarly, identification of the parasite on digestive biopsy samples stained by classic procedures is difficult for inexperienced observers. Immunological methods based on the use of monoclonal antibodies in immunofluorescent or immunoenzymatic techniques have thus been developed to identify *Cryptosporidium* in faeces [1, 67] or biopsy samples [8] and offer a valuable diagnostic alternative. An enzyme linked immunoassay using polyclonal sera was shown to detect cryptosporidial antigen in microscopy-negative stool specimens, suggesting that some patients may excrete only disintegrated organisms or secretory products [75]. This observation is consistent with a report that demonstrated a high asymptomatic carriage rate for *Cryptosporidium* [60]. Thus, detection of coproantigens and genomic amplification may be powerful tools for diagnostic purposes and for epidemiological studies. Serum antibodies to cryptosporidium oocysts or sporozoites have been detected in infected animals and humans by immunofluorescent assay [12, 74], ELISA [78] and Western blot [39, 46, 77]. However, most reports concern epidemiological studies or immune response analysis, and the possible value of serology in the clinical diagnosis of cryptosporidiosis has still to be determined.

Recent outbreaks of waterborne cryptosporidiosis have confirmed the importance of *Cryptosporidium* as a possible contaminant of drinking water supplies [65]. The prevalence of cryptosporidial oocysts in natural waters and the efficiency of purification procedures need to be investigated to appreciate the likely risk of human infection and to define effective water treatments. Such studies require sensitive and specific procedures to recover and identify cryptosporidial oocysts potentially pathogen for man. Monoclonal antibodies and DNA probes may provide useful tools for such accurate screening of environmental samples.

Treatment

None of the antimicrobial agents tested in animal and human cryptosporidiosis have been consistently effective in clearing the infection [17, 72]. The effectiveness of spiramycin treatment suggested by Portnoy *et al.* [54] was not confirmed by further studies. Thus, the search for a non-toxic anti-cryptosporidial molecule needs multiple complementary approaches, including clinical research and experimental screening systems *in vivo* and *in vitro*.

A few drugs recently tested in clinical practice demonstrated activity against *Cryptosporidium.* Halofuginone lactate cured experimental cryptosporidiosis in lambs and calves [50, 83]. Eflornithine, an inhibitor of ornithine decarboxylase, induced complete or partial clinical and/or parasitological response in some AIDS patients [61]. A cure of cryptosporidiosis has been reported in AIDS patients during treatment with azidothymidine [14]. Whether this effect is due to partial immune function recovery or to a direct anti-cryptosporidial activity of azidothymidine is unclear. The poor effect of this molecule on asexual growth of the parasite *in vitro* [43] is consistent with the former hypothesis. However, this observation does not rule out the possibility that azidothymidine may act later in the life cycle of *Cryptosporidium.*

The anti-cryptosporidial activity of 20 drugs has been evaluated *in vitro* on the asexual development of the parasite in a mouse fibroblast cell line. Except for monensin and halofuginone, none of the drugs tested could inhibit growth of the parasite at low concentrations [43]. Chemically synthesized amphipathic peptides, analogues of natural lytic peptides that occur in insects and frogs (magainin, cecropin), were shown to reduce sporozoite viability *in vitro* [2]. However, additional studies are needed to evaluate the local and systemic toxicity of these compounds to the host before their use in clinical practice is considered.

The difficulties encountered in conventional chemotherapy and the likely role of the immune system in the control of the disease have stimulated efforts to develop immunotherapeutic procedures. The observations that hyperimmune bovine colostrum may be effective in the treatment of cryptosporidiosis [79], and the protection of mouse infection by incubating sporozoites in immune serum, colostrum or monoclonal antibodies [53] led to investigation of the possible use of immunoglobulins. Sporozoite and merozoite antigens which induce neutralizing antibodies have been characterized [5, 58]. These molecules may be candidates for the development of a vaccine in cattle, and may provide effective immunogens for the recovery of immune colostrum containing potent neutralizing agents. In addition to immunoglobulins, a possible therapeutic effect of bovine dialysable leucocyte extract [42] and thymostimulin [4] has also been demonstrated by different groups.

Finally, a major problem regarding the treatment of cryptosporidiosis concerns water treatment processes. The usual chlorination systems have proved ineffective in clearing oocysts from contaminated water, and ozone is probably the only disinfectant likely to inactivate cryptosporidial oocysts in drinking water supplies [34].

6.2.7 Conclusion

Since the emergence of C. *parvum* as a significant human pathogen in the late 1970s, most studies have focused on the epidemiology and clinical features of cryptosporidiosis. The prevalence of the disease in domestic animals and humans, its clinical spectrum in immunocompetent and immunocompromised patients, and the mode of transmission of the parasite have thus been established [17, 72]. In the early 1990s, clinicians and researchers in this field are facing two major challenges. The first is the urgent requirement for a consistently effective treatment of cryptosporidiosis, at least in immunocompromised patients. The second concerns the potential for environmental transmission of *C. parvum* and the need for accurate surveys and effective disinfection procedures of drinking water supplies. No doubt, in coming years, molecular and cellular biology and immunology will provide useful information to improve prophylaxis and treatment of cryptosporidiosis, at both individual and environmental levels.

6.3 ISOSPORA

6.3.1 Introduction

Isospora belli infections have been reported mostly from tropical and subtropical regions, and sporadically from temperate countries [49]. The incidence of isosporiasis in immunocompetent populations is low, even in tropical areas. Thus, before the AIDS pandemic, between 1976 and 1980, *I. belli* infection was found in only 0.17% of 1139 refugees from southeast Asian countries [81]. In contrast, the frequency of isosporiasis is higher in immunodeficient patients. In Haiti, chronic diarrhoea of at least 2 months duration was found in 85% of patients with AIDS. *I. belli* and *Cryptosporidium* were identified in the stools of 53% (*Cryptosporidium*: 41%; *I. belli*: 17%) of the patients with chronic diarrhoea [52]. Insofar as isosporiasis is associated with immunosuppressive states, high frequencies of *I. belli* infection in tropical areas might be used as epidemiological markers of immunodeficiency [69].

6.3.2 Life cycle

I. belli is a monoxenous human coccidian parasite that inhabits the distal duodenum and proximal jejunum. Ingested sporulated oocysts (Figure 6.2) release sporozoites which invade enterocytes and develop into trophozoites. Trophozoites grow and become meronts which, on division, generate merozoites. The merozoites can invade other enterocytes and develop into other meronts, or into sexual forms of macrogamonts, and microgamonts which in turn develop into macrogametes and microgametes, respectively. Microgametes fertilize macrogametes, producing oocysts. Only fertilized macrogametes are diploid, indicating that meiosis occurs during sporogony [16]. The oocysts, containing one or two cytoplasmic masses (or

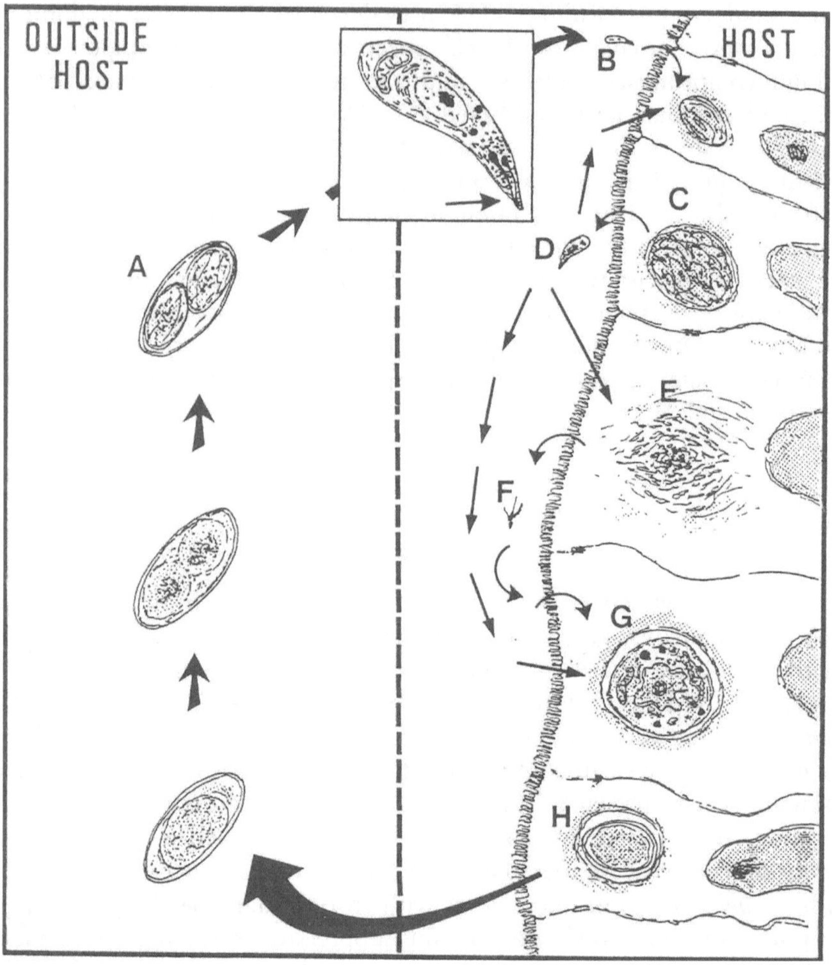

Figure 6.2 Life cycle of *Isospora belli*, sporulated oocysts (A); infective sporozoites (inset and B); meront (C); merozoites (D); microgamont (E); microgametes (F) and macrogametes (G); oocysts (H). Oocysts are excreted with the stools. Mature oocysts (A) have two sporcysts each containing four infective sporozoites. In the inset, rhoptries and micronemes are well visible at the apical pole (arrow) of a sporozoite.

sporoblasts) are excreted in the stools. Outside the body, during a 1–2 day period of ripening, the sporoblasts divide internally, developing into two sporocysts, each containing four infectious sporozoites. Humans contract *I. belli* infection by ingesting mature oocysts. Possible person-to-person sexual transmission of *I. belli* remains controversial. As the minimum sporulation time is 20–24 hours, transmission between homosexuals who practice aniliction seems possible [49]. The prepatent period appears to be about 10 days [36]. The incubation period has not been defined but is presumably 3–11 days [11].

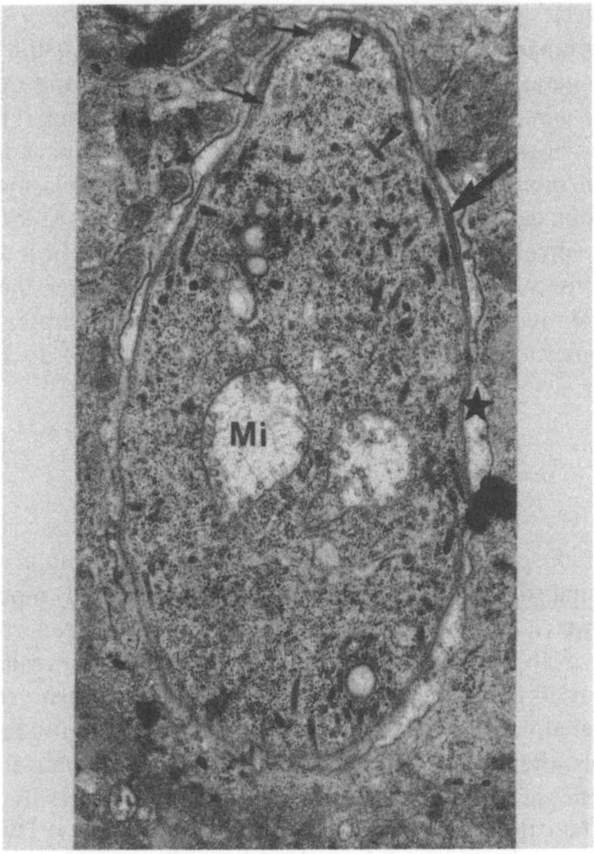

Figure 6.3 Intracellular *Isospora belli* merozoite parasitizing an enterocyte (small bowel biopsy of an AIDS patient). The organism is surrounded by the membrane of the parasitophorous vacuole (star) and shows the inner membrane complex (arrow), subpellicular microtubules (small arrows) and micronemes (arrowheads). Mitochondria (Mi) with tubular cristae are also seen (magnification ×20 400).

6.3.3 Ultrastructure

Ultrastructurally, the infective stages (sporozoites and merozoites) or zoites of *I. belli* show, as in other coccidia, a cell membrane lined by an inner membrane complex, subpellicular microtubules and a typical apical complex consisting of a polar ring, conoid, micronemes, rhoptries and dense granules (Figures 6.2, 6.3).

To our knowledge, no molecular studies have been made of the structure of *I. belli*. In other coccidia, particular attention has been given to structures involved in cell invasion. This process is probably identical for all intracellular sporozoans [25]. Infective organisms induce the formation of a parasitophorous vacuole and gain access to this intracellular compartment. The initial step of this process is the formation of a moving junction between the apical end of the zoite and the host cell plasmalemma, which becomes thicker on the contact zona. The moving junction glides backward on the zoite which enters the parasitophorous vacuole, the membrane of which is continuous with the plasmalemma of the host cell but depleted of intramembranous particles. Apical complex structures are involved in the invasion process of the host cell. Thus rhoptry proteins, which would be released during or after the host cell invasion, were recently immunolocalized in the membrane of the parasitophorous vacuole [25]. The role of micronemes and dense granules is still unclear.

6.3.4 Clinical and pathological aspects

I. belli is now known to cause chronic diarrhoea in immunocompromised patients [63]. Nonspecific radiographic findings include prominent mucosal folds, thickening of the intestinal wall and disordered motility [66]. Lactose intolerance and fat malabsorption have also been reported [66]. A careful study of an AIDS patient with isosporiasis [48] revealed marked perturbation of D-xylose and vitamin B_{12} absorption, steatorrhoea and increased faecal alpha 1 antitrypsin clearance. The disappearance of *I. belli* from the stools after treatment was associated with normalization of the absorption tests and the cessation of symptoms, suggesting that the parasite was the causative agent of the observed pathology. Furthermore, diarrhoea and malabsorption relapsed with recurrence of the parasitosis [48, 52].

The different intracellular parasitic stages of *I. belli* may be found in the epithelial intestinal cells in biopsy specimens, associated with marked shortening of the villi, hypertrophy of crypts and infiltration of the lamina propria with neutrophils, lymphocytes, plasma cells and eosinophils [22]. However, clinical signs may be present in infected hosts with minimal or undetectable histological abnormalities of the mucosa [10, 48].

Recently, a case of disseminated isosporiasis in a patient with AIDS was reported [57]. Intracellular and extracellular *I. belli* organisms were observed in the mucosae and lamina propriae of the small and large intestines as well as in the mesenteric and tracheo-bronchial lymph nodes.

6.3.5 Pathophysiology: immunological and molecular aspects

Although the pathophysiology of isosporiasis remains unexplained, some gut disturbances caused by *I. belli* and other human and animal coccidia are beginning to be understood [10]. Diarrhoea in coccidiosis could be explained by epithelial loss associated with villous atrophy, fusion and enterocyte sloughing. Cell mediated immunity may be an important factor in the villous atrophy [10]. As activated T-cells are mitogenic to enterocytes, crypt cell hyperplasia may be initiated by T-cell activation. Enterocyte proliferation produces crypt hypertrophy which leads to villous atrophy. Increased epithelial turnover may result in increased loss of epithelial cells, which is associated with villous fusion in infections with *Eimeria* and *Cryptosporidium*. Desmosomal junctions are observed between enterocytes of adjacent villi. Local T-cell activation also results in lymphokine release, increased expression of HLA-DR antigens on villous and crypt enterocytes and increased numbers of intraepithelial lymphocytes [10]. Intestinal mast cells could also be involved. In rats and lambs, the coccidian *Eimeria* induces a marked increase of mucosal mast cells, which respond to infection by releasing a collagen IV protease. This enzyme degrades the collagen of basement membranes [10].

Coccidian infection induces an increase in intestinal mucosal permeability, which causes loss of plasma proteins as reported by Modigliani *et al.* [48] in one patient infected with *I. belli*, two patients infected with *Cryptosporidium* and one patient infected with the microsporidian *Enterocytozoon bieneusi*. The cause of the increased permeability was not determined in coccidiosis. Other protozoan parasites, for instance *Entamoeba histolytica*, increase the permeability of the plasma membrane of the target cell by the release of pore forming proteins (PFPs). The PFPs break the cell permeability barrier, collapsing the electrical potential across the plasma membrane and provoking swelling and bursting of the target cell by water entry [51].

6.3.6 Laboratory diagnosis

The diagnosis of isosporiasis is generally made by examination of faeces, but it can also be made by direct detection of the parasite in intestinal aspirates or in small bowel biopsy. Faecal examination commonly reveals Charcot–Leyden crystals and may also reveal mucus. Blood and leucocytes are rarely observed [66]. Concentration methods should always be performed. Formalin–ether sedimentation is, according to our own experience, particularly useful to detect isosporal oocysts in small numbers. Although oocysts of *I. belli* are acid fast, wet direct smears or concentration smears are recommended, rather than permanent smears stained with modified acid-fast stains now widely used to detect *Cryptosporidium* [28].

The different intracellular parasitic stages of *I. belli* may be found in the intestinal epithelial cells in biopsy specimens [22]. *I. belli* may be detected in ultrastructural preparations of tissue samples. The different parasitic stages are identified by the presence of typical apical complex and trilaminar pellicle.

Diagnostic methods by *in situ* molecular hybridization or by genomic amplification have not yet been developed. Immunodiagnostic tests are also not currently available.

6.3.7 Treatment

Isosporiasis in patients with AIDS can be treated effectively with trimethoprim and sulphamethoxazole [52]. However, AIDS patients are especially susceptible to the development of hypersensitivity reactions to trimethoprim and sulphamethoxazole. Lyell syndrome was reported in an AIDS patient with *I. belli* infection and an alternative treatment could be metronidazole [62]. Recurrence is frequent and can be prevented by prophylaxis with either trimethoprim and sulphamethoxazole or pyrimethamine and sulfadoxine [52].

ACKNOWLEDGEMENT

The authors are indebted to J.F. Dubremetz for critical review of the manuscript.

REFERENCES

1. Anusz, K.Z., Mason, P.H., Riggs, M.W., *et al.* (1990) Detection of *Cryptosporidium parvum* oocysts in bovine feces by monoclonal antibody capture enzyme-linked immunosorbent assay. *J. Clin. Microbiol.*, **28**, 2770–4.
2. Arrowood, M.J., Jaynes, J.M., Healey, M.C. (1991) *In vitro* activities of lytic peptides against the sporozoites of *Cryptosporidium parvum. Antimicrob. Agents Chemother.*, **35**, 224–7.
3. Arrowood, M.J., Mead, J.R., Mahrt, J.L. *et al.* (1989) Effects of immune colostrum and orally administered antisporozoite monoclonal antibodies on the outcome of *Cryptosporidium parvum* infections in neonatal mice. *Infect. Immun.*, **57**, 2283–88.
4. Barbaro, G., Murri, G., Profenna, L.M., *et al.* (1990) Role of thymostimulin in the treatment of acute *Cryptosporidium* enteritis with erythromycin in patients with acquired immunodeficiency syndrome. *Curr. Ther. Res.*, **48**, 974–9.
5. Bjorneby, J.M., Riggs, M.W., Perryman, L.E. (1990) *Cryptosporidium parvum* merozoites share neutralization-sensitive epitopes with sporozoites. *J. Immunol.*, **145**, 298–304.
6. Bonnin, A., Dubremetz, J.F., Camerlynck, P. (1991) Characterization and

immunolocalization of an oocyst wall antigen of *Cryptosporidium parvum* (*Protozoa, Apicomplexa*). *Parasitology*, **103**, 171–7.

7. Bonnin, A., Dubremetz, J.F., Camerlynck, P. (1991) Characterization of microneme antigens of *Cryptosporidium parvum* (*Protozoa, Apicomplexa*). *Infect. Immun.*, **59**, 1703–8.
8. Bonnin, A., Petrella, T., Dubremetz, J.F., *et al.* (1990) Histopathological method for diagnosis of cryptosporidiosis using monoclonal antibodies. *Eur. J. Clin. Microbiol. Inf. Dis.*, **9**, 664–6.
9. Bonnin, A., Salimbeni, I., Dubremetz, J.F., *et al.* (1990) Mise au point d'un modèle expérimental de culture *in vitro* des stades asexués de *Cryptosporidium* spp. *Ann. Parasitol. Hum. Comp.*, **65**, 41–3.
10. Buret, A., Gall, D.G., Nation, P.N., *et al.* (1990) Intestinal protozoa and epithelial cell kinetics, structure and function. *Parasitol. Today*, **6**, 375–80.
11. Casemore D.P. (1990) Foodborne protozoal infection. *Lancet*, **336**, 1427–32.
12. Casemore, D.P. (1987) The antibody response to *Cryptosporidium*: development of a serological test and its use in a study of immunologically normal persons. *J. Infect.*, **14**, 125–34.
13. Center for Disease Control. (1987) Revision of the CDC surveillance case definition for acquired immunodeficiency syndrome. *M.M.W.R.*, (**Suppl. 1S**), 36.
14. Connolly, G.M., Dryden, M.S., Shanson, D.C., *et al.* (1988) Cryptosporidial diarrhoea in AIDS and its treatment. *Gut*, **29**, 593–7.
15. Cook, D.J., Kelton, J.G., Stanisz A.M., *et al.* (1988) Somatostatin treatment for cryptosporidial diarrhoea in a patient with the acquired immunodeficiency syndrome (AIDS). *Ann. Intern. Med.*, **108**, 708–9.
16. Cornelissen, W.C.A., Overdulve, J.P., Van Der Ploeg, M. (1984) Determination of nuclear DNA of five eucoccidian parasites, *Isospora* (*Toxoplasma*) *gondii*, *Sarcocystis cruzi*, *Eimeria tenella*, *E. acervulina* and *Plasmodium berghei*, with special reference to gamontogenesis and meiosis in *T. gondii*. *Parasitology*, **88**, 531–53.
17. Current, W.L. (1988) The biology of *Cryptosporidium*. *ASM News*, **54**, 605–11.
18. Current, W.L., Haynes, T.B. (1984) Complete development of *Cryptosporidium* in cell culture. *Science*, **224**, 603–5.
19. Current, W.L., Long, P.L. (1983) Development of human and calf *Cryptosporidium* in chicken embryos. *J. Infect. Dis.*, **148**, 1108–13.
20. Current, W.L., Reese, N.C. (1986) A comparison of endogenous development of three isolates of *Cryptosporidium* in suckling mice. *J. Protozool.*, 33, 98–108.
21. Current, W.L., Upton, S.J., Haynes, T.B. (1986) The life cycle of *Cryptosporidium baileyi n.* sp. (*Apicomplexa, Cryptosporidiidae*). infecting chickens. *J. Protozool.*, **33**, 289–96.
22. Dammin, G.J., Dooley, J.R. (1976) Coccidiosis, in *Pathology of Tropical and Extraordinary Diseases*, Vol. 1, (eds C.H. Binford, D.H. Connor). A.F.I.P., Washington, pp. 332–5.

23. Datry, A., Danis, M., Gentilini, M. (1989) Developpement complet de *Cryptosporidium* en culture cellulaire: applications. *Med./Sci.*, 5, 762–6.
24. Ditrich, O., Palkovic, L., Sterba, J., *et al.* (1991) The first finding of *Cryptosporidium baileyi* in man. *Parasitol. Res.*, 77, 44–7.
25. Dubremetz, J.F. (1990) Caractéristiques communes de l'invasion de la cellule hôte par les sporozoaires. *Ann. Parasitol. Hum. Comp.*, **65** (**Suppl. 1**), 23–5.
26. Fayer, R., Leek, R.G. (1984) The effects of reducing conditions, medium, pH, temperature, and time on *in vitro* excystation of *Cryptosporidium*. *J. Protozool.*, **31**, 567–9.
27. Fayer, R., Perryman, L.E., Riggs, M.W. (1989) Hyperimmune bovine colostrum neutralizes *Cryptosporidium* sporozoites and protects mice against oocyst challenge. *J. Parasitol.*, **75**, 151–3.
28. Garcia, L.S., Bruckner, D.A. (1988) *Diagnostic Medical Parasitology*, Elsevier, New York.
29. Garza, D.H., Fedorak, R.N., Soave, R. (1986) Enterotoxin-like activity in cultured cryptosporidia: role in diarrhoea. *Gastroenterology*, **90**, 1424 (Abstract).
30. Gentile, G., Baldassari, L., Caprioli, A., *et al.* (1987) Colonic vascular invasion as a possible route of extraintestinal cryptosporidiosis. *Am. J. Med.*, **82**, 574–5.
31. Hill, B.D., Blewett, D.A., Dawson, A.M., *et al.* (1990) Analysis of the kinetics, isotype and specificity of serum and coproantibody in lambs infected with *Cryptosporidium parvum*. *Res. Vet. Sci.*, **48**, 76–81.
32. Jacyna, M.R., Parkin, J., Goldin, R., *et al.* (1990) Protracted enteric cryptosporidial infection in selective immunoglobulin A and saccharomyces opsonin deficiencies. *Gut*, **31**, 714–6.
33. Johnson, A.M., Fielke, R., Lumb, R. *et al.* (1990) Phylogenetic relationships of *Cryptosporidium* determined by ribosomal RNA sequence comparison. *Int. J. Parasitol.*, **20**, 141–7.
34. Korich, D.G., Mead, J.R., Madore, M.S., *et al.* (1990) Effects of ozone, chlorine dioxide, chlorine, and monochloramine on *Cryptosporidium parvum* oocyst viability. *Appl. Env. Microbiol.*, **56**, 1423–8.
35. Lazo, A., Barriga, O.O., Redman, D.R., *et al.* (1986) Identification by transfer blot of antigens reactive in the enzyme-linked immunosorbent assay (ELISA) in rabbits immunized and a calf infected with *Cryptosporidium* spp. *Vet. Parasitol.*, **21**, 151–63.
36. Levine, N.D. (1973) *Protozoan Parasites of Domestic Animals and of Man*. Burgess Publishing Company, Minneapolis.
37. Levine, N.D. (1984) Taxonomy and review of the coccidian genus *Cryptosporidium* (*Protozoa, Apicomplexa*). *J. Protozool.*, **31**, 94–8.
38. Luft, B.J., Payne, D., Woodmansee, D., *et al.* (1987) Characterization of the *Cryptosporidium* antigens from sporulated oocysts of *Cryptosporidium parvum*. *Inf. Immun.*, **55**, 2436–41.
39. Lumb, R., Lanser, J.A., O'Donoghue P.J. (1988) Electrophoretic and immunoblot analysis of *Cryptosporidium oocysts*. *Immunol. Cell Biol.*, **66**, 369–76.

40. Lumb, R., Smith, K., O'Donoghue, P.J. *et al.* (1988) Ultrastructure of the attachment of *Cryptosporidium* sporozoites to tissue culture cells. *Parasitol. Res.*, **74**, 531–6.
41. Lumb, R., Smith, P.S., Davies, R., *et al.* (1989) Localization of a 23 000 MW antigen of *Cryptosporidium* by immunoelectron microscopy. *Immunol. Cell Biol.*, **67**, 267–70.
42. MacMeeking, A., Borkowsky, W., Klesius, P.H., *et al.* (1990) A controlled trial of bovine dialyzable leucocyte extract for cryptosporidiosis in patients with AIDS. *J. Infect. Dis.*, **161**, 108–12.
43. MacDonald, V., Stables, R., Warhurst, D.C. *et al.* (1990) *In vitro* cultivation of *Cryptosporidium parvum* and screening for anticryptosporidial drugs. *Antimicrob. Agents Chemother.*, **34**, 1498–500.
44. Marcial, M.A., Madara, J.L. (1986) *Cryptosporidium*: cellular localization, structural analysis of absorptive cell–parasite membrane–membrane interactions in guinea pigs, and suggestion of protozoan transport by M cells. *Gastroenterology*, **90**, 583–94.
45. Mead, J.R., Arrowood, M.J., Current, W.L., *et al.* (1988) Field inversion gel electrophoretic separation of *Cryptosporidium* spp. chromosome-sized DNA. *J. Parasitol.*, **74**, 366–9.
46. Mead, J.R., Arrowood, M.J., Sterling, C.R. (1988) Antigens of *Cryptosporidium* sporozoites recognized by immune sera of infected animals and humans. *J. Parasitol.*, **74**, 135–43.
47. Mead, J.R., Humphreys, R.C., Sammons, D.W., *et al.* (1990) Identification of isolate-specific sporozoite proteins of *Cryptosporidium parvum* by two dimensional gel electrophoresis. *Infect. Immun.*, **58**, 2071–5.
48. Modigliani, R., Bories, C., Le Charpentier, Y., *et al.* (1985) Diarrhoea and malabsorption in acquired immune deficiency syndrome: a study of four cases with special emphasis on opportunistic protozoan infestations. *Gut*, **26**, 179–87.
49. Morakote, N., Muangyimpong, Y., Somboon, P., *et al.* (1987) Acute human isosporiasis in Thailand: a case report. *Southeast Asian J. Trop. Med. Pub. Health*, **18**, 107–11.
50. Naciri, M., Yvoré, P. (1989) Efficacité du lactate d'halofuginone dans le traitement de la cryptosporidiose chez l'agneau. *Recueil Med. Vet.*, **165**, 823–6.
51. Ojcius, D.M., Young, D.E. (1990) A role for pore-forming proteins in the pathogenesis by parasites? *Parasitol. Today*, **6**, 163–5.
52. Pape, J.W., Verdier, R.I., Johnson, W.D. (1989) Treatment and prophylaxis of *Isospora belli* in patients with the acquired immunodeficiency syndrome. *N. Eng. J. Med.*, **320**, 1044–47.
53. Perryman, L.E., Riggs, M.W., Mason, P.H., *et al.* (1990) Kinetics of *Cryptosporidium parvum* sporozoite neutralization by monoclonal antibodies, immune bovine serum and immune bovine colostrum. *Infect. Immun.*, **58**, 257–9.
54. Portnoy, D., Whiteside, M.E., Buckley, E., *et al.* (1984) Treatment of intestinal cryptosporidiosis with spiramycin. *Ann. Intern. Med.*, **101**, 202–4.

55. Reduker, D.W., Speer, C.A. (1985) Factors influencing excystation in *Cryptosporidium* oocysts from cattle. *J. Parasitol.*, **71**, 112–5.
56. Reduker, D.W., Speer, C.A., Blixt, J.A. (1985) Ultrastructural changes in the oocyst wall during excystation of *Cryptosporidium parvum* (*Apicomplexa*; *Eucoccidiorida*). *Can. J. Zool.*, **63**, 1892–6.
57. Restrepo, C., Macher, A.M., Radany, E.H. (1987) Disseminated extraintestinal isosporiasis in a patient with acquired immune deficiency syndrome. *Am. J. Clin. Pathol.*, **87**, 536–42.
58. Riggs, M.W., McGuire, T.C., Mason P.H., *et al.* (1989) Neutralization-sensitive epitopes are exposed on the surface of infectious *Cryptosporidium parvum* sporozoites. *J. Immunol.*, **143**, 1340–5.
59. Riggs, M.W., Perryman, L.E. (1987) Infectivity and neutralization of *Cryptosporidium parvum* sporozoites. *Infect. Immun.*, **55**, 2081–7.
60. Roberts, W.G., Green, P.H.R., Ma, J., *et al.* (1989) Prevalence of cryptosporidiosis in patients undergoing endoscopy: evidence for an asymptomatic carrier state. *Am. J. Med.*, **87**, 537–9.
61. Rolston, K.V.I., Fainstein, V., Bodey, G.P. (1989) Intestinal cryptosporidiosis treated with eflornithine: A prospective study among patients with AIDS. *J. Acq. Immune Def. Synd.*, **2**, 426–30.
62. Romeu, J., Clotet, B., Tural, C., *et al.* (1989) Therapeutic challenge for *Isospora belli* enteritis in an AIDS patient who developed Lyell syndrome after cotrimoxazole therapy. *Am. J. Gastroenterol.*, **84**, 207–9.
63. Shaffer, R., Moore, L. (1989) Chronic travelers diarrhea in a normal host due to *Isospora belli. J. Infect. Dis.*, **159**, 596–7.
64. Sloper, K.S., Dourmashkin, R.R., Bird, R.B., *et al.* (1982) Chronic malabsorption due to cryptosporidiosis in a child with immunoglobulin deficiency. *Gut*, **23**, 80–2.
65. Smith, H.V., Rose, J.B. (1990) Waterborne cryptosporidiosis. *Parasitol. Today*, **6**, 8–12.
66. Soave, R., Johnson, W.D. (1988) *Cryptosporidium* and *Isospora belli* infections. *J. Infect. Dis.*, **157**, 225–9.
67. Sterling, C.R., Arrowood, M.J. (1986) Detection of *Cryptosporidium* sp. infections using a direct immunofluorescent assay. *Pediatr. Infect. Dis.*, **5**, S139–42.
68. Taghi-Kilani, R., Sekla, L., Hayglass, K.T. (1990) The role of humoral immunity in *Cryptosporidium* spp. infection. Studies with B cell-depleted mice. *J. Immunol.*, **145**, 1571–6.
69. Therizol-Ferly, M., Tagliante-Saracino, J., Kone, M., *et al.* (1989) Diarrhées chroniques et parasitoses chez des adultes suspectés de SIDA en Cote d'lvoire. *Bull. Soc. Pathol. Exot. Filiales*, **82**, 690–3.
70. Tilley, M., Fayer, R., Guidry, A., *et al.* (1990) *Cryptosporidium parvum* (*Apicomplexa*: *Cryptosporidiidae*) oocyst and sporozoite antigens recognized by bovine colostral antibodies. *Infect. Immun.*, **58**, 2966–71.
71. Tilley, M., Upton, S.J., Blagburn, B.L., *et al.* (1990) Identification of outer oocyst wall proteins of three *Cryptosporidium* (*Apicomplexa*: *Cryptosporidiidae*) species by ^{125}I surface labeling. *Infect. Immun.*, **58**, 252–3.
72. Tzipori, S. (1988) Cryptosporidiosis in perspective. *Adv. Parasitol.*, **27**, 63–129.

73. Tzipori, S., Angus, K.W., Campbell, I., *et al.* (1980) *Cryptosporidium*: evidence for a single-species genus. *Infect. Immun.*, **30**, 884–6.
74. Tzipori, S., Campbell, I. (1981) Prevalence of *Cryptosporidium* antibodies in 10 animal species. *J. Clin. Microbiol.*, **14**, 455–6.
75. Ungar, B.L.P. (1990) Enzyme-linked immunoassay for detection of *Cryptosporidium* antigens in fecal specimens. *J. Clin. Microbiol.*, **28**, 2491–5.
76. Ungar, B.L.P., Burris, J.A., Quinn, C.A., *et al.* (1990) New mouse models for chronic *Cryptosporidium* infection in immunodeficient hosts. *Infect. Immun.*, **58**, 961–9.
77. Ungar, B.L.P., Nash, T.E. (1986) Quantification of specific antibody response to *Cryptosporidium* antigens by laser densitometry. *Infect. Immun.*, **53**, 124–8.
78. Ungar, B.L.P., Soave, R., Fayer, R., *et al.* (1986) Enzyme immunoassay detection of immunoglobulin M and G antibodies to *Cryptosporidium* in immunocompetent and immunocompromised persons. *J. Infect. Dis.*, **153**, 570–8.
79. Ungar, B.L.P., Ward, D.J., Fayer, R., *et al.* (1990) Cessation of *Cryptosporidium* associated diarrhoea in an acquired immunodeficiency syndrome patient after treatment with hyperimmune bovine colostrum. *Gastroenterology*, **98**, 486–9.
80. Upton, S.J., Current, W.L. (1985) The species of *Cryptosporidium* (*Apicomplexa*: *Cryptosporidiidae*) infecting mammals. *J. Parasitol.*, **71**, 625–9.
81. Vernes, A., Abdellatifi, M., Dei-Cas, E., *et al.* (1982) Parasitoses et hémoglobinopathies chez 1170 réfugiés du Sud-Est asiatique. *Nouv. Presse Med.*, **11**, 2687–91.
82. Vetterling, J.M., Takeuchi, A., Madden, P.A. (1971) Ultrastructure of *Cryptosporidium wrairi* from the guinea pig. *J. Protozool.*, **18**, 248–60.
83. Villacorta, I., Peeters, J.E., Vanopdenbosch, E. (1991) Effect of halofuginone lactate against *Cryptosporidium parvum* in calves. *Antimicrob. Agents Chemother.*, **35**, 283–7.
84. Whitmire, W.M., Harp, J.A. (1990) *In vitro* murine lymphocyte blastogenic responses to *Cryptosporidium parvum*. *J. Parasitol.*, **76**, 450–2.

NOTE ADDED IN PROOF

Since submission of our manuscript, two important articles dealing with the molecular genetics of *Cryptosporidium parvum* have been published:

Laxer, M.A., Timblin, B.K. and Patel, R.J. (1991) DNA sequences for the specific detection of *Cryptosporidium parvum* by the polymerase chain reaction. *Am. J. Trop. Med. Hyg.*, 45, 688–94.

Kim, K., Goozé, L., and Petersen, C. (1992) Isolation, sequence and molecular karyotype analysis of the actin gene of *Cryptosporidium parvum*. *Mol. Biochem. Parasitol.*, 50, 105–14.

7

Toxoplasma

DEMETRIS SAVVA

7.1 INTRODUCTION

Toxoplasmosis is a zoonosis caused by the intracellular protozoan parasite *Toxoplasma gondii*. The organism was named in 1909 [63] following its first accurate description in 1908 by Nicolle and Manceaux who found it in Tunisia in the North American rodent *Ctenodactylus gondii*. The parasite is so called because it looks like a bow (from the Greek for bow, 'toxon'), while its species name comes from that of its first host. Infection is widespread throughout the world, occurring naturally in all warm blooded animals including man [17, 24, 51]. The only known final hosts where the life cycle of the parasite may be completed are members of the *Felidae*, including domestic and wild cats [42]. However, the importance of the disease in the medical and the veterinary fields is due to the unique ability of the parasite to become established in man and in a wide range of animals; in fact, *T. gondii* is ubiquitous in human beings (with an estimated 500 million infected people worldwide) and all species of mammals and birds [89]. Although toxoplasmosis is an extremely common infection, it causes disease only rarely. In human medicine, toxoplasmosis can cause complications and can be life threatening in three groups of individuals: firstly, the foetus when the mother experiences a primary infection during pregnancy [15, 22], secondly, patients being immunosuppressed for organ transplantation or for tumour therapy [25, 34, 72, 73], and thirdly, patients with the acquired immunodeficiency syndrome (AIDS) [29, 29a, 35, 55, 68, 89, 96]. The first reported case of toxoplasmosis in man dates back to 1923 when an autopsy on an infant

Molecular and Cell Biology of Opportunistic Infections in AIDS
Edited by Steven Myint and Alan Cann
Published in 1992 by Chapman & Hall, London ISBN 0 412 45330 4

showed the presence of the parasite in the eye [44]. In the veterinary field, toxoplasmosis is considered to be an important cause of abortion in sheep and goats resulting in considerable economic losses in many countries [18].

7.2 TAXONOMY AND LIFE CYCLE OF *TOXOPLASMA*

Taxonomically the organism belongs to the phylum *Apicomplexa*, class *Sporozoa*, sub-class *Coccidia*, order *Eimeriina*, family *Toxoplasmatidae*. It is thought that there is only one species of the organism but many strains, which vary in the degree of virulence [1]. The life cycle of the parasite (Figure 7.1) was elucidated about 20 years ago [23, 42] and has been reviewed thoroughly in recent years [17, 67], so only a short account will be presented here. There are two phases in the life cycle of *T. gondii*: the reproductive phase occurring in the intestine of the final hosts (members of the cat family), and the proliferative phase which takes place in all the intermediate hosts (all warm blooded animals including man). Three forms of the organism can be distinguished during the two phases of the life cycle: trophozoites or tachyzoites, bradyzoites and sporozoites. The trophozoites or tachyzoites (Figure 7.2) are vegetative and proliferative forms, which multiply rapidly (tachys means speedy in Greek), and are usually crescent shaped and about 5–8 μm in length and 2–5 μm in width; although no locomotion structures are present, attenuated forms tend to be motile. Bradyzoites are similar to tachyzoites, but they multiply slowly (bradys means slow in Greek) and are present in tissue cysts; the spherical cysts (15–100 μm in diameter) contain large numbers of bradyzoites within a highly resistant parasitic membrane. Sporozoites are the result of sporulation of oocysts which are 10–12 μm in diameter and are produced in the intestine of cats.

Cats become infected after eating either rodents and birds infected with *T. gondii*, or grass or soil contaminated with faecal oocysts. It is believed that infection occurs when cats are old enough to hunt for their own food [90]. Production and shedding of oocysts by cats is achieved most effectively following ingestion of tissue cysts although all three forms of the parasite can result in infection [26]. Once ingested by cats, tissue cysts are dissolved by digestive fluids in the stomach and small intestine, releasing the bradyzoites which penetrate the epithelium of the small intestine. Some bradyzoites invade the lamina propria of the intestine and multiply as tachyzoites, which are then distributed to extra-intestinal tissues in the lymph system and in the blood, just as in the case of the intermediate hosts (below). Following a number of asexual generations in the intestinal epithelium, the sexual phase begins 3–15 days after ingestion of infected material, resulting in the production of micro- and macrogametes (male and female gametes, respectively) which fuse to form the

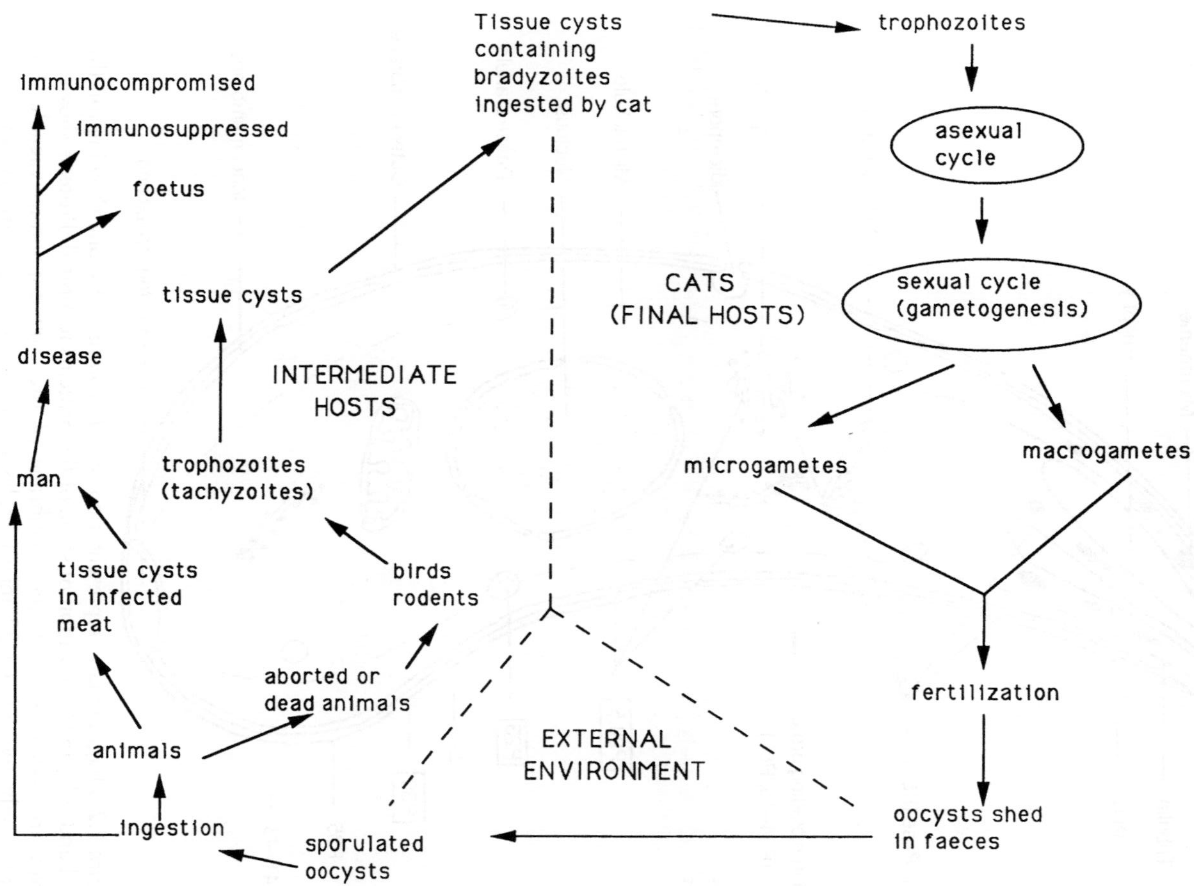

Figure 7.1 Life cycle and transmission of *T. gondii*.

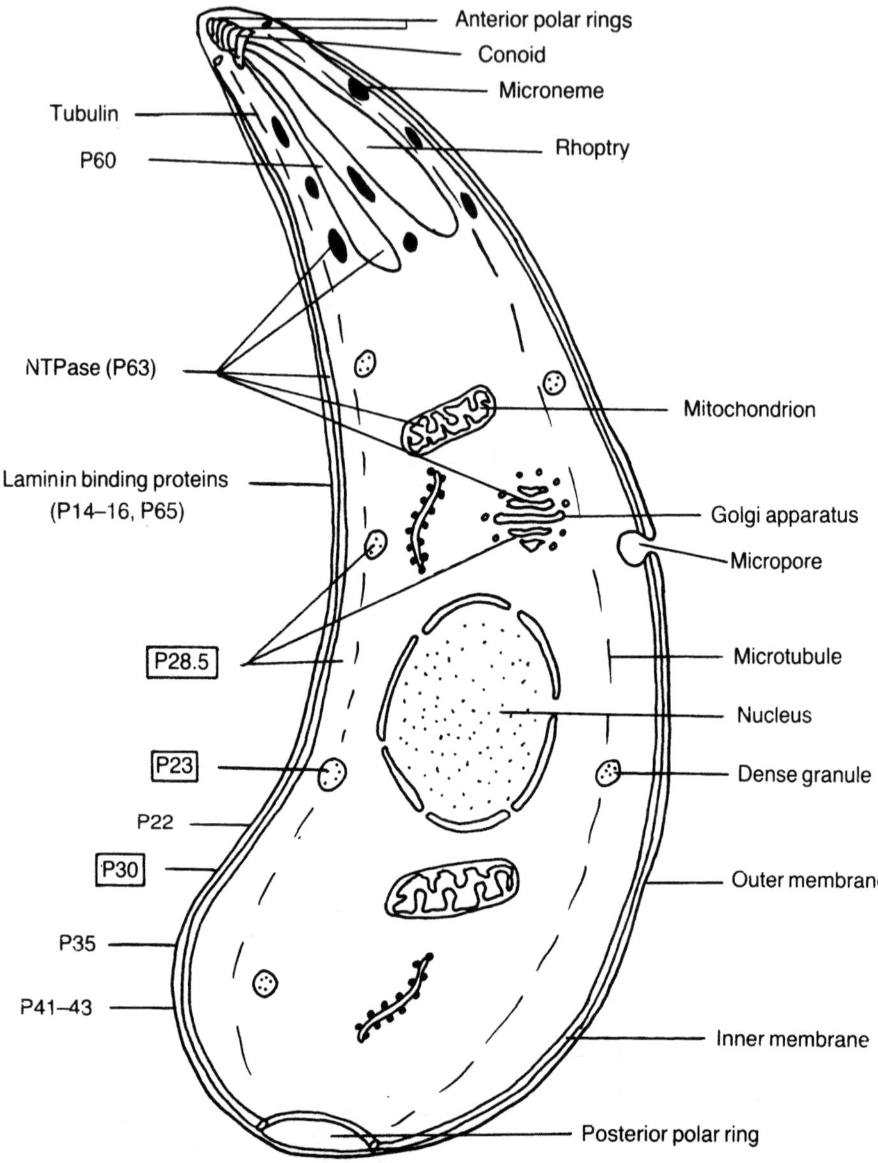

Figure 7.2 Diagrammatic representation of a *T. gondii* tachyzoite showing (on the left-hand side) the localization of some of the proteins identified. Proteins shown in boxes are also present on the membrane of the parasitophorous vacuole. For clarity, only two of the 22 microtubules of the cytoskeleton are shown.

oocysts. The peak of oocyst production occurs 5–8 days after ingestion. Oocysts are shed in the faeces following rupture of infected epithelial cells in the intestine. In primary infections, this occurs over a period of 1–3 weeks, with as many as 10^7 oocysts shed per day [27]. Once in the environment, an oocyst sporulates within 3 weeks, depending on factors such as humidity, temperature and oxygen availability. This process takes 2–4 days and results in the formation of two sporocysts each containing four sporozoites. Sporulated oocysts survive in the environment in soil and on plants and can remain infective for months [27]. They can be transmitted mechanically by rain and snow and by invertebrates such as cockroaches, flies and earthworms.

Food contaminated with sporulated oocysts may be eaten by man and by the large number of intermediate hosts as well as by cats. In the intermediate hosts the proteolytic enzymes in the digestive fluids cause the rupture of the oocysts releasing the sporozoites. These start dividing rapidly with the resulting tachyzoites being distributed throughout the body in the leucocytes, causing transient parasitaemia. Tachyzoites divide in enucleated cells every 5–12 hours until the cell is filled with parasites and ruptures. Tachyzoites then invade adjacent cells to repeat the cycle, creating microfoci of infection from which the host apparently recovers and the parasite enters the chronic stage of the infection. This involves the tachyzoites clinging together, dividing more slowly and encysting, thus forming tissue cysts which are packed tightly with bradyzoites. These cysts develop and remain in the cytoplasm of the host cell with young cysts containing as few as four bradyzoites, while older ones may contain several hundred. Cysts can persist for the lifetime of the host and are commonly found in the central nervous system (particularly the brain), in the eye and in skeletal and cardiac muscle.

7.3 ANTIGENIC AND GENOMIC ORGANIZATION OF *TOXOPLASMA*

Most laboratory studies on the structure of *T. gondii* are based on the use of trophozoites (tachyzoites) which can be maintained easily, either through cell culture or through serial passage through mice and rats. Although the life cycle of the parasite is known, a lot has yet to be discovered regarding the structure of its different forms, in particular the tissue cysts and the oocysts. The ultrastructure of *T. gondii* has been discussed in detail elsewhere [10, 17, 59, 67] and thus only recent work on its DNA and RNA and on the localization of some of the antigenic proteins will be mentioned here.

Analysis of the proteins of *T. gondii* (both native and denatured) by polyacrylamide gel electrophoresis and staining or by Western blotting using antitoxoplasma antibodies, indicates that the antigenic structure of the organism is complex [13, 67]. However, differences between various developmental stages of the parasite can be seen following polyacrylamide gel electrophoresis, as shown in Figure 7.3. Using human sera containing

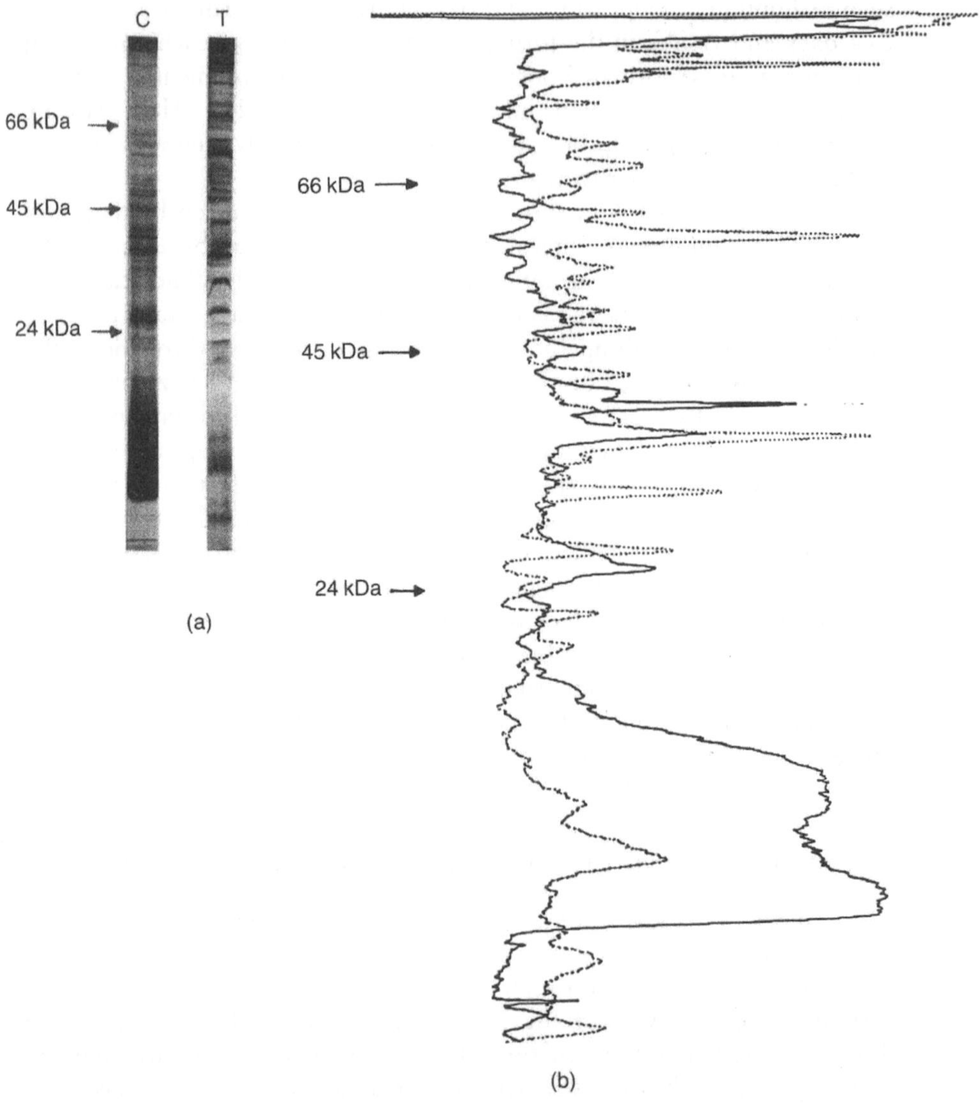

Figure 7.3 Protein profiles of tachyzoites and tissue cysts of *T. gondii*. (a) Proteins from RH strain tachyzoites (T) and human strain 18691 tissue cysts (C) were separated by electrophoresis on a discontinuous 12% polyacrylamide gel and visualized by silver staining (the photograph was kindly provided by Ms Gillian Woodison and Dr Judith Smith, Department of Pure and Applied Biology, University of Leeds). (b) Scan of (a) using laser densitometry with the solid and dotted lines representing tissue cysts and tachyzoites respectively.

antitoxoplasma antibodies, a number of strain-specific antigenic differences have been demonstrated [91, 97]. However, analysis of membrane bound antigens, obtained by treating tachyzoites with detergents, reveals a simpler picture with four major and a few minor proteins [17]. Although the precise molecular weights of these proteins differ slightly depending on the method of analysis, they are approximately 22, 30, 35 and 43 kDa in size and the proteins are designated P22, P30, P35 and P43 respectively. All are anchored to the membrane of the organism by glycophospholipids [60, 86] which form a membranous network within the paratisophorous vacuole following invasion of host cells [83]. Three dominant cyst antigens have also been identified and the molecular weights of these common antigens are about 25, 36 and 67 kDa [97]. Some of these antigens are present in all forms of the parasite but others are stage-specific; some of the antigens may be absent or down regulated in bradyzoites and sporozoites. For example, using monoclonal antibodies, P22 and P30 have been detected on the surface of tachyzoites but not on bradyzoites or sporozoites [50, 65, 97]. However, preliminary evidence from the author's laboratory using the polymerase chain reaction (PCR) to amplify toxoplasma RNA indicates that bradyzoites contain the mRNA for P22.

The genome of *T. gondii* is haploid except during two stages: first, during sexual division in the intestinal epithelium of the final host and, second, just before endodyogeny of the tachyzoites. More recent work indicates that the genome consists of a minimum of 10 linkage groups ranging in size from 2×10^6 bp to more than 6×10^6 bp [82a]. *T. gondii* DNA has a GC content of 55% and its haploid genome contains about 8×10^8 base pairs (bp) [12] none of which are methylated [45]. The circular mitochondrial DNA is about 36 kbp and contains a 10 kbp inverted repeat [4]. It is thought that the nucleus contains 12 chromosomes, although pulsed field gel electrophoresis has revealed only eight with the smallest about 2×10^6 bp and the largest more than 10^7 bp [59]. *T. gondii* RNA is similar to that from other eukaryotes in that the mRNA is polyadenylated at the 3′ end [71] and that the rRNA consists of the 28S and 18S types.

The advent of recombinant DNA procedures has resulted, over the last 6 years, in the molecular cloning and characterization of a number of *T. gondii* genes from genomic DNA and cDNA libraries. Introns have been demonstrated in the α- and β-tubulin genes which exist as single copy genes [64] and the B1 gene which is tandemly repeated 50 times [3]. A number of other genes have been shown to lack introns; these include the genes for P22 [70], the NTPase P63 [46], P30 [6] and P23 [7]. It is not clear whether the P28 gene has any introns [69] and a number of other cDNA and genomic DNA clones still remain to be characterized fully [3, 76, 98]. These molecular cloning studies have given an insight into the structure of some of the gene products. For example, the P30 gene codes for a mRNA of 1.5 kbp and a primary translation product of 34.7 or 33 kDa including a presumptive hydrophobic signal sequence of 47 or 30 amino acids depending on the ATG initiation codon used [6]. In the case of the P22 gene, the putative primary translation product has a theoretical molecular weight of 18 959 and a predicted *N*-terminal signal sequence of 26 amino acids [70]. Both the P22 and the P30 gene products have a C-terminal hydrophobic region characteristic of proteins undergoing post-translational cleavage and modification with a phospholipid anchor for membrane attachment.

A number of excreted–secreted antigens (ESAs) which enhance survival in mouse and rat models for toxoplasmosis have also been studied. The P21, P27 and P28.5 ESAs are localized in dense granules in tachyzoites, in the Golgi apparatus and in the parasitophorous vacuole [9] while the P23 ESA is a calcium binding protein localized in the dense granules and the parasitophorous vacuole [7].

7.4 ENTRY OF *TOXOPLASMA* INTO THE HOST

Although it is not yet clear how *T. gondii* enters into the cytoplasm of host cells, the evidence suggests that the most likely method of entry is by a complex process of active invasion which takes place in approximately 15 seconds [93]. Since *T. gondii* parasitizes many different cell types it must recognize a common component of the host cell membrane such as lipids and glycosyl residues; lipids are the most likely component, with cholesterol thought to be involved in the attachment process [67]. Also, there is some evidence that the laminin binding proteins of 14, 16 and 60 kDa on the outer membrane of the trachyzoites are involved in the attachment and penetration of host cells through interactions with members of the β1 integrin family [32]. It is known that contact with the host is through the parasite's anterior end which contains the conoidal region, rhoptries and micronemes. It is thought that these structures secrete a number of substances, including phospholipase A_2 and one termed penetration enhancing factor (PEF), which enhance penetration of *T. gondii* into the host cell (reviewed in [67]). During penetration, the rhoptries secrete their contents including dense granules [9] and proteins of 60.5 (which may be the PEF), 60, 59.5, 59 and 55 kDa [53]. These secretions enable the parasite to form a tight junction with the cell membrane. This junction is capped in a process involving transport of polyclonal antibodies and pellicular proteins from the anterior to the posterior end of the tachyzoite. Subsequent shedding of these causes forward motion which thrusts the organism into the cytoplasm of the host cell, creating a parasitophorous vacuole. When the junction reaches the posterior end of the tachyzoite, it breaks and the host cell membrane fuses spontaneously. Thus, the exterior surface of the host cell membrane becomes the interior of the parasitophorous vacuole, but in the process the Mg^+ ATPase and 5′-nucleotidase are lost [14]. This may be connected with the host's inability to acidify the parasitophorous vacuole [84] and with the failure of lysosomes and phagosomes to fuse to it [48].

In the parasitophorous vacuole, *T. gondii* multiplies by endodyogeny (reviewed in [67]) in which each tachyzoite gives rise to two daughter cells. Multiplication results in enlargement of the parasitophorous vacuole, and about 48 hours after invasion, the host cell ruptures to release 64–256 motile tachyzoites which proceed to enter other cells.

7.5 HOST–PARASITE INTERACTIONS

Much remains to be discovered regarding the interactions between *T. gondii* and host cells (reviewed in [67]). However, it is known that the processes of DNA replication, transcription and translation of the host cell are not required. It is also known that, although the parasite is capable of pyrimidine synthesis, it is incapable of purine synthesis and is therefore totally dependent on the supply of purines from the host [81]. Although nothing is known about lipid biosynthesis in *T. gondii*, it is thought that it is no different from that in other protozoa in the same taxonomic group (e.g. *Eimeria tenella* and *Plasmodium berghei*) which are incapable of cholesterol synthesis. ATP synthesis occurs using both glycolysis and oxidative phosphorylation, but the relative importance of each is not known.

Immunity to *T. gondii* is also not understood fully but both humoral and cell mediated immunity are known to be important (reviewed in [82]). Infection is first combatted by humoral immunity and then by cell mediated immunity. Antitoxoplasma IgM antibodies are detectable 7 days after infection and their concentration rises to a peak after a few weeks before declining gradually. IgG antibodies appear 1–2 weeks after infection and their levels peak after 6 weeks, declining only very slowly over a period of months to years, with some persisting for life. Normally, a recent infection is indicated by the presence of IgM antibodies or rising levels of IgG antibodies, while a previous infection is indicated by the presence of IgGs alone. The recent demonstration of IgMs in the absence of any *T. gondii* infection [31] is important in that it complicates diagnosis, as will be seen later.

Although the macrophages of healthy individuals appear to have some degree of resistance to *T. gondii* [43], the most important effector cells in host defence are activated macrophages which kill or inhibit the intracellular multiplication of the parasite. Macrophages are activated by soluble products from lymphocytes sensitized by parasite antigens. In mice, the major such activator is interferon gamma (IFN-γ) [49] which induces the transcription of specific genes [11]. Activated macrophages secrete tumour necrosis factor (TNF) [57], the role of which in parasitic disease is not fully understood; secretion of TNF has also been observed prior to lethal *T. gondii* infection [8]. It is possible that TNF and IFN-γ act synergistically to inhibit the multiplication of the parasite. Other cytokines implicated in immunity to *T. gondii* include IFN-β [80] and interleukin [8]. Also, killer cells may play a very important role in the early stages of infection [33].

7.6 TOXOPLASMOSIS IN MAN

As mentioned in the introduction, infection with *T. gondii* is extremely common but clinical disease is rare. The age at which infection occurs and the immune status of the infected person determine the course of toxoplasmosis, which can be described as congenital (transplacental intrauterine infection) or acquired (post-natal infection). Infection occurs either by ingestion of raw or undercooked meat containing tissue cysts or by ingestion of fruits, vegetables or soil contaminated with faecal oocysts; in rare cases infection may result following contamination by saliva, blood or milk containing trophozoites.

7.6.1 Infection and the immunocompetent individual

In immunocompetent people, acute acquired toxoplasmosis normally goes undiagnosed or is misdiagnosed as an influenza-like illness. It is estimated that approximately 500 million people worldwide are infected with *T. gondii*. The importance of the disease varies in different countries, depending on a range of social, economic and medical factors. For example, in France 80–90% of people are infected, in the USA 50–60% and in the UK 20–40% [2, 21]. Following infection, the parasite is distributed by the blood and lymphatic systems to a variety of tissues where it forms tissue cysts which can persist for life or rupture occasionally to release trophozoites. The incubation period is variable and can be up to 3 weeks. The most common symptoms of disease are fever, lymphadenopathy, headache, malaise, anorexia, sore throat, eye pain and abdominal pain but the chronic phase of the infection is asymptomatic. Severe forms of the disease are very rare but in such cases manifestations include hepatitis, encephalitis and pneumonia.

7.6.2 Infection and pregnancy

Congenital toxoplasmosis is the result of *T. gondii* crossing the placental barrier. This does not occur if a woman is immune before pregnancy. It is generally accepted that transmission of *T. gondii* to the foetus may occur when the mother is infected during pregnancy and that in subsequent pregnancies the foetus is not at risk [15]. Initial infection of pregnant females may result in the formation of cysts in the placenta where necrotic lesions may develop. The parasite may then spill into the circulation and be disseminated, producing one of several degrees of disease. Congenital toxoplasmosis may manifest itself as severe thrombocytopenia, hepatosplenomegaly, cerebral calcifications, chorioretinitis, recurrent retinochoroiditis, lymphadenopathy, hydrocephalus, disturbance of head size, pneumonia, retinal scars or blindness. The severity of the disease depends on the virulence, site and time of the infection *in utero* as well as on the immunological state of the mother and the foetus. In general, maternal infections occurring in early pregnancy result in greater damage to the foetus and may lead to stillbirths. Maternal infections after the first trimester of pregnancy tend to result in mild or subclinical disease of the foetus [52]. The risk of congenital transmission increases from 25% in the first trimester to 65% in the third trimester of pregnancy [15]. The incidence of congenital toxoplasmosis amongst live births has been estimated to be 0.01% in the UK [22] and 0.1% in France [15]. It has been estimated that 50–90% of the exposed foetuses become infected and that only 10–20% of these show any symptoms at birth, with about 3% dying in the neonatal period [15, 95]. Although the majority of infants appear to be normal at birth, more than 85% of them will subsequently develop chorioretinitis [95].

7.6.3 Infection and the immunodeficient host

During the past 20 years it has become increasingly evident that patients being immunosuppressed for tumour therapy or for organ transplantation may develop generalized toxoplasmosis which can be serious or even fatal [25, 34, 72, 73]. Patients with Hodgkin's disease, non-Hodgkin's lymphoma and those receiving organ transplants are especially vulnerable. Since such patients generally have antitoxoplasma antibodies before immunosuppression, it is thought that generalized toxoplasmosis represents recrudescent infections resulting from the spread of quiescent, encysted *T. gondii* as a result of the breakdown of the patient's immunity status. In the case of organ transplants, an additional explanation may be the transplantation of an infected organ or transfusion of infected packed leucocytes into a non-infected (thus non-immune) recipient, giving rise to generalized toxoplasmosis due to the immunosuppression. It appears that transplantation of an organ from an infected to a non-infected individual is more common than that from a non-infected to an infected one. In transplant recipients, since clinical disease is common in those who seroconvert during immunosuppression but rare in seropositive patients

[56], it appears that the major cause of disease is primary acute infection acquired from the donated organ (especially the heart). These infections often involve the central nervous system (CNS) but until recently CNS involvement was extremely rare; for example, in Boston between 1969 and 1979 less than 2% of CNS infections in chronically immunosuppressed patients were caused by *T. gondii* [41].

Recent findings have demonstrated that patients with the acquired immunodeficiency syndrome (AIDS) constitute a further group prone to severe disseminated toxoplasmosis which results from reactivation of latent disease rather than from primary infection [56, 89]. In this group, 30–70% of patients show neurological symptoms [29] and evidence suggests that *T. gondii* may be the commonest cause of focal lesions [29] as well as the most prevalent opportunistic infection [89]. The incidence of CNS toxoplasmosis in AIDS patients is not certain, but published estimates range from 2.6% to 80% [29, 35, 55, 68, 96]. Toxoplasmosis was found to be the cause of 25–30% of CNS lesions in AIDS sufferers [20, 54] and it has been estimated that up to 30% of AIDS sufferers with positive toxoplasma serology may develop CNS infections [62]. In AIDS patients *T. gondii* has also been shown to infect the lungs, heart, adrenal gland, colon, testis and stomach [87], and the eyes [66]. Toxoplasmosis, therefore, constitutes a life threatening opportunistic infection in AIDS patients.

7.7 DIAGNOSIS OF TOXOPLASMOSIS

As mentioned in the previous section, toxoplasmosis can manifest itself with any or several of a multitude of symptoms and it can remain latent for years. Consequently, laboratory testing is necessary, since clinical diagnosis is difficult or impossible. The laboratory diagnosis of toxoplasmosis can be performed in a number of ways [22, 58]. Methods used only in exceptional cases include:

1. The histological examination of lymph nodes and other biopsy (usually post-mortem) tissues; although the histological picture is characteristic, many cases may be missed due to non-specific results [22], especially in AIDS patients.
2. Isolation and identification of the parasite following intraperitoneal injection of the suspect specimen into infection free mice. As with histological examination, this may also fail to detect some cases and, in addition, it is time consuming, requiring 6–8 weeks. Also, in the case of suspect specimens from AIDS patients, isolation of the parasite cannot

be performed in mice and has to be carried out by infection of cell monolayers which have a higher failure rate, and in any case, will not distinguish current, active infection from the presence of tissue cysts.

Currently, routine laboratory diagnosis of *T. gondii* infections is carried out using a range of serological methods which rely on the fact that toxoplasma antigens react with antitoxoplasma antibodies present in the serum of infected individuals. Following infection, the first antitoxoplasma antibodies to appear are of the IgM class and these are soon followed by IgG, which persists in the bloodstream for life. The serological test used most commonly to detect antitoxoplasma antibodies is the dye test (DT) of Sabin and Feldman [74] which is generally accepted to be the most sensitive and most specific of the serological methods. In this test, live *T. gondii* reacts with any antitoxoplasma antibodies and an accessory factory (complement-like) present in the serum specimen. This results in the lysis of the parasite which therefore fails to stain with alkaline methylene blue and can be seen in a stained smear under a microscope. The DT detects mainly the IgG antibodies which react with the parasite's membrane antigens. A DT titre of 1024 or more is described as marked elevation in the levels of IgG and is characteristic of an active infection. However, AIDS patients rarely show DT titres of this magnitude, despite disseminated toxoplasmosis. The test is used by the World Health Organization to standardize IgG reference sera but it is difficult to perform and is thus carried out in specialized laboratories. Due to the special requirements of the DT and the need for specialized laboratories, a number of other serological methods, which can be used in peripheral laboratories, have been developed.

1. The indirect fluorescent antibody test (IFAT), in which killed *T. gondii* is dried on a slide, incubated with the serum sample, and the antigen–antibody complexes are detected using fluorescein-labelled anti-human IgG or IgM. Thus, the test can detect both IgGs and IgMs directed primarily against membrane proteins but some non-specific IgM binding does occur. IgG determinations agree well with those obtained with the DT.
2. The indirect haemagglutination test (IHAT) uses sheep red blood cells sensitized by the chemical attachment of *T. gondii* antigen (mostly cytoplasmic proteins) and incubated with the serum protein. Sensitized cells agglutinate in the presence of antitoxoplasma antibodies [85]. The test detects IgG antibodies but IgM can be demonstrated indirectly if serum samples are treated with 2-mercaptoethanol to destroy this class of antibody. With cytoplasmic antigens, the test is useful for detecting past exposure; with membrane antigens it could allow early detection but it is more sensitive to non-specific antibodies.

3. The complement fixation test (CFT) uses a haemolytic system of sheep red blood cells and anti-sheep haemolysin. Antigen (cytoplasmic proteins from *T. gondii*) reacts with antitoxoplasma antibodies in the serum sample and the complex fixes complement thus preventing haemolysis. In the absence of specific antibodies haemolysis occurs. This test will detect both IgGs (except IgG_4) and IgMs which fix complement, but these appear later in infection than those detected by the DT. Although the CFT is useful for the diagnosis of active infections, its reproducibility is poor and its sensitivity is low. It is also very important to ensure that the complement being used in the test is free of any antitoxoplasma antibodies.
4. The direct agglutination test (DAT) in its original form [28], whole, formalin-treated *T. gondii* was incubated with the serum specimen to bring about agglutination of the cells when antitoxoplasma antibodies are present. The test detects both IgG and IgM directed against membrane antigens. Being sensitive to IgM, the test is useful for diagnosis of congenital and active acquired disease but it must be remembered that some non-specific IgM also reacts. In a modification of this test [16], the antigen is prepared by enzymatic treatment of whole, killed *T. gondii* and the sera are first treated with 2-mercaptoethanol to destroy IgM, thus allowing detection of IgG only. The results obtained by the modified DAT agree well with those from the DT.
5. The latex agglutination test (LAT) uses latex particles sensitized by the attachment of toxoplasma antigens, incubated with the untreated serum specimen. They agglutinate if the serum contains antitoxoplasma IgG or IgM. Although the LAT results for IgG are in good agreement with those from the DT and the IFAT, a number of false positive results are obtained due to non-specific IgM reactions taking place [37].
6. Enzyme-linked immunosorbent assays (ELISAs) use soluble toxoplasma antigens (mostly cytoplasmic proteins) fixed to a plastic microtitre plate. They are incubated with the suspected serum and any antigen–antibody complexes are reacted with an enzyme-conjugated anti-human IgG or IgM. These complexes are detected by colorimetry following hydrolysis of an appropriate substrate. Depending on the nature of the enzyme conjugate used, ELISAs detect IgG or IgM but the IgG detected appears later than that detected by the DT. The test is sensitive and is therefore suitable for the detection of IgM early in the infection. However, specific IgM is rarely detected during episodes of disseminated toxoplasmosis in AIDS sufferers. A modified ELISA to detect IgM antibodies (IgM-ELISA) has also been described [61]. In this method, the walls of microtitre plates are coated with anti-IgM antibodies which then capture the IgM in the serum specimen. The

presence of antitoxoplasma IgM is detected following reaction with toxoplasma antigen and then an enzyme-conjugated antitoxoplasma antiserum and substrate. This assay is not affected by any IgG present in the serum as this is washed away.

7. The immunosorbent agglutination assay (ISAGA) is a procedure for detecting antitoxoplasma IgM. Microtitre plates coated with anti-human IgM monoclonal antibody are employed to capture IgM in the serum sample; antitoxoplasma IgM is revealed by adding toxoplasma tachyzoites which either agglutinate when IgM is present or sediment at the bottom of the wells when IgM is absent. The IgM–ISAGA has been shown to detect antitoxoplasma IgM earlier than other methods [36] but again is of limited value in detecting disseminated toxoplasma infection in AIDS patients because IgM is produced only rarely.

In addition to the above serological procedures, which detect anti-toxoplasma antibodies, there are ELISA methods for detection of toxoplasma antigens. In such methods, the tissues under investigation are homogenized, centrifuged and the supernatant is incubated in microtitre plates coated with antitoxoplasma IgG. The antigen–antibody complexes are then revealed with an enzyme-conjugated antitoxoplasma antiserum and an appropriate substrate.

None of the methods currently used can be relied upon to give good results in patients with AIDS or in other immunosuppressed individuals. In some cases computed tomographic (CT) scans, magnetic resonance imaging (MRI) and lumbar punctures may provide further information. Also, serological procedures do not always offer a reliable diagnosis of toxoplasmosis in foetal specimens. Thus, there is a need for quicker and more sensitive diagnostic procedures for the early detection of toxoplasmosis.

7.8 DNA PROBES

The detection of a parasite's nucleic acids in clinical samples may provide a good alternative to serological procedures. During the last decade a number of bacterial, viral and protozoan pathogens have been detected in clinical specimens using DNA probes, and more recently such tests, based on gene amplification by the polymerase chain reaction (PCR) [75], have become available for the detection of *T. gondii* [5, 19, 77–79, 91a]. These tests rely on the amplification of the *T. gondii* B1 gene [3, 5] or the P30 gene [6, 19, 77–79]. DNA isolated from the tissue under investigation is subjected to PCR and the reaction products are analysed with either a radioactive or a non-radioactive DNA probe following gel electrophoresis and Southern hybridization [5, 19, 77–79]. Alternatively the PCR product is re-amplified with a second set of internal oligonucleotide primers and is then visualized directly following gel electrophoresis [77–79]. This 'nested' PCR assay is shown diagrammatically in Figure 7.4. The B1-based PCR assay can detect as few as 10 parasites in the presence of 10^5 leucocytes and, although subjective judgement has to be applied to the results, the test provides correct diagnosis of *T. gondii* infection if they are interpreted carefully [30]. In the case of the P30-based double PCR assay, in which the result is clearly positive or clearly negative, and which requires no subjective judgement, a single organism can be detected even in the presence of 10^6 human cells [79]. The P30 double PCR assay has been used to examine clinical specimens from AIDS and from transplant patients [38–40] and from pregnant women and their foetuses and infants [39, 47]. It has also been used to detect *Toxoplasma* following abortion in sheep [77, 88b, 94] and in the placentas of horses [88, 88a]. In common with the DT, these tests for the detection of toxoplasmosis require specialized laboratories. Also, the extreme sensitivity of the PCR can sometimes present problems, and consequently great care must be taken to

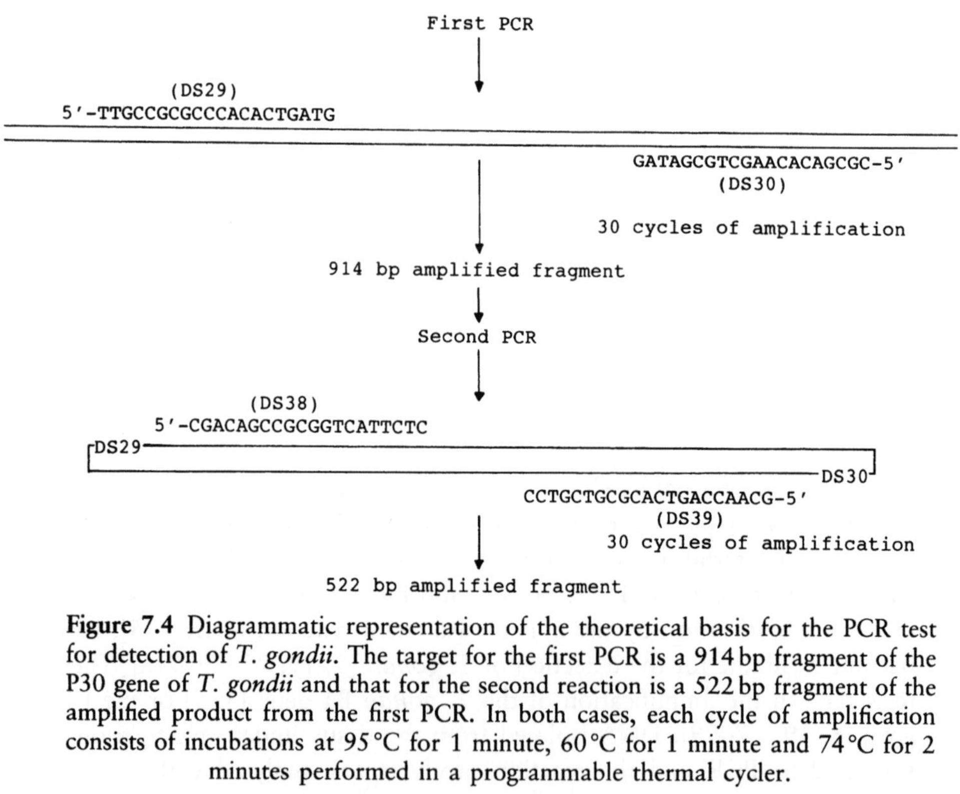

Figure 7.4 Diagrammatic representation of the theoretical basis for the PCR test for detection of *T. gondii*. The target for the first PCR is a 914 bp fragment of the P30 gene of *T. gondii* and that for the second reaction is a 522 bp fragment of the amplified product from the first PCR. In both cases, each cycle of amplification consists of incubations at 95 °C for 1 minute, 60 °C for 1 minute and 74 °C for 2 minutes performed in a programmable thermal cycler.

avoid cross contamination of samples (especially with amplified products from earlier reactions) before amplification.

7.9 MANAGEMENT OF TOXOPLASMOSIS

Although treatment of *T. gondii* infections with drugs slows down the multiplication of tachyzoites, it does nothing to eliminate tissue cysts or to prevent their formation. In immunocompetent individuals, toxoplasmic lymphadenopathy is a self-limiting condition and drug treatment is not required unless the patient is seriously ill with severe and persistent symptoms. However, pregnant women with primary *T. gondii* infection, infants with congenital toxoplasmosis, patients with chorioretinitis or myocarditis, and immunosuppressed or immunocompromised individuals with disseminated toxoplasmosis are normally given drug treatment. The most effective drug treatment in current use is a combination of pyrimethamine and sulphadiazine, but, since both drugs are folic acid antagonists and may cause bone marrow depression, folinic acid or yeast supplements must be used to prevent this side effect [22]. Pyrimethamine has also been used on its own for the treatment of AIDS patients, but it has a high failure rate [29a, 62]. Another regimen is to use sulphamethoxazole in combination with trimethoprim but this also has a high failure rate in AIDS patients [29a]. In AIDS patients with an adverse response to sulphonamides an alternative treatment has been a combination of pyrimethamine and clindamycin [29a, 62]. During pregnancy (especially during the first trimester), spiramycin has been used extensively to treat primary toxoplasma infections because of its lower toxicity [22].

ACKNOWLEDGEMENTS

Thanks are due to Dr Rick Holliman and Mrs Julie Johnson (Public Health Laboratory Service, St George's Hospital, Tooting), Dr Judith Smith (Department of Pure and Applied Biology, University of Leeds) and Miss Ruth Wheeler and Mr Saleem Mohammed (Department of Biochemistry and Physiology, University of Reading) for very useful comments during long discussions. Thanks are also due to Ms Gillian Woodison and Dr Judith Smith (Department of Pure and Applied Biology, University of Leeds) for providing the photograph used in Figure 7.3. Original work in the author's laboratory is funded by grants from the Medical Research Council, the Royal Society and the SmithKline (1982) Foundation, and postgraduate studentships from the Medical Research Council and the Science and Engineering Research Council in collaboration with the Public Health Laboratory Service.

REFERENCES

1. Baker, J.R. (1973) *Parasitic protozoa*, 2nd edn, Hutchinson University Library, London, pp. 127–35.
2. Beattie, C.P. (1982) The ecology of toxoplasmosis. *Ecol. Dis.*, **1**, 13–20.
3. Boothroyd, J.C., Burg, J.L., Nagel, S.D., et al. (1987) Antigen and tubulin genes of *Toxoplasma gondii*, in *Molecular Strategies of Parasitic Invasion*, (eds N. Agabian, H. Goodman, N. Nogueira), Liss, New York, pp. 237–50.
4. Borst, P., Overdulve, J.P., Weijers, P.J., Fase-Fowler, F., Vandenberg, M. (1984) DNA circles with cruciforms from Isospora *Toxoplasma gondii*. *Biochim. Biophys. Acta*, **781**, 100.
5. Burg, J.L., Grover, C.M., Pouletty, P., Boothroyd, J.C. (1989) Direct and sensitive detection of a pathogenic protozoan, *Toxoplasma gondii*, by polymerase chain reaction. *J. Clin. Microbiol.*, **27**, 1787–92.
6. Burg, J.L., Perelman, D., Kasper, L.H., Ware, P.L., Boothroyd, J.C. (1988) Molecular analysis of the gene encoding the major surface antigen of *Toxoplasma gondii*. *J. Immunol.*, **141**, 3584–91.
7. Cesbron-Delauw, M.F., Guy, B., Torpier, G., *et al.* (1989) Molecular characterization of a 23-kilodalton major antigen secreted by *Toxoplasma gondii*. *Proc. Natl. Acad. Sci. U.S.A.*, **86**, 7537–41.
8. Chang, H.R., Grau, G.E., Pechere, J-C. (1990) Role of tumour necrosis factor and interleukin-1 in infections with *Toxoplasma gondii*. *Immunology*, **69**, 33–7.
9. Charif, H., Darcy, F., Torpier, G., Cesbron-Delauw, M.F., Capron, A. (1990) *Toxoplasma gondii*: characterization and localization of antigens secreted from tachyzoites. *Exp. Parasitol.*, **71**, 114–24.
10. Chobotar, B., Schotyseck, E. (1982) Ultrastructure, in *The Biology of the Coccidia*, (ed. P.L. Long), University Park Press, Baltimore, pp. 101–65.
11. Collard, M., Belin, D., Vassalli, J.-D., De Kossodo, S., Vassalli, P. (1986) Gamma interferon enhances macrophage transcription of the tumour necrosis factor/cachectin, interleukin-1 and urokinase genes which are controlled by short-lived repressors. *J. Exp. Med.*, **164**, 2113–8.
12. Cornelissen, A.W.C.A., Overdulve, J.P., Van Der Ploeg, M. (1984) Determination of nuclear DNA of five eucoccidian parasites, *Isospora* (*Toxoplasma*) *gondii*, *Sarcocystis cruzi*, *Eimeria tenella*, *E. acervulina* and *Plasmodium berghei*, with special reference to gamontogenesis and meiosis in *I.* (*T.*).*gondii*. *Parasitology*, **88**, 531–53.
13. Couvreur, G., Sadak, A., Fortier, B., Dubremetz, J.F. (1988) Surface antigens of *Toxoplasma gondii*. *Parasitology*, **97**, 1–10.
14. De Carvalho, L., De Souza, W. (1990) Internalization of surface anionic sites and phagosome–lysosome fusion during interaction of *Toxoplasma gondii* with macrophages. *Eur. J. Cell. Biol.*, **51**, 211–9.
15. Desmonts, G., Couvreur, J. (1974) Congenital toxoplasmosis: a prospective study of 378 pregnancies. *N. Engl. J. Med.*, **290**, 1110–6.
16. Desmonts, G., Remington, J.S. (1980) Direct agglutination test for diagnosis of *Toxoplasma* infection. *J. Clin. Microbiol.*, **11**, 562–8.

17. Dubey, J.P., Beattie, C.P. (1988) *Toxoplasmosis in Animals and Man*, CRC Press, Boca Raton.
18. Dubey, J.P., Towle, A. (1986) *Toxoplasmosis in Sheep*. C.A.B. International, Slough.
19. Dupouy-Camet, J., Lavareda De Souza, S., Bougnox, M.E., *et al.* (1990) Preventing congenital toxoplasmosis. *Lancet*, **336**, 1018.
20. Elkin, C.M., Leon, E., Grenell, S.L., Leeds, N.E. (1985) Intracranial lesions in the acquired immunodeficiency syndrome: Radiological (computed tomographic) features. *JAMA*, 253, 393–6.
21. Feldman, H.A. (1974) Toxoplasmosis: an overview. *Bull. N.Y. Acad. Med.*, **50**, 110–27.
22. Fleck, D.G., Kwantes, W. (1980) *The Laboratory Diagnosis of Toxoplasmosis, Public Health Laboratory Service Monograph Series, No. 13, HMSO, London.*
23. *Frenkel, J.K. (1973) Toxoplasma in and around us. Bioscience*, **23**, 343–52.
24. Frenkel, J.K. (1990) Toxoplasmosis in human beings. *J. Am. Vet. Med. Assoc.*, **196**, 240–8.
25. Frenkel, J.K., Amare, M., Larsen, W. (1978). Immune competence in a patient with Hodgkin's disease and relapsing toxoplasmosis. *Infection*, **6**, 84–91.
26. Frenkel, J.K., Dubey, J.P., Miller, N.L. (1970) *Toxoplasma gondii* in cats: fecal stages identified as coccidian oocysts. *Science*, **167**, 896..
27. Freyre, A., Dubey, J.P., Smith, D.D., Frenkel, J.K. (1989) Oocysts-induced *Toxoplasma gondii* infections in cats. *J. Parasitol.*, 75, 750–5.
28. Fulton, J.D., Turk, J.L. (1959) Direct agglutination test for *Toxoplasma gondii. Lancet*, **ii**, 1068–9.
29. Gapen, P. (1982) Neurological complications now characterizing many AIDS victims. *JAMA*, **248**, 2941–2.
29a. Gellin, B.G., Soave, R. (1992) Coccidian infections in AIDS. *Medical Clinics of North America*, **76**, 205–34.
30. Grover, C.M., Thulliez, P., Remington, J.S., Boothroyd, J.C. (1990) Rapid prenatal diagnosis of congenital toxoplasma infection by using polymerase chain reaction and amniotic fluid. *J. Clin. Microbiol.*, **28**, 2297–301.
31. Gusetti, N., D'Elia, R. (1990) Natural immunoglobulin M antibodies against *Toxoplasma gondii* during pregnancy. *Am. J. Obstet. Gynecol.*, **162**, 1359–60.
32. Hall, B.F., Joiner, K.A. (1991) Strategies for obligate intracellular parasites for evading host defences, in *Immunoparasitology Today*, (eds C. Ash, R.B. Gallagher), Elsevier Trends Journals, Cambridge, pp. A22–27.
33. Hauser, W.E., Tsai, V. (1986) Acute toxoplasma infection of mice induces spleen NK cells that are cytotoxic for *T. gondii in vitro. J. Immunol.*, **136**, 313–9.
34. Hirsch, R., Burke, B.A., Kersey, J.H. (1984) Toxoplasmosis in bone marrow transplant recipients. *J. Paediatr.*, **105**, 426–8.
35. Holliman, R.E. (1986) Toxoplasmosis and the acquired immunedeficiency syndrome. *J. Infection*, **16**, 121–8.
36. Holliman, R.E., Johnson, J.D. (1989) The postnatal serodiagnosis of congenital toxoplasmosis. *Serodiag. Immunother. Infect. Dis.*, **3**, 323–7.

37. Holliman, R.E., Johnson, J., Duffy, K., New, L. (1989) Discrepant toxoplasma latex agglutination test results. *J. Clin. Pathol.*, **42**, 200–3.
38. Holliman, R.E., Johnson, J.D., Savva, D. (1990) Diagnosis of cerebral toxoplasmosis in association with AIDS using the polymerase chain reaction. *Scand. J. Infect. Dis.*, **22**, 243–4.
39. Holliman, R.E., Johnson, J.D., Constantine, G., Bissenden, J.G., Nicolaides, K., Savva, D. (1991) Difficulties in the diagnosis of congenital toxoplasmosis by cordocentesis: a case report. *Brit. J. Obstet. Gynecol.*, **98**, 832–4.
40. Holliman, R.E., Johnson, J.D., Gillespie, S.H., Johnson, M.A., Squire, S.B., Savva, D. (1991) New methods in the diagnosis and management of cerebral toxoplasmosis associated with the acquired immune deficiency syndrome. *J. Infection*, **22**, 281–5.
41. Hooper, D.L., Pruitt, A., Rubin, A.H. (1982) Central nervous system infection in the chronically immunosuppressed. *Medicine*, **61**, 166–88.
42. Hutchinson, W.M., Dunachie, J.F., Work, K., Sim, J.C. (1970) The life cycle of the coccidian parasite *Toxoplasma gondii* in the domestic cat. *Trans. R. Soc. Trop. Med. Hyg.*, **65**, 380–99.
43. Israelski, S.D., Araujo, F.G., Wachtel, J.S., Heinrichs, H.L., Remington, J.S. (1990) Differences in microbiocidal activities of human macrophages against *Toxoplasma gondii* and *Trypanosoma cruzi*. *Infect. Immun.*, **58**, 263–5.
44. Janku, J. (1923) Pathogenesis and pathologic anatomy of colomboma of macula lutea in eye of normal dimensions and in microphthalmic eye, with parasites in retina. *Cas. Lek. Cesk.*, **62**, 1021–7.
45. Johnson, A.M., Illana, S., Dubey, J.P., Dame, J.B. (1987) *Toxoplasma gondii* and *Hammondia hammondii*: DNA comparison using cloned rRNA gene probes. *Exp. Parasitol.*, **63**, 272–8.
46. Johnson, A.M., Illana, S., McDonald, P.J., Asai, T. (1989) Cloning, expression and nucleotide sequence of the genes fragment encoding an antigenic portion of the nucleotide triphosphate hydrolase of *Toxoplasma gondii*. *Gene*, **85**, 215–20.
47. Johnson, J.D., Holliman, R.E., Savva, D. (1990) Detection of *Toxoplasma gondii* using the polymerase chain reaction. *Biochem. Soc. Trans.*, **18**, 665.
48. Joiner, K.A., Fuhrman, S.A., Miettinen, H.M., Kasper, L.H., Mellman, I. (1990) *Toxoplasma gondii*: Fusion competence of parasitophorous vacuoles in Fc receptor-transferred fibroblasts. *Science*, **249**, 641–6.
49. Jones, T.C., Alkan, S., Erb, P. (1986) Spleen and lymph node cell populations *in vitro*, cell proliferation and interferon gamma production during the primary immune reponse to *Toxoplasma gondii*. *Parasite Immunol.*, **8**, 619–29.
50. Kasper, L.H. (1989) Identification of stage-specific antigens of *Toxoplasma gondii*. *Infect. Immun.*, **57**, 668–72.
51. Kwantes, W. (1987) Toxoplasmosis, in *Oxford Textbook of Medicine*, 2nd edn, (eds D.J. Weatherall, J.G.G. Ledingham, D.A. Warrell), Oxford University Press, Oxford, pp. 5506–9.
52. Lee, R.V. (1988) Parasites and pregnancy: The problem of malaria and toxoplasmosis. *Clin. Perinatol.*, **15**, 351–63.

53. Leriche, A.M., Dubremetz, J.F. (1991) Characterization of the protein contents of rhoptries and dense granules of *Toxoplasma gondii* tachyzoites by subcellular fractionation and monoclonal antibodies. *Mol. Biochem. Parasitol.*, **45**, 249–60.
54. Levy, R.M., Pons, V.G., Rosenblum, M.L. (1984) Central nervous system mass lesions in the acquired immunodeficiency syndrome (AIDS). *J. Neurosurg.*, **61**, 9–16.
55. Luft, B.J., Brooks, R.G., Conley, F.K., McCabe, R.E., Remington, J.S. (1984) Toxoplasma encephalitis in patients with acquired immune deficiency syndrome. *JAMA*, **252**, 913–7.
56. Luft, B.J., Naot, Y., Araujo, F.G., Stinson, E.B., Remington, J.S. (1983) Primary and reactivated toxoplasma infection in patients with cardiac transplants: clinical spectrum and problems in diagnosis in a defined population. *Ann. Intern. Med.*, **99**, 27–31.
57. Matthews, N., Riley, H.C., Neale, M.L. (1980) Tumour necrosis factor from the rabbit IV: Purification and chemical characterisation. *Br. J. Cancer*, **42**, 416–22.
58. McCabe, R.E., Remington, J.S. (1983) The diagnosis and treatment of toxoplasmosis. *Eur. J. Clin. Microbiol.*, **2**, 95–104.
59. McLeod, R., Mack, D., Brown, C. (1991) *Toxoplasma gondii*: New advances in cellular and molecular biology. *Exp. Parasitol.*, **72**, 109–21.
60. Nagel, S.D., Boothroyd, J.C. (1989) The major surface antigen, P30, of *Toxoplasma gondii* is anchored by a glycolipid. *J. Biol. Chem.*, **264**, 5569–74.
61. Naot, Y., Remington, J.S. (1980) An enzyme-linked immunosorbent assay for detection of IgM antibodies to *T. gondii*: use for diagnosis of acute acquired toxoplasmosis. *J. Infect. Dis.*, **142**, 757–66.
62. Navia, B.A., Petito, C.K., Gold, J.W.M., Cho, E.S., Jordon, B.D., Price, R.W. (1986) Cerebral toxoplasmosis complicating the acquired immune deficiency syndrome, clinical and neurological findings in 27 patients. *Ann. Neurol.*, **19**, 224–38.
63. Nicole, C., Manceaux, L. (1909) Sur un protozoaire nuveau du Gondi. *Compte Rendu Hebdomadaire des Seances de l'Academie des Sciences, Paris*, **148**, 369–72.
64. Nigel, S.D., Boothroyd, J.C. (1988) The α- and β-tubulins of *Toxoplasma gondii* are encoded by single copy genes containing multiple introns. *Mol. Cell. Parasitol.*, **29**, 261–73.
65. Omata, Y., Igarashi, M., Ramos, M.I., Nakabayashi, T. (1989) *Toxoplasma gondii*: antigenic differences between endozoites and cystozoites defined by monoclonal antibodies. *Parasitol. Res.*, **75**, 189–93.
66. Parke, D.W., Font, R.L. (1986). Diffuse toxoplasmic retinochoroiditis in a patient with AIDS. *Arch. Ophthalmol.*, **104**, 571–5.
67. Pfefferkorn, E.R. (1990) Cell biology of *Toxoplasma gondii*, in *Modern Parasite Biology: Cellular, Immunological and Molecular Aspects*, (ed. D.J. Wyler), Freeman, New York, pp. 26–50.
68. Pitchenik, A.E., Fischl, M.A., Walls, K.W. (1983) Evaluation of cerebral mass lesions in acquired immunodeficiency syndrome. *N. Engl. J. Med.*, **308**, 1099.

69. Prince, J.B., Araujo, F.G., Remington, J.S., Burg, J.L., Boothroyd, J.C., Sharma, S.D. (1989) Cloning of cDNAs encoding a 28 kilodalton antigen of *Toxoplasma gondii. Mol. Biochem. Parasitol.*, **34**, 3–14.
70. Prince, J.B., Auer, K.L., Huskinson, J., Parmley, S.F., Araujo, F.G. Remington, J.S. (1990) Cloning, expression, and cDNA sequence of surface antigen P22 from *Toxoplasma gondii. Mol. Biochem. Parasitol.*, **43**, 97–106.
71. Prince, J.B., Kovan-Quinn, M.A., Remington, J.S., Sharma, S.D. (1985) Cell free synthesis of *Toxoplasma gondii* antigens. *Mol. Biochem. Parasitol.*, **17**, 163–70.
72. Remington, J.S. (1974) Toxoplasmosis in the adult. *Bull. N.Y. Acad. Med.*, **50**, 211–27.
73. Ruskin, J., Remington, J.S. (1976) Toxoplasmosis in the immunocompromised. *Ann. Intern. Med.*, **84**, 193–9.
74. Sabin, A.B., Feldman, H.A. (1948) Dyes as microchemical indicators of a new immunity phenomenon affecting a protozoan parasite (toxoplasma). *Science*, **108**, 660–3.
75. Saiki, R.K., Scharf, S., Faloona, F., *et al.* (1985). Enzymatic amplification of β-globin genomic sequences and restriction site analysis for diagnosis of sickle cell anemia. *Science*, **230**, 1350–4.
76. Savva, D. (1989) Isolation of a potential DNA probe for *Toxoplasma gondii. Microbios*, **58**, 165–72.
77. Savva, D., Holliman, R.E. (1989) Diagnosis of ovine toxoplasmosis: an alternative to serology. *Vet. Rec.*, **125**, 212.
78. Savva, D., Holliman, R.E. (1990) Diagnosis of toxoplasmosis using DNA probes. *J. Clin. Pathol.*, **43**, 260–1.
79. Savva, D., Morris, J.C., Johnson, J.D., Holliman, R.E. (1990) Polymerase chain reaction for detection of *Toxoplasma gondii. J. Med. Microbiol.*, **32**, 25–31.
80. Schmitz, J.L., Carlin, J.M., Borden, E.C., Byrne, G.I. (1989) Beta interferon inhibits *Toxoplasma gondii* growth in human monocyte derived macrophages. *Infect. Immun.*, **57**, 3254–6.
81. Schwartzman, J.D., Pfefferkorn, E.R. (1982) *Toxoplasma gondii*: Purine synthesis and salvage in mutant host cells and parasites. *Exp. Parasitol.*, **53**, 77–86.
82. Sharma, S.D. (1990) Immunology of toxoplasmosis, in *Modern Parasite Biology: Cellular, Immunological and Molecular Aspects*, (ed. D.J. Wyler), Freeman, New York, pp. 184–99.
82a. Sibley, L.D., Boothroyd, J.C. (1992) Construction of a molecular karyotype for *Toxoplasma gondii. Molec. Biochem. Parasitol.*, **51**, 291–300.
83. Sibley, L.D., Krahenbuhl, J.L. (1988) Modification of host cell phagosomes by *Toxoplasma gondii* involves redistribution of surface proteins and secretion of a 32-kDa protein. *Eur. J. Cell Biol.*, **47**, 81–7.
84. Sibley, L.D., Weidner, E., Krahenbuhl, J.L. (1985) Phagosome acidification blocked by intracellular *Toxoplasma gondii. Nature*, **315**, 215–22.
85. Thorburn, H., Williams, H. (1972) A stable haemagglutinating antigen for detecting toxoplasma antibodies. *J. Clin. Pathol.*, **25**, 762–7.
86. Tomavo, S., Schwarz, R.T., Dubremetz, J.F. (1989) Evidence for glyco-

sylphosphatidyl-inositol anchoring of *Toxoplasma gondii* major surface antigens. *Mol. Cell. Biol.*, **9**, 4576–80.

87. Tschirhart, D., Klatt, E.C. (1988) Disseminated toxoplasmosis in the acquired immunodeficiency syndrome. *Arch. Pathol. Lab. Med.*, **112**, 1237–41.
88. Turner, C.B., Savva, D. (1990) Evidence of *Toxoplasma gondii* in an equine placenta. *Vet. Rec.*, **127**, 96.

88a. Turner, C.B., Savva, D. (1991) Detection of *Toxoplasma gondii* in equine eyes. *Vet. Rec.*, **129**, 128.

88b. Turner, C.B., Mohammed, S., Savva, D. (1991) Detection of *Toxoplasma gondii* in ovine samples. *Vet. Rec.*, **129**, 436.

89. Velimirovic, B. (1984) Toxoplasmosis in immunosuppression and AIDS. *Infection*, **12**, 315–7.
90. Wallace, G.D. (1971) Isolation of *Toxoplasma gondii* from the faeces of infected cats. *J. Infect. Dis.*, **124**, 227–8.
91. Ware, P.L., Kasper, L.H. (1987) Strain-specific antigens of *Toxoplasma gondii. Infect. Immun.*, **55**, 778–83.

91a. Weiss, L.M., Udem, S.A., Salgo, M., Tanowitz, H.B., Wittner, M. (1992) Sensitive and specific detection of *Toxoplasma* DNA in an experimental model: use of *Toxoplasma gondii*-specific cDNA and polymerase chain reaction. *J Infect. Dis.*, **163**, 180–6.

92. Weiss, L.M. Udem, S.A., Tanowitz, H., Wittner, M. (1988) Western blot analysis of the antibody response of patients with AIDS and toxoplasma encephalitis: Antigenic diversity among *Toxoplasma* strains. *J. Infect. Dis.*, **157**, 7–13.
93. Werk, R. (1985) How does *Toxoplasma gondii* enter host cells? *Rev. Infect. Dis.*, **7**, 449–57.
94. Wheeler, R., Wilmore, H., Savva, D., Turner, C.B. (1990) Diagnosis of ovine toxoplasmosis using PCR. *Vet. Rec.*, **126**, 249.
95. Wilson, C.B., Remington, J.S., Stagno, S. (1980) Development of adverse sequalae in children born with subclinical congenital toxoplasma infection. *Pediatrics*, **66**, 767–74.
96. Wong, B., Gold, J.M.W., Brown, A.E., *et al.* (1984) Central nervous system toxoplasmosis in homosexual men and parenteral drug abusers. *Ann. Intern. Med.*, **100**, 36–42.
97. Woodison, G., Smith, J.E. (1990) Identification of the dominant cyst antigens of *Toxoplasma gondii. Parasitology*, **100**, 389–92.
98. Xia, A.-D., Gu, Y.-Z., Xu, S.-J., *et al.* (1992) Construction of a genomic DNA library of the *Toxoplasma gondii* ZS_2 strain, screening of specific clones, and DNA diagnosis of toxoplasmosis. *Am. J. Trop. Med. Hyg.*, **46**, 50–6.

8

Pneumocystis carinii

JULIAN M. HOPKIN and ANN E. WAKEFIELD

8.1 INTRODUCTION

Pneumocystis carinii is the principle cause of severe pneumonia in the immunosuppressed, particularly those with T-lymphocyte immune deficiency, including extremely malnourished premature infants, leukaemics treated with chemotherapy, organ transplant recipients and those with HIV infection causing AIDS [16]. The disease takes the form of a diffuse pneumonitis with bilateral chest X-ray change and causes marked hypoxaemia which is the ultimate cause of death if treatment is not promptly instituted [17]. Histopathology shows filling of scattered alveolar spaces with a foamy debris that includes the two forms of the parasite. The trophozoite, a thin-walled amoeboid phase of 2–6 µm has the main interaction with alveolar tissue, an adherence to the flat type I pneumocyte which lines the alveolar space (Figure 8.1). The thick-walled cyst containing up to 8 intracystic bodies, or sporozoites, is outnumbered by the trophozoite by a ratio of 100 : 1 and is believed to be a replicative phase, ultimately leading to the release of the intracystic bodies as trophozoites [55]. Morphological studies, based principally on electron-microscopy, have left doubts about the taxonomy of the organism; is it fungal or protozoal [43]?

Though clinical disease is not recognized in non-immunosuppressed individuals, serological studies indicate that 80% or more of normal children are exposed, without ill effect, before the age of 10 years [28, 48]. One assumption has been that dormant foci of organisms, the legacy of this childhood infection, are the source of pneumonia in those who may become immunosuppressed for various reasons during later life. The

Molecular and Cell Biology of Opportunistic Infections in AIDS
Edited by Steven Myint and Alan Cann
Published in 1992 by Chapman & Hall, London ISBN 0 412 45330 4

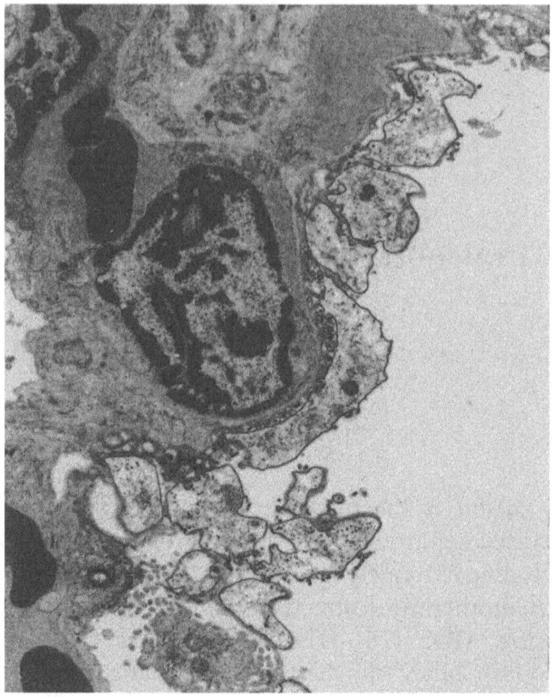

Figure 8.1 Electron micrograph of *P. carinii* infected alveolar space showing many thin walled trophozoites, most adherent to the alveolar pneumocytes.

testing of this idea, as well as establishing the diagnosis of clinical disease, has been restricted by failure to efficiently culture the organism *in vitro* [34] and by the relatively low sensitivity and specificity of the staining methods used to identify the organism by microscopy [36]. The clinical diagnosis has therefore often been dependent upon obtaining deep lung samplings by the invasive and potentially hazardous procedure of bronchoscopy [15]. Practical difficulties in prevention and treatment also remain, due to the toxicity of the agents used in chemoprophylaxis and treatment, particularly in AIDS patients [53].

Molecular approaches are now providing greater insights into the biology of *P. carinii* and therefore offering hopes for improved diagnosis, treatment and prevention of disease.

8.2 BIOCHEMISTRY

Early studies by Bruns [3] showed that the *P. carinii* cell wall was composed principally of mucopolysaccharide. ATPase was demonstrated by histochemistry in the pellicle of the trophozoite but not the cyst [2], suggesting the presence of an active electrolyte transport system which might be coupled to the inward transport of simple nutrients such as glucose. Enzymological study of selected dehydrogenase enzymes by Mazer and colleagues [27] was performed on tissue from the rat model. Light microscope histoenzymology, taking advantage of tetrazolium dyes that allow topographic localization of enzymatic activity, was used to show the presence of lactate dehydrogenase (which allows glycolysis under anaerobic conditions), succinate dehydrogenase (which functions in the Krebs tricarboxylic acid cycle), and glutamate dehydrogenase (which forms the link between the Krebs cycle and the amino acid cycle). Pesanti [30] used conventional biochemical techniques on highly enriched *P. carinii* from rat lungs (10^8 organisms/ml with less than 0.5% contamination with rat cells and fewer than 500 Gram-negative bacteria/ml); changes in oxygen tension were monitored by polarographic electrode, superoxide dismutase and glutathione oxidase activities by spectrophotometric assay, and the viability of the organisms was assessed by metabolism of radiolabelled glucose. Organisms utilized glucose and oxygen and produced carbon dioxide; this metabolism was cyanide sensitive, and presumably cytochrome mediated. Later, Pesanti used starch-gel electrophoresis of *P. carinii* lysates, compared with rat macrophage and erythrocyte lysates as controls, for *in situ* metabolic studies that allowed for more confident detection of *P. carinii* specific enzymes [31]. The presence of superoxide dismutase and catalase were confirmed, as well as glucose-6-phosphate dehydrogenase.

The successful empirical treatment of *P. carinii* pneumonia (*PCP*) with drug combinations that included either trimethoprim or pyrimethamine plus sulphonamide made the folate pathway of the parasite of interest. Allegra *et al.* [1] using radiolabelling studies in whole *P. carinii* organisms showed that the parasite did not absorb compounds with the classic folate structure, indicating that it probably lacks a specific active transport system for folates to cross cellular membranes.

In further studies, Kovacs *et al.* [18], used whole harvested organisms from rat lungs, and maintained on a human lung fibroblast cell line, for studies that utilized radiolabelling of para-aminobenzoic acid (PABA), and measured uptake and metabolism by combined scintillation counting and high performance liquid chromatographic elution profiles of folates produced. The studies demonstrated that *P. carinii* does synthesize folate *de novo*, and that this process could be inhibited by the sulphonamide, sulphamethoxazole and dapsone.

The use of difluoromethylornithine (DFMO) as an anti-*P. carinii* agent prompted the study of polyamine metabolism in this organism [23]. DFMO is a specific inhibitor of ornithine decarboxylase, the rate-limiting enzyme in the biosynthesis of polyamines, which are required for growth and differentiation. It was found that the rat-derived parasites contained putrescine and spermidine and were able to incorporate exogenous ornithine and arginine into these polyamines, but not into spermine. However, contrary to expectation, catalytically active ornithine decarboxylase could not be demonstrated in cell-free lysates of *P. carinii*.

Different studies on the cell wall of the cyst have emphasized an important carbohydrate component [55]. Using methenamine silver and aniline blue dyes to selectively stain β-glucan, Matsumoto [26] showed positive staining of the wall of both *P. carinii* and *Saccharomyces cerevisiae*; treatment with zymolyase (the major component of which is β-1,3-glucan laminari pentahydrolase) produced lysis of the outer two layers of the cyst wall with loss of staining. Involvement of the β-glucan component of the parasite cell wall was also demonstrated in an *in vivo* study using the rat model of *PCP*. Compounds from two classes of 1,3-β-glucan synthesis inhibitors, the echinocandins and the papulacandins, were found to show some effect in reducing parasite numbers [37]. Studies using lectins as probes for different sugars (fluorescent-labelled for light microscopy or gold-labelled for ultrastructural studies) show specific binding to the outer wall of the trophozoite and outer and middle layers of the cyst [4, 56]. Immunocytochemical studies show that chitin is present on the surface of both cyst and trophozoite [49]. Observations on the triggering of the classical complement cascade by glucan/mannan residues, without requirement for specific interacting antibody [41], may explain the prominent, apparently ineffective interstitial inflammatory cell reaction seen in the severely immunosuppressed human and the experimental animal with *PCP*.

8.3 ANTIGEN STUDIES

Though the many studies by different groups on the major surface antigens of *P. carinii* have used many methods, a coherent picture is emerging [14]. The development of monoclonal antibodies has provided the opportunity for studying the parasite's antigenicity, for comparing *P. carinii* isolates from different hosts, and also for characterising *P. carinii*-specific proteins [11, 14, 19, 20, 21]. Gigliotti developed monoclonal antibodies to surface proteins of rat-derived *P. carinii* and showed that some monoclonal antibodies were reactive to the organism from only certain hosts, whilst another antibody identified a protein common to *P. carinii* from different host species [14]. This 90–95 kDa protein is the same as that identified as 110–116 kDa, under different conditions by another monoclonal antibody developed by Graves *et al.* [12]. This protein was originally only sporadically demonstrated in human preparations, perhaps because fresh human tissue is hardly ever available for study; but more recently Walzer and Linke [50] have shown its presence in human-derived tissue, and this has also been documented by Kovacs *et al.* [19]. Gigliotti has shown that the monoclonal antibody reactive with this antigen confers some protection against *PCP* in animal models, and also, by lectin binding studies [10], that this antigen is a glycoprotein which specifically includes mannose and *N*-acetyl glucosamine residues. Further analysis of this major surface protein shows that its antigenicity in immunoblotting experiments is disturbed by pretreatment with the proteinase, trypsin, or an agent, such as metaperiodate, that oxidizes the hydroxyl radicals of carbohydrates, all indicating its glycoprotein nature [14]. Tanabe *et al.* [42] used isoelectric focusing to show that this 116 kDa glycoprotein comprises six major forms, each sharing a common peptide map, suggesting an identical core protein but differing in the degree of their glycosylation, which could account for up to 5 kDa of the glycoprotein's molecular weight. Further studies have demonstrated differences in the carbohydrate composition of this glycoprotein between rat-derived and human-derived *P. carinii.* Glucose, mannose, galactose and glucosamine occurred in approximately equimolar ratios in the human *P. carinii* protein, whereas glucose and mannose were the predominant sugars of the rat *P. carinii* protein [25]. The mannose moiety of this predominant surface glycoprotein may be involved in the uptake of *P. carinii* by alveolar macrophages. Binding and uptake of cultured rat *P. carinii* by human and rat alveolar macrophages was shown to be competitively reduced by inhibitors of the macrophage mannose receptor activity. Using a human lung fibroblast cell line transfected with cDNA coding for the human macrophage mannose receptor, this study suggested that the mannose receptor alone was sufficient for macrophage binding and ingestion of *P. carinii* [8].

Different groups have also identified bands at approximately 45 and 60 kDa. Recent studies [14, 22, 51] including comparative protein preparations from other forms of inflamed lung, provide data supporting the specificity of these bands to *P. carinii.* Other bands on human material have been regularly identified at 20 and 40 kDa. Differences have been identified in electrophoretic patterns of *P. carinii*-infected lungs from human and rat hosts, although immune serum from both species shows partial cross-reactivity for different bands on immunoblotting. Graves *et al.* [13] have shown that sera, from rats recovering from *P. carinii* and from normal humans, react with the same bands; therefore *P. carinii* infecting these different hosts must share some major antigenic determinants. Differences observed in monoclonal antibody reactions with *P. carinii* isolates from different hosts suggest, but do not necessarily prove, genotype differences between these isolates.

8.4 DNA STUDIES

The cloning of *P. carinii* DNA was impeded by the lack of an *in vitro* means of culture and the resulting difficulties in purifying the parasite from host lung tissue, but it has now been achieved using fairly coarse enrichment procedures, followed by cloning and selection procedures. Wakefield *et al.* [44] selected clones of interest by exclusion of those clones which hybridized to host (rat) genomic DNA, and then testing the putative *P. carinii* clones by an *in situ* hybridization assay. Cloned DNA sequences of interest annealed to *P. carinii* in the infected lung of both rat and man, with a distribution limited to the alveolar spaces and with a morphology characteristic of the arrangement of the nuclei within the cyst and the trophozoite phase of the parasite (Figure 8.2). Using these methods, chromosomal and mitochondrial gene sequences have been isolated including those encoding apocytochrome B, NADH dehydrogenase subunits 1, 2, 3, and 6, cytochrome oxidase subunit 11, and the small subunit of ribosomal RNA [35] (Figure 8.3). A number of preliminary studies on chromosomal DNA from highly purified parasite, using pulsed

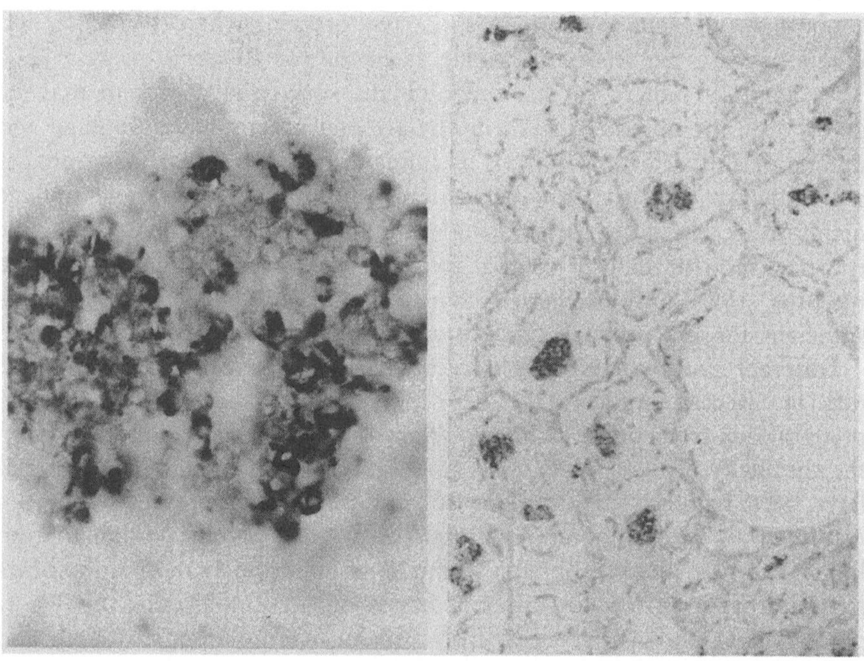

Figure 8.2 A photomicrograph of *in situ* hybridization with biotin labelled cloned *P. carinii* DNA to infected human lung tissue showing: (a) at low power, the parasite restricted to the alveolar spaces; (b) at high power, the characteristic annealing with the nuclear DNA of the parasite, forming scattered 'dot signals' from the trophozoites and 'ring signals' from the cluster of intracystic sporozoites.

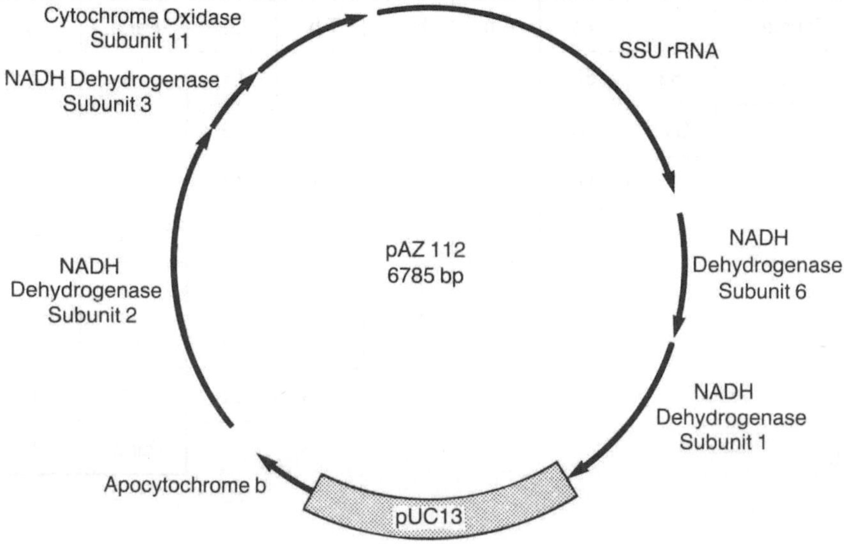

Figure 8.3 A map of pAZ112, a recombinant plasmid containing cloned *P. carinii* DNA, showing the arrangement of seven *P. carinii* mitochondrial genes.

field gel electrophoresis, have suggested a genome size for *P. carinii* ranging from 7 to 16 × 10^6 base pairs, with estimates of the number of chromosomes ranging from 13 to 20 [9, 24, 54].

Edman and colleagues [6] have screened cloned DNA from *P. carinii*-infected lung tissue with *Entamoeba histolytica* 16S-like rRNA genes, thereby providing opportunity for identifying and sequencing this 'ubiquitous' gene from *P. carinii* and comparing it with sequences from other species to provide further information on the taxonomy of the organism. Contrary to some expectations, genetic similarity was greatest to the fungi and not the protozoa. Other groups using RNA sequencing have also found homology with fungal sequences for fragments of the *P. carinii*-derived, small ribosomal RNA sequences [40], and 5S ribosomal RNA [52].

However, defining taxonomy, by constructing phylogenetic trees based on DNA sequencing, properly requires the comparison of many DNA sequences. Comparative sequencing analysis of six contiguous mitochrondrial gene sequences from *P. carinii* strongly supports a fungal classification for the parasite [35] (Figure 8.4, Table 8.1). Taxonomic deliberation can descend into sterile debate, but these DNA studies, with

(a)

Source	S.c.	N.c.	A.n.	H.s.	T.b.	L.t.	Par.a.	P.y.
P.c.	53	57	54	36	17	15	9	7
	S.c.	56	61	31	13	12	15	4
		N.c.	70	37	11	10	15	7
			A.n.	32	10	8	15	8
				H.s.	12	12	9	8
					T.b.	76	11	6
						L.t.	9	8
							Par.a.	8

(b)

Source	N.c.	P.a.	S.c.	A.n.	H.s.	T.b.	L.t.	Par.a.
P.c.	65	63	60	58	43	25	23	11
	N.c.	92	59	71	41	24	25	12
		P.a.	59	68	41	24	25	12
			S.c.	54	38	27	25	16
				A.n.	37	26	24	7
					H.s.	26	24	8
						T.b.	81	6
							L.t.	6

Figure 8.4 Similarities (%) of amino acids sequences, derived from primary DNA sequences, amongst a range of organisms: (a) part of apocytochrome B; (b) cytochrome oxidase sub-unit 11. *Pneumocystis carinii* (P.c.); *Homo sapiens* (H.s.); the fungi *Neurospora crassa* (N.c.); *Aspergillus nidulans* (A.n.), *Saccharomyces cerivisiae* (S.c.), *Popspora anserina* (P.a.); and the protozoans *Trypanosoma brucei* (T.b.), *Leishmania tarentolae* (L.t.), *Paramecium aurelia* (Para.a), and *Plasmodium yeolii* (P.y.).

Table 8.1 Similarity (%) of derived amino acid sequence from six contiguous mitochondrial genes of *P. carinii* compared with fungal, protozoal and human sequences

	N.c	*S.c*	*A.n*	*H.s*	*T.b*	*L.t*
Cytochrome oxidase 11	64.8	60.1	57.9	42.9	24.7	22.8
Apocytochrome b	58.4	52.9	53.9	36.0	21.8	18.4
NADH1	51.6	NI	53.4	42.5	24.6	22.4
NADH2	39.6	NI	34.8	26.1	NP	NP
NADH3	38.8	NI	38.3	22.1	NP	NP
NADH6	NP	NI	35.0	12.1	NP	NP

NP, not published; NI, not identified; N.c, *neurospora crassa*; S.c, *Saccharomyces cerevisiae*; A.n, *Aspergillus nidulans*; H.s, *Homo sapiens*; T.b, *Trypanosoma brucei*; L.t, *Leishmania tarentolae.*

their consistent suggestion of fungal homology, must lead to re-evaluation of perspectives regarding *in vitro* culture and choice of candidate chemotherapeutic agents for the parasite, and also the epidemiology of infection.

Molecular genetic techniques are also being applied to clone specific *P. carinii* genes and express their protein products, thus providing the opportunity for studies on pathogenic mechanisms and the development of pharmacological inhibitors for key biochemical pathways. Edman and colleagues [5, 7] have cloned the *P. carinii* genes for thymidylate synthase, a protein of interest since it catalyses *de novo* synthesis of one essential DNA base (dTMP) while reconverting tetrahydrofolate to dihydrofolate and dihydrofolate reductase. Expression in bacterial vectors has been achieved, making purified enzymic products available for study.

Careful selection of oligonucleotide primers, based on a mitochondrial ribosomal RNA sequence of *P. carinii*, has allowed the development of efficient DNA amplification, using the polymerase chain reaction (PCR), for these *P. carinii* sequences [47]. The technique has been applied to diagnostic work and epidemiological studies. DNA amplification, using these primers, has allowed comparison of sequences of DNA from *P. carinii* derived from different hosts. Small differences in this region of the ribosomal gene are consistently observed between the sequences derived from rat and human hosts, finally demonstrating that the parasites are genetically distinct (Figure 8.5) [38]. This implies that different mammals, infected with *P. carinii*, are not the source of infection for man; this is consistent with the observation that exposure to animals in the home is not a risk factor for the development of *PCP*, as documented in the study of childhood leukaemics [17].

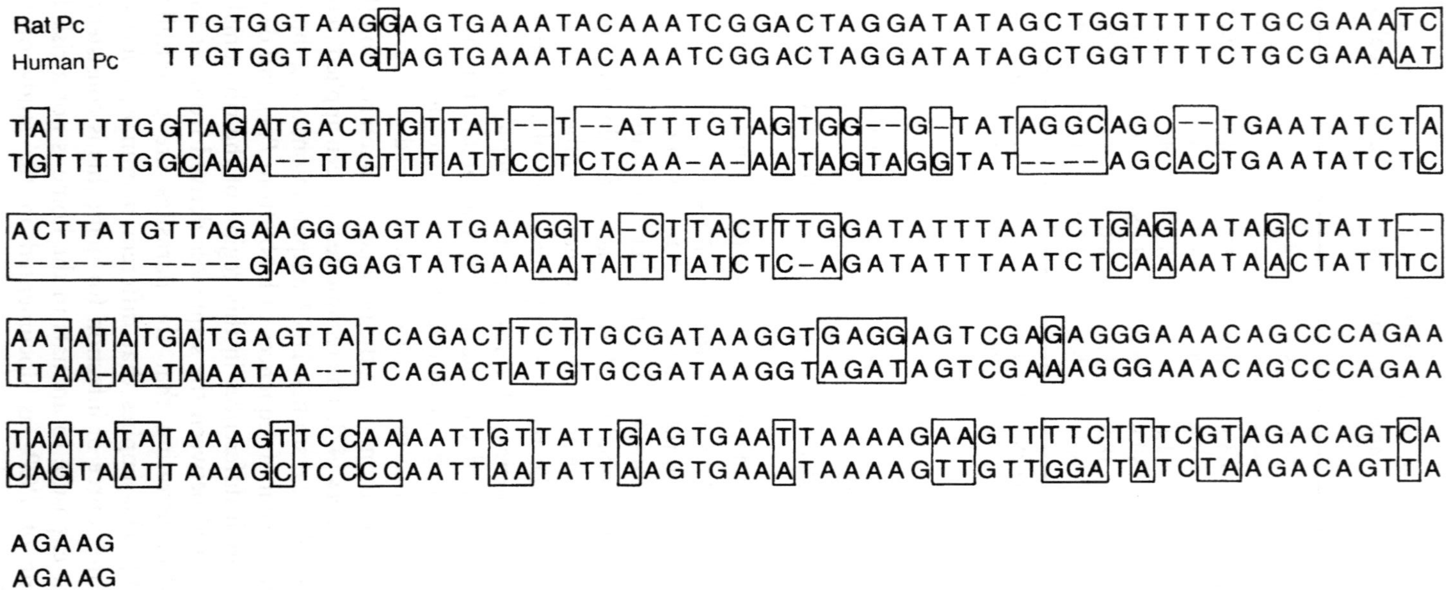

Figure 8.5 Comparative DNA sequencing of *Pneumocystis carinii* DNA derived from the rat and human host.

The use of these oligonucleotide primers for the specific amplification of *P. carinii* DNA has been described in two clinical studies [45, 46]. The recognition of the specific amplified DNA was made by direct visualization under UV irradiation after simple ethidium bromide staining of gel electrophoresed products, and subsequent internal oligonucleotide hybridization and autoradiography of Southern transferred fragments (Figure 8.6). The second step increases sensitivity of the DNA assay 100-fold; in experiments using rat-derived *P. carinii* for seeding clinical samples, it has been shown that the band detected on an ethidium stained gel (strong signal) indicates the presence of 100 organisms or more, whereas identification of the product by oligonucleotide hybridization alone (weak signal) indicates fewer organisms, which are detectable to a limit to 1–2 organisms in the PCR sample [32].

These methods were applied to both bronchoscopic alveolar lavage samples and non-invasive induced sputum samples [45, 46]. The patient groups and results are summarized in Tables 8.2 and 8.3. The results from the two studies can be summarized: DNA amplification is a significantly more sensitive method for identifying *PCP* than silver staining and reliably identifies the condition in 95% of instances whether the sampling is lavage or induced sputum; simple calibration of the test into strong and weak signals, known to correspond to 100 organisms and more or fewer organisms down to a detectable level of 1–2 in a PCR sample also allows diagnostic specificity; and although 20% of immunosuppressed patients may show weak signals in diagnostic samples (indicative of subclinical infection), a strong signal indicates *PCP* with a high probability (~95% based on these data).

Although the method offers a highly sensitive and specific diagnostic technique, no *P. carinii* was detected in any of the clinical samples from immunocompetent subjects, either by ethidium staining or subsequent oligonucleotide hybridization. This observation, together with other reports [29], seriously questions assumptions made that *PCP* necessarily represents recrudescence of dormant pulmonary infection, the legacy of childhood exposure. To test this further, DNA amplification has been applied directly to post-mortem lung samplings from immunocompetent subjects ranging from 15–70 years of age [33]. *P. carinii*-specific DNA was not detected in 45 DNA amplification reactions from 15 lungs. The results do not favour the concept of carriage of *P. carinii* in normal lungs and make fresh infection of the immunosuppressed, causing PCP, a more likely alternative. Reported case clusters of *PCP* in different immunosuppressed groups suggest that horizontal transmission may be significant [39] whilst the recent demonstration of the fungal origins of *P. carinii* emphasize the possibility of an, as yet unidentified, environmental source of the parasite and infectious particles.

Table 8.2 One hundred subjects studied by DNA amplification and silver staining for *P. carinii* in bronchoalveolar lavage samples

	Clinical PCP	*Silver staining*	*DNA amplification*	
			Gel	*Blot*
Immunosuppressed (n = 90)				
Clinical PCP and silver positive (n = 30)	30	30	30	30
Clinical PCP but silver negative (n = 14)	14	0	10	10
Not PCP silver negative (n = 46)	0	0	1[a]	11
Immunocompetent (n = 10)	0	0	0	0

PCP, *P. carinii* pneumonia.
[a] developed PCP within 10 weeks of study.

Table 8.3 Thirty-eight immunosuppressed subjects studied by DNA amplification and silver staining for *P. carinii* in paired bronchoalveolar lavage and induced sputum samples [45]

		Lavage			*Sputum*		
			DNA			*DNA*	
	Clinical PCP	*Silver staining*	*Gel*	*Blot*	*Silver staining*	*Gel*	*Blot*
Clinical PCP and silver positive (n = 15)	15	14[a]	15	15	7[a]	15	15
Clinical PCP but silver negative (n = 6)	6	0	4	4	0	3	4
Not PCP silver negative (n = 17)	0	0	1[b]	4	0	1[b]	1

PCP, *P. carinii* pneumonia.
[a] One patient was silver positive on sputum but not lavage.
[b] One patient developed PCP within 10 weeks of study.

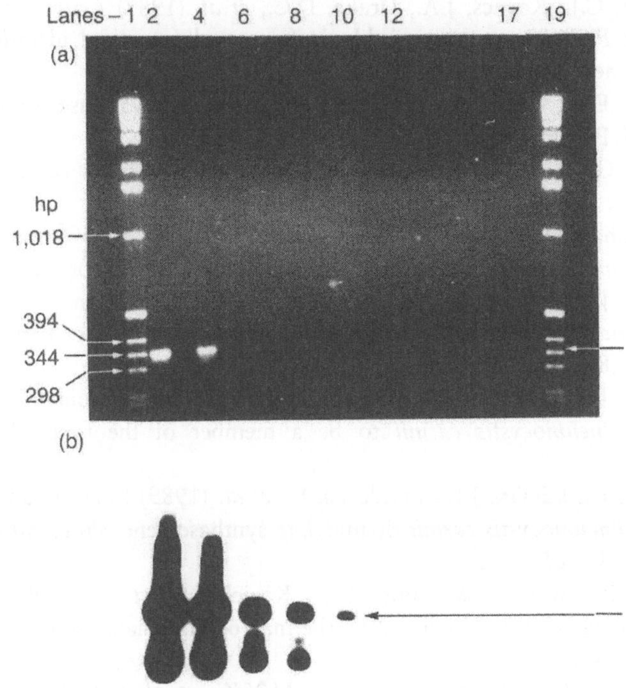

Figure 8.6 DNA amplification for *P. carinii* specific sequences. Sensitivity of detection of amplification products using serially ten-fold diluted DNA template prepared from bronchoalveolar lavage from a patient with confirmed *P. carinii* infection. (a) Ethidium bromide stained gel. (b) Autoradiograph of Southern blot probed with an internal oligonucleotide. The asterisk indicates the position of the *P. carinii* specific amplification product.

8.5 CONCLUSIONS

Study of the molecular biology of *P. carinii* has just begun. A clearer understanding of the epidemiology of infection has quickly emerged. *P. carinii* is a fungal organism and *PCP* in the immunosuppressed individual develops as the result of fresh infection. The source of infectious particles may be a free-living environmental form of the parasite. Various genes in the organism have now been cloned and protein produced by recombinant genetic methods; this progress provides the hope for the development of novel chemotherapeutic agents valuable for chemoprophylaxis and treatment, and may ultimately to lead to an understanding of the pathogenicity of the organism.

REFERENCES

1. Allegra, C.J., Kovacs, J.A., Drake, D.C., *et al.* (1987) Activity of antifolates against *Pneumocystis carinii* dihydrofolate reductase and identification of a potent new agent. *J. Exp. Med.*, **165**, 926–31.
2. Barton, E.G., Campbell, W.G. (1967) Further observations on the ultrastructure of Pneumocystis. *Arch. Pathol.*, **83**, 527–34.
3. Bruns, G. (1956) Der Foormwechsel bei *Pneumocystis carinii. Acta Histochem.*, **2**, 222.
4. De Stephano, J.A., Cushion, M.T., Trinkle, L.S., *et al.* (1989) Lectins as probes to *Pneumocystis carinii* surface glycocomplexes. *J. Protozool.*, **36**, 65–6.
5. Edman, J.C., Edman, U., Cao, M., *et al.* (1989) Isolation and expression of the *Pneumocystis carinii* dihydrofolate reductase gene. *Proc. Natl. Acad. Sci. U.S.A.*, **86**, 8625–9.
6. Edman, J.C., Kovacs, J.A., Masur, H., *et al.* (1988) Ribosomal RNA sequence shows *Pneumocystis carinii* to be a member of the fungi. *Nature*, **334**, 519–22.
7. Edman, U., Edman, J.C., Lundgren, B., *et al.* (1989) Isolation and expression of the *Pneumocystis carinii* thymidylate synthase gene. *Proc. Natl. Acad. Sci. U.S.A.*, **86**, 6503–7.
8. Ezekowitz, R.A.B., Williams, D.J., Koziel, H., *et al.* (1991) Uptake of *Pneumocystis carinii* mediated by the macrophage mannose receptor. *Nature*, **351**, 155–8.
9. Fishman, J.A., Ullu, E., Armstrong, M.Y.K., *et al.* (1989) Organization of DNA and RNA from rat *Pneumocystis carinii. J. Protozool.*, **36**, 4–5.
10. Gigliotti, F., Hughes, W.T. (1988) Passive immunoprophylaxis with specific monoclonal antibody confers partial protection against *Pneumocystis carinii* pneumonia in animal models. *J. Clin. Invest.*, **81**, 1666–8.
11. Gigliotti, F., Stokes, D.C., Cheatham, A.B., *et al.* (1986) Development of murine monoclonal antibodies to *Pneumocystis carinii. J. Infect. Dis.*, **154**, 315–22.
12. Graves, D.C., McNabb, S.J.N., Ivey, M.H., *et al.* (1986) Development and characterisation of monoclonal antibodies to *Pneumocystis carinii. Infect. Immun.*, **51**, 125–33.
13. Graves, D.C., McNabb, S.J.N., Worley, M.A., *et al.* (1986) Analysis of rat *Pneumocystis carinii* antigens recognised by human and rat antibodies by using Western immunoblotting. *Infect. Immun.*, **54**, 96–103.
14. Graves, D.C. (1989) Immunological studies of *Pneumocystis carinii. J. Protozool.*, **36**, 60–9.
15. Hopkin, J.M., Turney, J.H., Young, J.A., *et al.* (1983) Rapid diagnosis of obscure pneumonia in immunosuppressed renal patients by cytology of alveolar lavage fluid. *Lancet*, **ii**, 299–301.
16. Hopkin, J.M. (1991) *Pneumocystis carinii.* Oxford University Press, Oxford.
17. Hughes, W.T., Price, R.A., Ho-Kyun, K., *et al.* (1973) *Pneumocystis carinii* pneumonitis in children with malignancies. *J. Paediatr.*, **82**, 404–15.

18. Kovacs, J.A., Allegra, C.J., Beaver, J., *et al.* (1989) Characterization of *de novo* folate synthesis in *Pneumocystis carinii* and *Toxoplasma gondii*: potential for screening therapeutic agents. *J. Infect. Dis.*, **160**, 312–20.
19. Kovacs, J.A., Halpern, J.L., Swan, J.C., *et al.* (1988) Identification of antigens and antibodies specific for *Pneumocystis carinii. J. Immunol.*, **140**, 2023–31.
20. Lee, C., Bolinger, C.D., Bartlett, M.S., *et al.* (1986) Production of monoclonal antibody against *Pneumocystis carinii* by using a hybrid of rat spleen and mouse myeloma cells. *J. Clin. Microbiol.*, **23**, 505–8.
21. Linder, E., Lundin, L., Vorma, H. (1987) Detection of *Pneumocystis carinii* in lung-derived samples using monoclonal antibodies to an 82 kD parasite component. *J. Immunol. Meth.*, **98**, 57–62.
22. Linke, M.J., Cushion, M.T., Walzer, P.D. (1989) Properties of the major antigens of rat and human *Pneumocystis carinii. Infect. Immun.*, **57**, 1547–55.
23. Lipschik, G.Y., Masur, H., Kovacs, J.A. (1991) Polyamine metabolism in *Pneumocystis carinii. J. Infect. Dis.*, **163**, 1121–7.
24. Lundgren, B., Cotton R., Lundgren, J.D., *et al.* (1990) Identification of *Pneumocystis carinii* chromosomes and mapping of five genes. *Infect. Immun.*, **58**, 1705–10.
25. Lundgren, B., Lipschik, G.Y., Kovacs, J.A. (1991) Purification and characterization of a major human *Pneumocystis carinii* surface antigen. *J. Clin. Invest.*, **87**, 163–70.
26. Masumoto, Y., Matsuda, S., Tegoshi, T. (1989) Yeast glucan in the cyst wall of *Pneumocystis carinii. J. Protozool.*, **36**, 21–3.
27. Mazer, M.A., Kovacs, J.A., Swan, J.C., *et al.* (1987) Histoenzymological study of selected dehydrogenase enzymes in *Pneumocystis carinii. Infect. Immun.*, **55**, 727–30.
28. Meuwissen, J.H.E.T., Tauber, I., Leevwenberg, A.D.E.M., *et al.* (1977) Parasitologic and serologic observations of infection with *Pneumocystis* in humans. *J. Infect. Dis.*, **136**, 43–9.
29. Millard, P.R., Heryet, A.R. (1988) Observations favouring *Pneumocystis carinii* pneumonia as a primary infection: a monoclonal antibody study on paraffin sections. *J. Pathol.*, **154**, 365–70.
30. Pesanti, E.L. (1984) *Pneumocystis carinii*: oxygen uptake, antioxidant enzymes and susceptibility to oxygen-mediated damage. *Infect. Immun.*, **44**, 7–11.
31. Pesanti, E.L. (1987) Phospholipid profile of *Pneumocystis carinii* and its interaction with alveolar type II epithelial cells. *Infect. Immun.*, **55**, 736–41.
32. Peters, S.E., Wakefield, A.E., Bannerji, S., *et al.* (1992) Quantification of DNA amplification for *Pneumocystis carinii. Mol. Cell. Probes*, **6**, 115–17.
33. Peters, S.E., Wakefield, A.E., Sinclair, K., *et al.* (1992) A search for latent *Pneumocystis carinii* infection in post-mortem lungs by DNA amplification. *J. Pathol*, **166**, 195–8.
34. Pifer, L.L., Hughes, W.T., Murphy, M.J. (1977) Propagation of *Pneumocystis carinii in vitro. Pediatr. Res.*, **11**, 305–16.
35. Pixley, F.J., Wakefield, A.E., Banerji, S., *et al.* (1991) Mitochondrial DNA

sequence demonstrates fungal homology of *Pneumocystis carinii*. *Molec. Microbiol.*, **5**, 1347.

36. Reinhardt, D.J., Kaplan, W., Chandler, F.W. (1977) Morphologic resemblance of zygomycete spores to *Pneumocystis carinii* cysts in tissue. *Am. Rev. Respir. Med.*, **115**, 170–2.
37. Schmatz, D.M., Romancheck, M.A., Pittarelli, L.A., *et al.* (1990) Treatment of *Pneumocystis carinii* pneumonia with 1,3β-glucan synthesis inhibitors. *Proc. Natl. Acad. Sci. U.S.A.*, **87**, 5950–4.
38. Sinclair, K., Wakefield, A.E., Bannerji, S., *et al.* (1991) *Pneumocystis carinii* organisms derived from rat and human hosts are genetically distinct. *Mol. Biochem. Parasitol.*, **45**, 183–4.
39. Singer, C., Armstrong, D., Rosen, P.P., *et al.* (1975) *Pneumocystis carinii* pneumonia: a cluster of 11 cases. *Ann. Intern. Med.*, **82**, 772–7.
40. Stringer, S.L., Stringer, J.R., Blase, M.A., *et al.* (1989) *Pneumocystis carinii*: sequence from ribosomal RNA implies a close relationship with fungi. *Exp. Parasitol.*, **68**, 450–61.
41. Super, M., Thee, S., Lu, J., *et al.* (1989) Association of low levels of mannan-binding protein with a common defect of opsonization. *Lancet*, **ii**, 1236–9.
42. Tanabe, K., Takasaki, S., Watanabe. J., *et al.* (1989) Glycoproteins composed of major surface immunodeterminants of *Pneumocystis carinii*. *Infect. Immun.*, **57**, 1363–8.
43. Ul Haque, A., Plattner, S.B., Cook, R.T., *et al.* (1987) *Pneumocystis carinii* taxonomy as viewed by electron microscopy. *Am. J. Clin. Pathol.*, **87**, 504–10.
44. Wakefield, A.E., Hopkin, J.M., Burns, J., *et al.* (1988) Cloning of DNA from *Pneumocystis carinii*. *J. Infect. Dis.*, **158**, 859–62.
45. Wakefield, A.E., Miller, R.M., Gwiver, L., *et al.* (1991) DNA amplification for the diagnosis of *Pneumocystis carinii* pneumonia from induced sputum. *Lancet*, **i**, 1378–9.
46. Wakefield, A.E., Pixley, F.J., Bannerji, S., *et al.* (1990) Detection of *Pneumocystis carinii* by DNA amplification. *Lancet*, **ii**, 451–3.
47. Wakefield, A.E., Pixley, F.J., Sinclair, K., *et al.* (1990) Amplification of mitochondrial ribosomal RNA sequences from *Pneumocystis carinii* DNA of rat and human origin. *Mol. Biochem. Parasitol.*, **43**, 69–76.
48. Wakefield, A.E., Stewart, T.J., Moxon, E.R., *et al.* (1990) Infection with *Pneumocystis carinii* is prevalent in healthy Gambian children. *Trans. R. Soc. Trop. Med. Hyg.*, **84**, 800–2.
49. Walker, A.N., Garner, R.E., Horst, M.N. (1990) Immunocytochemical detection of chitin in *Pneumocystis carinii*. *Infect. Immun.*, **58**, 412–5.
50. Walzer, P.D., Linke, M.J. (1987) A comparison of the antigenic characteristics of rat and human *Pneumocystis carinii* by immunoblotting. *J. Immunol.*, **138**, 2257–65.
51. Walzer, P.D., Stranforth, D., Linke, M.J., *et al.* (1987) *Pneumocystis carinii*: immunoblotting and immunofluorescent analyses of serum antibodies during experimental rat infection and recovery. *Exp. Parasitol.*, **63**, 319–28.
52. Watanabe, J., Hori, H., Tanabe, K., *et al.* (1989) Phylogenetic association of

Pneumocystis carinii with the 'Rhizopoda/Myxomycota/Zygomycota group' indicated by comparison of 5S ribosomal RNA sequences. *Mol. Biochem. Parasitol.*, **32**, 163–8.

53. Wharton, J.M., Coleman, D.L., Wofsy, C.B., *et al.* (1986) Trimethoprim-sulfamethoxazole or pentamidine for *Pneumocystis carinii* pneumonia in the acquired immunodeficiency syndrome. *Ann. Intern. Med.*, **105**, 37–44.
54. Yoganathan, T., Lin, H., Buck, G.A. (1989) An electrophoretic karyotype and assignment of ribosomal genes to resolved chromosomes of *Pneumocystis carinii. Mol. Microbiol.*, **3**, 1473–80.
55. Yoshida, Y. (1989) Ultrastructural studies of *Pneumocystis carinii. J. Protozool.*, **36**, 53–60.
56. Yoshikawa, H., Tegoshi, T., Yoshida, Y. (1989) Detection of surface carbohydrates of *Pneumocystis carinii. J. Protozool.*, **36**, 63–4.

9

Candida

RUTH MATTHEWS

9.1 INTRODUCTION

Oral and oesophageal candidosis (candidiasis) are common in patients with the acquired immunodeficiency syndrome (AIDS). Indeed, unexplained oral thrush is often the initial manifestation of the syndrome [18, 54]. Oesophageal candidosis is a diagnostic criterion of 'full-blown' AIDS [17]. Candidal enteritis has also been described in AIDS patients but dissemination beyond the gastrointestinal tract is unusual [106]. There have been reports of pulmonary and cerebral abscesses due to *Candida albicans* [124], but widespread visceral candidosis is much less frequent than in other immunosuppressed individuals who do not have AIDS [109]. It may be that the incidence of systemic candidosis in AIDS will increase as improved management lengthens the survival time of patients with very severe, advanced immunodeficiencies.

In this chapter, the interaction of *Candida* with HIV will be examined. This may increase our understanding of the pathogenesis of candidal infections in individuals who do not have AIDS as well as in those who do. In healthy adults *Candida albicans* is part of the normal flora of the oral cavity and gastrointestinal tract. In immunocompetent individuals it is usually non-pathogenic but it may become invasive, giving rise to oral or vulvovaginal candidosis. Oral candidosis is prevalent in the very young or elderly or the terminally ill [31]. Otherwise it is unusual and likely to be symptomatic of an underlying condition such as AIDS, diabetes mellitus or malignancies. In contrast, vaginal candidosis is common, and approximately ten times more frequent than candidal infections of the penis [36, 94].

Molecular and Cell Biology of Opportunistic Infections in AIDS
Edited by Steven Myint and Alan Cann
Published in 1992 by Chapman & Hall, London ISBN 0 412 45330 4

Cutaneous candidosis, including napkin dermatitis, is associated with repeated wetting and occlusion of the skin. Candidal infections of the nail tend to arise as an occupational disease, for example in fruit canners. Rarely, persistent candidal infections arise simultaneously at multiple superficial sites, a condition referred to as chronic mucocutaneous candidosis (CMC). This is not a single disease entity. It is often, but not always, associated with defects in cell-mediated immunity and can be familial. Many different defects have been incriminated, including T-cell dysfunction and complement deficiencies [26, 53]. Some have selective antibody deficiencies, particularly secretory IgA, but serum antibodies to *Candida* are usually high [19]. In immunocompromised, non-AIDS patients the most important form of candidal infection is systemic candidosis. This is because of its high mortality, of up to 75% [42]. Most commonly this takes the form of disseminated candidosis, with spread via the bloodstream to multiple organs, but it may be localized to a single organ [63, 107].

C. albicans is by far the commonest species of *Candida* to cause these infections. First the molecular biology and genetics of this organism will be examined; this has led to the development of new typing systems, from which we can gain further insight into the epidemiology of these infections. New ideas regarding the pathogenesis of candidal infections will be discussed, including the role of heat shock proteins. The clinical implications of these findings, concerning diagnosis and treatment, will be explored.

9.2 MOLECULAR BIOLOGY, GENETICS AND STRUCTURE

C. albicans is a dimorphic fungus. *In vivo* it is seen as a mixture of yeasts and hyphae; and the latter were at one time thought to be indicative of the 'invasive' form of the infection. *In vitro* studies indicate that the hyphal form is better at adherence and more active in the initial stages of tissue penetration, but overall animal models suggest that preformed *Candida* hyphae are no more virulent than yeast cells. This subject has been comprehensively reviewed by Odds [95] who concludes that there is no real proof that either yeasts or hyphae play a superior role in pathogenesis.

C. albicans reproduces asexually. A unicellular yeast cell (blastospore) produces a new cellular outgrowth (bud) at one of its poles, which enlarges, undergoes mitosis, and a septum is laid down between parent and bud. Hyphae arise as branches from existing hyphae or yeast cells (germ tubes). Mitotic cell division occurs in the growing hyphae and each cell unit is separated by septa, laid down within the hyphae. The hyphae may give rise to hyphal branches or bud laterally to give new blastospores. Narrow, elongated 'pseudohyphae' arise from yeast cells or hyphae when the buds remain attached to the parent. Chlamydospores are large, thick-walled, refractile cells which are produced *in vitro* under conditions of nutrient depletion. Since *C. albicans* is the only species of *Candida* to produce these specialized structures they are useful for identification purposes.

Environmental conditions, such as pH, temperature, initial blastospore concentration, and the presence or absence of serum, influence the transition between yeast and hyphae. The regulatory processes involved are not yet known at a molecular level. Another phenotypic change, which is only just beginning to be understood, is the tendency of *C. albicans* to give rise to variant colonial forms. The ability to give rise to 'rough' colonies, was first described by Negroni in 1935 [92]. These colonies contain abundant hyphae. Varieties of colony types have been described by Di Menna [25] and Slutsky's group [122], the latter generated by UV treatment, which increased the frequency with which variants arose from the parental strain. Soll *et al.* [125] showed that 9 of 11 isolates from patients with vaginal candidosis were in a high frequency (10^{-2}–10^{-3}) mode of switching between different colony morphotypes. This may be the result of transposable genetic elements activating a range of silent genes [126]. It was suggested that this plasticity might help the yeast to develop resistance to antifungal agents, invade different tissues and evade the immune system by changing antigenicity. They subsequently characterized the strains isolated from a patient with recurrent vaginal candidosis [128] and demonstrated that a new switch phenotype arose from the same strain after the first drug treatment. However, this new switch phenotype did not show increased resistance to the antifungal drug clotrimazole.

Until recently, an understanding of the genetics of *C. albicans* was hampered by the fact that it is diploid and reproduction is asexual [136]. Generation of mutants in a diploid requires two steps, mutation in the gene of interest and a mitotic recombination event to render the mutation homozygous. The frequency of both events occurring can be made much greater by a two step process, such as nitrous acid followed by UV irradiation, rather than the latter alone [48]. To achieve exchange of genetic information between cells, in the absence of a mating system, techniques have been developed for fusing spheroblasts with polyethylene glycol [49, 106]. More recently, major advances have been made by cloning genes from *C. albicans*, transformation, site directed mutagenesis and electrophoretic karyotyping. Several genomic libraries have been prepared from *C. albicans* DNA, including a *Saccharomyces cerevisiae* library with the plasmid YEP13 (J. Gorman of Smith, Kline and French Research Laboratories) and λ libraries in *Escherichia coli* (S. Scherer, University of Minnesota; R. Matthews, University of Manchester). The mitochondrial genome has been mapped, cloned and used to analyse strains of *C. albicans* for mitochondrial DNA polymorphism [100]. Transformation in *C. albicans* was achieved by Kurtz *et al.* [56]. Directed mutagenesis has been widely used in elucidating gene function in *S. cerevisiae*, and similar techniques are being applied to *C. albicans* [51]. Since *C. albicans* is diploid, it has to be made homozygous for the gene disruption and this can be induced by UV irradiation.

The number of size of *C. albicans* chromosomes has been examined by pulsed field gel electrophoresis. Chromosome karyotypes from different strains show considerable variation in the migration distances of some of the chromosomes, indicating there could be differences in the genetic maps of these strains [69]. Lasker *et al.* [60] separated 8 bands of chromosome-sized DNA molecules by orthogonal field alternation gel electrophoresis (OFAGE). Six bands were observed by field inversion gel electrophoresis (FIGE). Southern blotting with a collection of cloned DNA probes specific for each of the 8 observed bands by OFAGE, showed that the 6 FIGE bands consisted of 4 singlets and 2 comigrating doublets. A mitochondrial DNA probe hybridized to material in the well but not to any band, suggesting all bands were due to nuclear DNA. They suggested that the *C. albicans* haploid genome contains a minimum of 8 chromosomes. This conclusion has also been reached by Iwaguchi *et al.* [47] who examined 27 strains. Magee *et al.* [70] suggested that *C. albicans* has 7 chromosomes, on the basis of using OFAGE, FIGE and contour-clamped homogenous field (CHEF) gel electrophoresis combined with Southern blotting. The largest chromosome (no. 1) which they identified in their study by CHEF gel electrophoresis, appears to have been separated into two distinct chromosomes by OFAGE [60]. The interpretation of these results is complicated by use of different methods of pulsed field gel electrophoresis, different running conditions and what appears to be the intrinsically variable karyotype of different strains of *C. albicans* [69].

Pulsed field gel electrophoresis has been applied to another asexual, diploid yeast, *Candida tropicalis* [132]. When applied to morphological mutants induced by UV irradiation of the parent strain, it was found that some of the mutants had changed their karyotype, suggesting that chromosomal rearrangements can be associated with morphological mutants. It may be that some of the variation in chromosomal karyotypes observed with different strains of *C. albicans* is in fact due to chromosomal rearrangement within individual strains.

9.3 PATHOGENESIS OF CANDIDOSIS

The species of *Candida* capable of causing human infections differ markedly in their inherent pathogenicity. In animal studies, *C. albicans* and *C. tropicalis* consistently carry the highest mortality and morbidity, with *C. albicans* usually being the more virulent [38, 44, 45, 67]. *C. albicans* is by far the commonest cause of superficial candidal infections, but accounts for only 50–70% of the commensal yeast flora in superficial sites [96]. In disseminated candidosis, *C. albicans* is identified as the causative agent in about 58–80% of cases, followed by *C. tropicalis* (10–13%) and *C. parapsilosis* (5–26%) [27, 101].

A prerequisite to invasion is the ability of the organism to adhere to the host's surfaces. *C. albicans* commonly colonizes the oral cavity and intestinal tract of man and other warm blooded animals. Indigenous bacterial flora may help to suppress colonization, so that treatment with antibacterial drugs enhances yeast colonization [24, 41]. In contrast, *S. cerevisiae*, which is generally believed to be non-pathogenic for mammals, is rapidly eliminated from the alimentary tract, even after administration in large doses [106]. King *et al.* [52] showed that *C. albicans* adhered to vaginal and buccal epithelial cells in considerably higher numbers than any of six other *Candida* species. Of the other six, only *C. tropicalis* and *C. stellatoidea* and, to a lesser extent, *C. parapsilosis*, showed significant adhesion. Therefore there appears to be some correlation between ability to adhere and colonize and subsequent invasion.

C. albicans produces a number of hydrolytic enzymes, expressed at its cell surface, which could damage host cells *in vivo* but incontrovertible evidence of this is hard to obtain. One of the strongest candidates as a virulence factor is the inducible acid proteinase. Two studies show significant reductions in the lethality of mutant strains deficient in this enzyme [57, 65]. The enzyme is produced *in vivo* [66]. Against the proteinase being a major virulence factor is the lack of correlation between gross proteinase production, as detected by plate tests, and strain virulence [119]. Indeed the strain responsible for a major outbreak of systemic candidosis at a London teaching hospital was proteinase-negative on biotyping [11]. Monoclonal antibodies have now been raised against this enzyme [6].

It has been suggested that *C. albicans* might produce toxins analogous to bacterial endotoxins. The toxic effect of *C. albicans* extracts has been studied particularly by Cutler *et al.* [21]. They found that the cell wall glycoprotein had some toxic properties, including lethality for mice and pyrogenicity for rabbits. Its potency is less than that of bacterial endotoxins. The finding that *C. albicans* contains some components with low, but detectable, potential toxicity is compatible with the organism being an opportunistic pathogen. If it released highly toxic substances it is unlikely that it would so frequently exist as an asymptomatic commensal. The pathological lesions observed at foci of invasive candidal infections are more likely to be the result of the host inflammatory response than a *Candida* toxin [96].

C. albicans and *C. stellatoidea* have cell surface receptors for the complement fragments C3d and iC3b, whereas these receptors appear to be absent from less pathogenic *Candida* species, such as *C. tropicalis*, *C. parapsilosis* and *C. krusei* [40]. Some, but not all, studies suggest these receptors are on pseudohyphae and hyphae rather than yeast cells [15, 28]. The receptor was isolated and characterized as a glycoprotein of about 60 kDa [64]. It may aid the fungus in evading host defence mechanisms in several ways. Gilmore *et al.* [35] have shown that they inhibit phagocytosis of *C. albicans* by human polymorphonuclear leucocytes by non-covalently binding iC3b.

9.4 EPIDEMIOLOGY AND TYPING METHODS

There have been many reports of inter-strain variation in virulence of *C. albicans* [2, 9, 86, 110, 111]. It has also recently become apparent that, although candidal infections usually arise by autoinfection by yeasts colonizing the patient's bowel or intravenous catheters [29, 55, 129], cross-infection can occur between susceptible individuals, due to a single outbreak strain [11, 14]. Therefore one possible explanation of the relative rarity of disseminated candidosis in AIDS patients is that cross-infection has occurred with strains capable of causing superficial, but not systemic, candidosis. To gain insight into the epidemiology of candidosis in AIDS patients it was first necessary to develop a reliable typing system.

Early typing systems for *C. albicans* were based on phenotypic differences. They included serotyping [37], morphotyping [7, 104], resistotyping [81, 105], biotyping [98, 99] and immunoblot fingerprinting [61]. Agglutination reactions essentially distinguish just two serotypes, A and B [38], with a third serotype, C, introduced by Muller and Kirchoff [88] to cover rare isolates that do not react clearly as type A or B. Morphotyping carries the disadvantage that the majority of isolates are the same morphotype, A1, which has smooth fringeless colonies [7]. Although modifications can be introduced which reduce the prevalence of false A1 morphotypes, such as taking care not to overheat the media during preparation [106], there is still some doubt over its reproducibility and whether it is affected by phenotypic switching [43]. Resistotyping [81] and biotyping [99] are both simple to perform but suffer from technical difficulties in standardization, and both intra- and interlaboratory variations in results are evident [97].

In using these methods to identify an outbreak, it was found necessary to use a combination of serotyping, morphotyping and biotyping to achieve sufficient discrimination between outbreak and non-outbreak isolates [11]. Subsequently, immunoblot fingerprinting was applied to *C. albicans* and successfully used to identify four further outbreaks [14]. This technique had the advantage that all isolates could be rapidly fingerprinted, and discrimination was much greater than with morphotyping or serotyping. It lacked the phenotypic variability observed with the biotyping procedure when applied to the outbreak isolates. Unfortunately gel-to-gel variation necessitated sequential comparison of isolates, precluding its use in large scale epidemiological studies. As with so many typing systems for *C. albicans*, one particular type predominated. Type 1 accounted for 43% of isolates [61].

More recently, typing systems have been developed based on genotypic differences between strains. DNA fingerprinting was first applied to *C. albicans* by Scherer and Stevens [118] and Magee *et al.* [68]. Scherer and Stevens [117, 118] found that EcoRI produced the largest number of distinct bands. The method was reproducible and the DNA fingerprints produced remained stable through multiple generations of the organism.

This technique proved to be by far the most useful to date in the rapid identification of outbreaks of systemic candidosis [76]. Applied to isolates from 5 distinct outbreaks, it gave highly reproducible results within a few days. Only when the suspected outbreak strain fingerprinted as type 2 was it necessary to use additional typing systems because type 2 was common among non-outbreak isolates (63%). Other DNA types accounted for 6% or fewer of the non-outbreak isolates. Smith *et al.* [123] have found the restriction enzyme *Hinf*I a useful alternative to EcoRI for the generation of restriction fragment length polymorphisms (RFLP).

Both DNA fingerprinting and immunoblot fingerprinting were applied to 51 oral isolates of *C. albicans* from patients with AIDS or the AIDS-related complex (ARC) [73]. This showed that the strains responsible for oral and oesophageal candidosis in AIDS and ARC patients were heterogeneous (Figure 9.1) and some were indistinguishable from those capable of causing systemic infections. A further study using *Eco*RI-generated RFLP showed similarity in the strains isolated from AIDS patients and those present in healthy individuals [137]. They went on to demonstrate that *C. albicans* isolates collected from different sites in the same patient (oral cavity and oesophagus) and at different times over a 5-month period showed identical DNA fingerprints. In contrast, Bruatto *et al.* [8] found frequent changes in the biotypes of *C. albicans* isolated from HIV-antibody positive patients with recurrent oral candidosis. They biotyped isolates on the basis of differences in susceptibility to yeast killer factors [106]. Candidosis in HIV-antibody patients is often recalcitrant to treatment [4, 103]. Bruatto's group noted at least one recurrence in 26 patients studied. They differentiated 21 biotypes among 66 strains isolated. There was no significant clustering of biotypes and no correlation

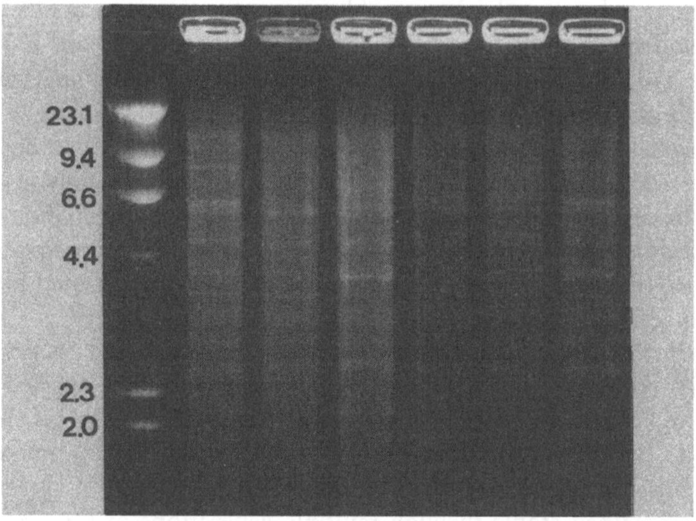

Figure 9.1 DNA fingerprints of *C. albicans* isolates from AIDS patients, with *Hind*III-cut markers on left-hand side. *Eco*Ri-generated RFLP were prepared as described in Matthews and Burnie (1989) *Br. Med. J.*, **298**, 354–7.

was observed between biotype, the clinical features of the oral candidosis, the clinical stage of the HIV-1 infection or CD4+ cell number. There was a significant correlation between the clinical features of the fungal lesion, the CD4+ cell number and the time of clinical disappearance of the oral lesions. Recurrence of the candidosis was almost always accompanied by a change in biotype of the strain isolated. The origin of these different biotypes is uncertain. They could be the result of exogenous reinfection or relapse, the different biotypes reflecting different phenotypic expressions of the same genotype.

Among isolates from AIDS patients and controls, 51% [137] to 61% [73] have identical RFLP following *Eco*RI digestion. One possible means of generating greater discrimination between strains is by the use of cloned probes and Southern blotting [100, 118]. Mitochondrial DNA is believed to evolve more rapidly than nuclear DNA in a variety of taxa. The 40 kb mitochondrial genome of *C. albicans* has been cloned as a series of fragments generated by *Eco*RI digestion. Detailed restriction maps of these fragments have been established [138]. Olivo's group described clearly discernable differences in the restriction patterns of mitochondrial DNA from a variety of isolates, demonstrating its epidemiological potential. They digested the *C. albicans* DNA with *Hae* III, electrophoresed in a 5% polyacrylamide gel, electroblotted and hybridized with ^{32}P-labelled cloned fragments of *C. albicans* mitochondrial DNA.

An alternative approach to further refining genetic typing of *C. albicans* may lie in pulsed field gel electrophoresis. As already discussed, considerable inter-strain variation occurs in the migration distances of several of the *C. albicans* chromosomes.

9.5 HOST DEFENCE AGAINST CANDIDOSIS

Animal models of systemic candidosis generally suggest that humoral immunity is more important than cell mediated immunity in protection against disseminated infection [112]. Congenitally athymic (nude) mice appear to be more resistant to intravenous *C. albicans* than their phenotypically normal littermates [22, 113]. Nude mice reconstituted with a syngeneic thymus graft become as susceptible as the normal littermate mice [113]. This may be due to the reconstitution of T-cell suppressor activity which inhibits phagocytic activity or B-lymphocytes.

Similarly, Giger *et al.* [32] have shown that thymectomized, irradiated and bone marrow reconstituted mice are no more susceptible than normal (euthymic) mice to disseminated candidosis. Others have found that active immunization of mice against *C. albicans* [39, 87, 125] or passive transfer of serum from hyperimmune vaccinated mice to non-immune mice [87, 100] confers protection. Pearsall *et al.* [100] demonstrated that resistance to candidosis was greater in mice injected with immune serum than those given normal serum. Transfer of immune lymphocytes had no effect on resistance to infection even though a state of *Candida*-specific cutaneous delayed hypersensitivity was conferred by such treatment.

Al-Doory [1] showed that serum taken from a baboon which died of candidosis could be used to passively immunize baboons so that they resisted challenge from a lethal dose of *C. albicans*. When a similar procedure was applied to rabbits the serum failed to protect or lengthen survival time over that of control animals. However, candidal antibodies induced by active immunization seemed to protect rabbits against the development of candidal endocarditis [116].

Zunino and Hudig [141] investigated the interaction of human natural killer (NK) lymphocytes with *C. albicans in vitro*. NK cells are non-phagocytic, large, granular lymphocytes which are cytotoxic *in vitro* to certain tumour cell lines and to virally infected cells. Mouse NK cells have been shown to inhibit the *in vitro* growth of another yeast, *Cryptococcus neoformans*, but it was found that human NK cells did not kill *Candida albicans in vitro*. However, *C. albicans* did block NK lysis of the tumour cell line K562.

Neutrophils can ingest and kill *C. albicans* bastospores and small hyphal forms (less than 12 μm) in the presence of complement *in vitro*, provided small inocula are used [23]. Large inocula overwhelm the phagocytes. Hyphae too large to be engulfed can be killed by an extracellular mechanism provided they are less than 200 μm in size. Three distinct types of neutrophil candidacidal mechanisms have been described. One is dependent on the production of hydrogen peroxide and potentiated with myeloperoxidase and halide ions. Myeloperoxidase deficiency is associated with systemic candidosis [62]. The second mechanism is mediated by a family of intrinsically microbicidal peptides which have been isolated from rabbit and human granulocytes [120]. A third mechanism has recently been described involving the release of an inhibitory substance from dying neutrophils [82]. This factor, with an estimated molecular weight of 30 kDa, differs from the above in that it is present in the cell sap rather than granules.

The role of neutrophils in man has not been established. Systemic candidosis occurs in two different high risk groups: neutropenic, haematological patients, usually following cytotoxic chemotherapy or as a result of their underlying lymphoproliferative disorder, and non-haematological, surgical or medical patients who are not neutropenic, in whom it may follow a variety of predisposing factors including major bowel surgery, broad spectrum antibiotics and urinary or vascular catheterization. There is no significant difference in the survival rate from systemic candidosis for neturopenic and non-neutropenic patients [72], nor is it known whether granulocyte transfusions have any value in the treatment of systemic candidosis.

The alternative pathway of complement is actived by *C. albicans* [108]. Complement deficiencies in man have sometimes been associated with chronic mucocutaneous candidosis [26] and, in mice, increased susceptibility to fatal infections [85]. Calderone *et al.* [15] identified receptors for C3d, a fragment of the complement component C3, on *C. albicans* pseudohyphae. Using a monoclonal antibody to affinity purify the receptor, it was identified as a 60 kDa cell surface glycoprotein [64]. It was not found on the yeast forms of *C. albicans*. The role of this receptor in the pathogenesis of candidosis is not yet known but it could compete with receptors for complement on phagocytic cells and it may degrade bound complement, lessening the chance of immune elimination.

Serial sera were examine from 92 patients with proven systemic candidosis, diagnosed by positive histology and culture from a deep biopsy taken ante-mortem or post-mortem, or two positive blood cultures taken from different sites at least 24 hours apart. Of the 74 patients with detectable antibodies, 92% had antibody to the 47 kDa component of *C. albicans*, 40% had antibody to the 60 kDa band and 27% had antibody to the 92 kDa band [72]. All survivors (30% of patients) were found to have a major antibody response to the 47 kDa antigen, whereas fatal cases produced little or no antibody, or initially had antibody which faded as their condition deteriorated [71, 72]. Eight of 21 superficially infected and colonized patients had antibody to the 47 kDa antigen. Antibody detection could not therefore be used diagnostically, but its absence is now regarded, in our laboratory, as a bad prognostic factor, whereas a well maintained antibody response to the 47 kDa antigen is a good prognostic sign.

The presence of an immunodominant 47 kDa antigen has now been independently confirmed by three further groups of workers [3, 30, 91]. It appears to be distinct from the 48–52 kDa enolase antigen described by Buckley's group [9], which has a different pattern of antibody reactivity [115, 130, 131]. Antibodies raised the 47 kDa antigen cross-react with the 92 kDa antigen and DNA sequencing has revealed it to a member of the HSP 90 family [77].

C. albicans HSP 90 and its breakdown products, including the 47 kDa antigen, circulate in the sera of systemically infected patients [72]. They have formed the basis of a series of serodiagnostic tests for systemic candidosis.

9.6 SEROLOGICAL TESTS FOR DIAGNOSIS

Some of the most widely used serological tests for systemic candidosis have been based on reverse passive latex agglutination (RPLA). The latex particles are sensitized with antiserum raised in rabbits against heat-killed candidal yeast cells [33] or a pressate of *C. albicans* [12]. The 47 kDa subcomponent of *C. albicans* HSP 90 is probably a dominant component of the circulating antigens measured by this assay system [13].

Subsequently a 47 kDa antigen-specific antibody probe was developed by affinity purification of the rabbit candidal antiserum against an immunoblot of this antigen [75]. This was compared, in a dot immunobinding assay, with the unfractionated candidal antiserum and also with RPLA. Use of the 47 kDa antigen-specific probe increased both the sensitivity and specificity of the assay system. Patients with systemic *C. albicans* infections were detected earlier on in the course of the infection. The rate of detection, in neutropenic patients, was 77%, compared to 55% with the unfractionated rabbit candidal antiserum and 29% with RPLA. In non-neutropenic patients, all three assay systems were positive in over 73% of the infected patients. One disadvantage of the 47 kDa antigen-specific probe is that it was relatively specific to *C. albicans* and was more likely to miss other species of yeast than the unfractionated antiserum.

Another disadvantage was the limited amounts of antibody probe to the 47 kDa antigen prepared by affinity purification. A way round this problem was to produce monoclonal antibodies to the 47 kDa antigen or to raise a rabbit antiserum against a synthetic peptide epitope of the antigen. The latter was achieved by epitope mapping and cloned and sequenced portion of *C. albicans* HSP 90 [79]. Using Geyson's method [34] an epitope was identified which was common to all patients infected with *C. albicans* who had antibody to the 47 kDa antigen. This epitope was also recognized by pooled sera from HIV-1 antibody-positive individuals with antibody to the 47 kDa antigen. It was not recognized by uninfected controls' sera. A rabbit was successfully immunized against it by conjugating the synthetic peptide epitope (LKVIRKNIVKKMIE) to keyhole limpet haemocyanin (KLH).

In the dot immunobinding assay, the epitope-specific antibody probe was more sensitive than the unfractionated hyperimmune candidal antiserum but less sensitive than the affinity-purified antibody to the 47 kDa antigen, which recognized multiple epitopes. In patients with proven systemic *C. albicans* infections, 70% were positive, 18% weakly positive and 12% gave false negatives with the epitope-specific probe. There were no false positives. This compared with 22% false negatives with the unfractionated hyperimmune candidal antiserum, and 8% false negatives with the 47 kDa antigen affinity-purified antiserum. The latter gave 7% false positives, though usually only weakly positive. The epitope-specific probe, like the 47 kDa antigen affinity-purified antiserum, successfully detected a systemic *C. parapsilosis* infection, but not infections due to other yeast species. With both probes, changes in antigen titres correlated with response to therapy, rising in those who died and falling in patients who recovered.

To optimize this test system it will be necessary to incorporate polyclonal antisera or monoclonal antibodies to other epitopes on *C. albicans* HSP 90. These might be against both conserved epitopes, present on other yeast species, and species-specific epitopes, so that *C. albicans* infections can be distinguished from other yeasts. It would also be helpful to develop a serological test for the diagnosis of candidal infections in AIDS patients. Such patients rarely develop systemic candidosis, but candidal oesophagitis can present a diagnostic problem since there are other causes of this condition in AIDS patients.

9.7 IMMUNOLOGY OF CANDIDOSIS IN AIDS

Sera was examined, by immunoblotting against *C. albicans*, from 76 HIV-1 antibody-positive patients, 39 having AIDS, 26 ARC and 11 being asymptomatic [73]. All the AIDS patients had IgM to the 47 kDa antigen with the exception of one patient with pulmonary candidosis, who died. Among AIDS patients with oesophageal candidosis 84% had IgG to the 47 kDa antigen, compared to 59% of AIDS patients with oral thrush only and 50% of AIDS patients with no evidence of *Candida* clinically. Among patients with ARC and oral thrush, 95% had IgM and 50% IgG to the

47 kDa antigen. This fell to 67% with IgM and 33% with IgG in ARC patients with no evidence of *Candida* clinically. In asymptomatic HIV-1 antibody positive patients with lymphadenopathy 50% had IgM to the 47 kDa antigen and 33% had IgG. Among those without lymphadenopathy, 40% had IgM and 20% had IgG. The presence of serum IgA to the 47 kDa antigen generally correlated with IgM.

The prevalence of antibody, especially IgM, to the 47 kDa antigen of *C. albicans* is therefore much higher in AIDS patients than in other groups of immunocompromised individuals with superficial candidosis. This antibody is present in about 30% of neutropenic patients with severe oral or oesophageal candidosis and in 38% of superficially infected surgical patients [72]. It is detectable in less than 5% of hospital inpatients with no clinical or laboratory evidence of candidosis. We were also able to detect either IgM or IgA antibody to the 47 kDa antigen in sera from three patients with chronic mucocutaneous candidosis [10, 73].

Clinical observation, combined with the data from animal models, suggests that T-lymphocyte abnormalities are associated with superficial candidosis but do not predispose to systemic candidosis. *In vitro* studies demonstrate the role of phagocytes in removing *C. albicans*, but in neutropenic patients, as in non-neutropenic patients, there is a close correlation between producing a good IgG or IgM response to the 47 kDa antigen and recovery from systemic candidosis. The frequency of IgM antibody to the 47 kDa antigen in AIDS patients is compatible with the hypothesis that antibody to this antigen plays a role in defence against disseminated candidosis. The much lower prevalence of this antibody in non-AIDS immunocompromised patients would then account for their greater susceptibility to systemic candidosis.

Humoral immunity is still functional early on in the course of HIV-1 infection. Hypergammaglobulinaemia, raised antiviral antibody titres and follicular B-cell hyperplasia are features of this stage. This is probably due to polyclonal B-cell activation by HIV or Epstein–Barr virus, or to chronic antigen stimulation, and might account for the exceptionally high fequency of this candidal antibody in AIDS patients. Lack of T-cell help could account for the higher frequency of IgM than IgG antibody. Perhaps the severe cell mediated immunodeficiency associated with HIV-1 infection is responsible for the very common and often recalcitrant superficial candidosis frequently seen as the presenting feature of AIDS. This could be responsible for activation of B-lymphocytes by the 47 kDa antigen and the subsequent creation of memory B-cells to this antigen. Polyclonal B-cell activation could then initiate the production of antibody to the 47 kDa antigen.

Antibody to *C. albicans* HSP 90 may help to protect AIDS patients with superficial candidosis from developing systemic spread [73, 74]. However, it clearly does not protect against oral and oesophageal candidosis. As a result, the AIDS patient's immune system becomes chronically overstimulated by this high antigenic load. Indeed, this may contribute to the host's inability to eliminate other infectious agents [121]. Both mannan [93] and a cell wall glycoprotein of *C. albicans* [16] can suppress the cellular immune response, so candidosis may itself aggravate a pre-existing T-lymphocyte deficiency.

9.8 THE ROLE OF HSP 90 AND AUTOANTIBODIES

Epitope mapping revealed eight continuous epitopes along the sequenced carboxy portion of *C. albicans* HSP 90 [79]. Three of these were recognized by the majority of patients, including HIV-1 antibody positive patients, with antibody to the 47 kDa antigen. Epitope LKVIRK, situated 317–322 amino acid residues from the carboxy terminus, was recognized by all patients but none of four rabbit hyperimmune candidal antisera. It is conserved between human and *C. albicans* HSP 90 and was immunogenic in rabbits when conjugated to KLH. The second most commonly recognized epitope was LSREM, 329–333 amino acid residues from the carboxy terminus. This showed a high degree of homology between human and yeast HSP 90. It was not immunogenic in rabbits and, by analogy to human HSP 90, it may be a carbohydrate binding site [140]. The third most commonly recognized epitope, was the *C. albicans* specific carboxy terminal epitope DEPAGE, recognized by about 50% of patients with antibody to the 47 kDa antigen. It was also recognized by all four hyperimmune rabbit candidal antisera. This epitope was only 8–13 amino acid residues from the carboxy terminus. When conjugated to KLH it was immunogenic in rabbits.

Serum taken from two patients with antibody to *C. albicans* HSP 90, prior to receiving antifungal chemotherapy, was given prophylactically to mice challenged with a lethal dose of the yeast [80]. In each case mortality was halved. This protective effect was found to be mediated by the immunoglobulin fraction of the serum, and not observed with normal serum. In one of these patients, the only detectable antibodies on immunoblotting, were to HSP 90 and its breakdown products, the 47 kDa and 40 kDa bands. The other patient also had antibodies to bands at 48 kDa and 60 kDa. The latter may correspond to *Saccharomyces cerevisiae* HSP 60, which shows extensive sequence homology to the HSP 65 family [114]. Members of the HSP 65 family are major targets for the immune response in many infections. Elevated antibodies to these proteins have been demonstrated in patients with superficial candidal infections [46, 135]. However, a monoclonal antibody to the 60 kDa antigen of *C. albicans* was not protective in mice [78]. Previously, attempts to induce protective immunity in mice against leprosy and tuberculosis by immunization with mycobacterial HSP 65 also failed [133].

T-cells reactive to mycobacterial HSP 65 have been linked to autoimmune diseases; this has been reviewed by Kaufman [50] and Cohen [20]. However, T-cells responsive to self epitopes of HSP 65 have been shown to exist in normal individuals [59, 90]. Increased antibodies to HSP have been observed in rheumatoid arthritis (HSP 65), ankylosing spondylitis (HSP 90) and systemic lupus erythematosus (HSP 70, HSP 90 and ubiquitin) [5, 58, 83, 84, 89, 133]. We now know that autoantibodies to conserved epitopes of HSP 90 are commonly produced in patients recovering from systemic candidosis, without any clinical evidence of autoimmune sequelae. Despite the overlap between the association of HSPs with autoimmunity, and their dominance in the immune response to many infections, the occurrence of symptomatic autoimmune disease following overt infection is a relatively rare event.

In order to explore further the possibility that autoantibodies to *C. albicans* HSP 90 might contribute to the protective effect of patients' sera in systemic candidosis, a monoclonal antibody was raised against the synthesized peptide epitope LKVIRK [80]. When given to mice challenged with a lethal dose of *C. albicans*, mortality at 24 hours fell to 66%. This was compared to a mortality of 100% in a group given an irrelevant control monoclonal antibody. When given to mice challenged with a tenfold lower dose of *C. albicans*, mortality at 24 hours fell from 60% in the control group, to 20% in the protected group.

Four hyperimmune rabbit candidal antisera were found to contain antibodies to many candidal antigens, including HSP 90, but not to the LKVIRK epitope. They had very little protective effect in the mouse model. There appears to be some species-dependent variability in the immunogenicity of different epitopes of HSP 90. This could explain some of the conflicting results which have followed studies on the protective effects of rabbit candidal antiserum [1, 139].

9.9 CONCLUSION

C. albicans is a common human commensal and many arms of the immune system are involved in defence against this opportunistic pathogen. This is reflected in the range of possible clinical manifestations, each indicative of a deficiency in one or more host defence mechanisms. Thus, AIDS patients with severe cell-mediated immunodeficiency, are particularly prone to recalcitrant oral and oesophageal candidosis, but rarely develop disseminated candidosis. Similarly, patients with chronic mucocutaneous candidosis, often associated with defects in cell-mediated immunity, suffer from multiple superficial candidal infections but not systemic candidosis.

In contrast, systemic candidosis is associated with failure to produce antibody to *C. albicans* HSP 90. The molecular biology of *C. albicans* is also complicated by the fact that it is diploid and there is considerable karyotic variability between different strains. There may be a correlation between this karyotic polymorphism and phenotypic variation. However, there has been great interest in the application of molecular biology to the study of *C. albicans* and, as a result, there have been significant advances in our understanding of the epidemiology and immunology of this yeast.

REFERENCES

1. Al-Doory, Y. (1970) An immune factor in baboon anti-candida serum. *Sabouraudia*, **8**, 41–7.
2. Allen, C.M., Beck, F.M. (1983) Strain-related differences in the pathogenicity of *Candida albicans* for oral mucosa. *J. Infect. Dis.*, **147**, 1036–40.
3. Au-Young, J.K., Troy, F.A., Goldstein, E. (1985) Serological analysis of antigen-specific reactivity in patients with systemic candidiasis. *Diag. Microbiol. Infect. Dis.*, **3**, 419–32.
4. Babajews, A., Poswillo, D.E., Griffin, G.E. (1985) Acquired immunodeficiency syndrome presenting as recalcitrant *Candida. Br. Dent. J.*, **159**, 106–8.
5. Bahr, G.M., Rook, G.A.W., Al-Saffar, M., Van Embden, J., Stanford, J.L., Behbehani, K. (1988) Antibody levels to mycobacteria in relation to HLA type: evidence for non-HLA-linked high levels of antibody to the 65 kD heat shock protein of *M. bovis* in rheumatoid arthritis. *Clin. Exp. Immunol.*, **74**, 211–5.
6. Borg, M., Walters, D., Reich, B., Ruchel, R. (1988) Production and characterization of monoclonal antibodies against secretory proteinase of *Candida albicans* CBS 2730. *Zentralbl. Bakteriol. Mikrobiol. Hyg. (A)*, **268**, 62–73.
7. Brown-Thomsen, J. (1968) Variability in *Candida albicans* Robin Berkhout I. Studies on morphology and biochemical activity. *Hereditas*, **60**, 355–98.
8. Bruatto, M., Vidotto, V., Marinuzzi, G., Raiteri, R., Sinicco, A. (1991) *Candida albicans* biotypes in human immunodeficiency virus type 1-infected patients with oral candidiasis before and after anti-fungal therapy. *J. Clin. Microbiol.*, **29**, 726–30.
9. Buckley, H.R., Price, M.R., Daneo-Moore, L. (1982) Isolation of a variant of *Candida albicans. Infect. Immun.*, **37**, 1209–17.
10. Burford-Mason, A.P., Matthews, R.C., Williams, J.R.B. (1987) Transient abrogation of immunosuppression in a patient with chronic mucocutaneous candidiasis following vaccination with *Candida albicans. J. Infect.*, **14**, 147–57.
11. Burnie, J.P., Odds, F.C., Lee, W., Webster, C., Williams, J.D. (1985) Outbreak of systemic *Candida albicans* in intensive care unit caused by cross infection. *Br. Med. J.*, **290**, 746–8.

12. Burnie, J. (1985) A reverse passive latex agglutination test for the diagnosis of systemic candidosis. *J. Immunol. Methods*, **82**, 267–80.
13. Burnie, J.P., Matthews, R.C., Featherbe, D.I., Tabaqchali, S. (1985) 47 kDa antigen of *Candida albicans*. *Lancet*, i, 1155.
14. Burnie, J.P., Matthews, R., Lee, W., *et al.* (1987) Four outbreaks of nosocomial systemic candidiasis. *Epidemiol. Infect.*, **99**, 201–11.
15. Calderone, R.A., Linehan, L., Wadsworth, E., Sandberg, A.L. (1988) Identification of C3d receptors on *Candida albicans*. *Infect. Immun.*, **56**, 252–8.
16. Carrow, E.W., Damer, J.E. (1985) Immunoregulation in experimental murine candidiasis, specific suppression induced by *Candida albicans* cell wall glycoprotein. *Infect. Immun.*, **49**, 172–81.
17. Centers for Disease Control. (1987) Revision of the CDC surveillance case definition for acquired immunodeficiency syndrome. *MMWR*, **36**, 1S.
18. Chandrasekar, P.H., Molinari, J.A. (1985) Oral candidosis: forerunner of acquired immunodeficiency syndrome (AIDS)? *Oral Surg.*, **60**, 532–4.
19. Chilgren, R.A., Quie, P.G., Meuwissen, H.J., Hong, R. (1967) Chronic mucocutaneous candidiasis, deficiency of delayed hypersensitivity and selective antibody defect. *Lancet*, ii, 688–93.
20. Cohen, I.R. (1990) Microbial heat-shock protein and autoimmunity, in *New Antibacterial Strategies*, (ed. H.C. Neu), Churchill Livingstone, New York, pp. 263–8.
21. Cutler, J.E., Friedman, L., Milner, K.C. (1972) Biological and chemical characterization of toxic substances from *Candida albicans*. *Infect. Immun.*, **6**, 616–27.
22. Cutler, J.E. (1976) Acute systemic candidiasis in normal and congenitally thymic-deficient (nude) mice. *J. Reticuloendothel. Soc.*, **19**, 121–4.
23. Davies, R.R., Denning, T.J.V. (1972) *Candida albicans* and the fungicidal activity of the blood. *Sabouraudia*, **10**, 301–12.
24. De Maria, A., Buckley, H., Von Lichtenberg, F. (1976) Gastrointestinal candidiasis in rats treated with antibiotics, cortisone and azathioprine. *Infect. Immun.*, **13**, 1761–70.
25. Di Menna, M.E. (1952) Natural occurrence of rough variant of a yeast: *Candida albicans*. *Nature*, **169**, 550–1.
26. Drew, J.H. (1973) Chronic mucocutaneous candidiasis with abnormal function of serum complement. *Med. J. Aust.*, **2**, 77–80.
27. Dyess, D.L., Garrison, R.N., Fry, D.E. (1985) Candida sepsis. *Arch. Surg.*, **120**, 345–8.
28. Edwards, J.E. Jr, Gaither, T.A., O'Shea, J.J., Rotrosen, D., Lawley, T.L. (1986) Expression of specific binding sites on *Candida* with functional and antigenic characteristics of human complement receptors. *J. Immunol.*, **137**, 3577–83.
29. Evans, E.G.V., Forster, R.A. (1976) Antibodies to *Candida* after operations on the heart. *J. Med. Microbiol.*, **9**, 303–8.
30. Ferreira, R.P., Yu, B., Niki, J., Armstrong. D. (1990) Detection of candida antigenuria in disseminated candidiasis by immunoblotting. *J. Clin. Microbiol.*, **28**, 1075–8.

31. Finlay, I.G. (1986) Oral symptoms and *Candida* in the terminally ill. *Br. Med. J.*, **292**, 592–3.
32. Giger, D.J., Domer, J., Moser, S., McQuitty, J. (1978) Experimental murine candidiasis: pathological and immune responses in T lymphocyte depleted mice. *Infect. Immun.*, **21**, 729–37.
33. Gentry, L.O., Wilkinson, I.A., Lea, A.S., Price, M.F. (1983) Latex agglutination test for detection of candidal antigen in patients with disseminated disease. *Eur. J. Clin. Microbiol.*, **76**, 122–8.
34. Geysen, H.M., Rodda, S.J., Mason, T.J., Tribbick, G., Schoofs, P.G. (1987) Strategies for epitope analysis using peptide synthesis. *J. Immunol. Methods.*, **102**, 259–74.
35. Gilmore, B.J., Retsinas, E.M., Lorenz, J.S., Hostetter, M.K. (1988) An iC3b receptor on *Candida albicans*: structure, function and correlates with pathogenicity. *J. Infect. Dis.*, **157**, 38–46.
36. Gough, P.M., Warnock, D.W., Turner, A., Richardson, M.D., Johnson, E.M. (1985) Candidosis of the genital tract in non-pregnant women. *Eur. J. Obstet. Gynecol. Reprod. Biol.*, **19**, 237–46.
37. Hasenclever, H.F., Mitchell, W.O. (1961) Antigenic studies of *Candida*. I. Observation of two antigenic groups in *Candida albicans*. *J. Bacteriol.*, **82**, 570–3.
38. Hasenclever, H.F., Mitchell, W.O. (1962) Pathogenicity of *Candida albicans* and *Candida tropicalis*. *Sabouraudia*, **1**, 16–21.
39. Hasenclever, H.F., Mitchell, W.O. (1963) Acquired immunity to candidosis in mice. *J. Bacteriol.*, **86**, 401–6.
40. Heidenreich, F., Dierich, M.P. (1985) *Candida albicans* and *Candida stellatoidea*, in contrast to other *Candida* species, bind iC3b and C3d but not C3b. *Infect. Immun.*, **50**, 598–600.
41. Helstrom, P.B., Balish, E. (1979) Effect of oral tetracycline, the microbial flora, and the athymic state on gastrointestinal colonization and infection of Balb/c mice with *Candida albicans*. *Infect. Immun.*, **23**, 764–74.
42. Horn, R.B., Wong, B., Kiehn, T.E., Armstrong, D. (1985) Fungemia in a cancer hospital: changing frequency, earlier onset and results of therapy. *Rev. Infect. Dis.*, **7**, 646–55.
43. Hunter, P.R., Fraser, C.A.M., Mackenzie, D.W.R. (1989) Morphotype markers of virulence in human candidal infections. *J. Med. Microbiol.*, **28**, 85–91.
44. Hurley, R. (1966) Pathogenicity of the genus *Candida*, in *Symposium on Candida Infections*, (eds H.I. Winner, R. Hurley), Livingstone, Edinburgh, pp 13–25.
45. Hurley, R., de Louvois, J., Mulhall, A. (1987) Yeasts as human and animal pathogens, in *The Yeasts*, Vol I, (eds A.H. Rose, J.S. Harrison), Academic Press, London, pp. 201–81.
46. Ivanyi, L., Ivanyi, J. (1990) Elevated antibody levels to mycobacterial 65 kDa stress protein in patients with superficial candidiasis. *J. Infect. Dis.*, **162**, 519–22.
47. Iwaguchi, S.I., Homma, M., Tanaka, K. (1990) Variation in the electrophoretic

karyotype analysed by the assignment of DNA probes in *Candida albicans*. *J. Gen. Microbiol.*, **136**, 2433–42.
48. Kakar, S.N., Partridge, R.M., Magee, P.T. (1983) A genetic analysis of *Candida albicans*: isolation of wide variety of auxotrophs and demonstration of linkage and complementation. *Genetics*, **104**, 241–55.
49. Kakar, S.N., Magee, R.T. (1982) Genetic analysis of *Candida albicans*: identification of different isoleucine-valine, methionine and arginine alleles by complementation. *J. Bacteriol.*, **151**, 1247–52.
50. Kaufmann, S.H.E. (1990) Heat shock proteins and the immune response. *Immunol. Today*, **11**, 129–36.
51. Kelly, R., Miller, S., Kurtz, M., Kirsch, D. (1987) Directed mutagenesis in *Candida albicans*: one-step gene disruption to isolate ura3 mutants. *Mol. Cell. Biol.*, **7**, 199–207.
52. King, R.D., Lee, J.C., Morris, A.L. (1980) Adherence of *Candida albicans* and other *Candida* species to mucosal epithelial cells. *Infect. Immun.*, **27**, 667–74.
53. Kirkpatrick, C.H., Rich, R.R., Bennett, J.E. (1971) Chronic mucocutaneous candidiasis: model-building in cellular immunity. *Ann. Intern. Med.*, **74**, 955–78.
54. Klein, R.S., Harris, C.A., Butkins Small, C., Moll, B., Lesser, M., Friedland, G.H. (1984) Oral candidosis in high-risk patients as the initial manifestation of the acquired immunodeficiency syndrome. *N. Engl. J. Med.*, **311**, 354–7.
55. Krause, W., Matheis, H., Wulf, K. (1969) Fungaemia and funguria after oral administration of *Candida albicans*. *Lancet*, **i**, 598–9.
56. Kurtz, M.B., Cortelyon, M.W., Kirsch, D.R. (1986) Integrative transformation of *Candida albicans*, using a cloned *Candida* ADE 2 gene. *Mol. Cell. Biol.*, **6**, 142–9.
57. Kwon-Chung, K.J., Lehman, D., Good, C., Magee, P.T. (1985) Genetic evidence for role of extracellular proteinase in virulence of *Candida albicans*. *Infect. Immun.*, **49**, 571–5.
58. Lakomek, H.J., Will, H., Zech, M., Krauskemper, H.L. (1984) A new serological marker in ankylosing spondylitis. *Arthritis Rheum.*, **27**, 961–7.
59. Lamb, J.R., Bal, V., Mendez-Samperio, P., *et al.* (1989) Stress proteins may provide the link between the immune response to infection and autoimmunity. *Int. Immunol.*, **1**, 191–6.
60. Lasker, B.A., Carle, G.F., Kobayashi, G.S., Medoff, G. (1989) Comparison of the separation of *Candida albicans* chromosome-sized DNA by pulse-field gel electrophoresis techniques. *Nucleic Acids Res.*, **17**, 3783–93.
61. Lee, W., Burnie, J., Matthews, R. (1986) Fingerprinting *Candida albicans*. *J. Immunol. Methods*, **93**, 177–82.
62. Lehr, R.I., Cline, M.J. (1969) Leukocyte myeloperoxidase deficiency and disseminated candidiasis, the role of myeloperoxidase in resistance to candida infection. *J. Clin. Invest.*, **48**, 1478–88.
63. Levy, R.M., Pons, V.G., Rosenblum, M.L. (1983) Intracerebral mass lesions in the acquired immunodeficiency syndrome. *N. Engl. J. Med.*, **309**, 1454–5.
64. Linehan, L., Wadsworth, E., Calderone, R. (1988) *Candida albicans* C3d

receptor, isolated by using a monoclonal antibody. *Infect. Immun.*, **56**, 1981–6.

65. MacDonald, F., Odds, F.C. (1983) Virulence for mice of a proteinase-secreting strain of *Candida albicans* and a proteinase-deficient mutant. *J. Gen. Microbiol.*, **129**, 431–8.
66. MacDonald, F., Odds, F.C. (1980) Inducible proteinase of *Candida albicans* in diagnostic serology and in the pathogenesis of systemic candidosis. *J. Med. Microbiol.*, **13**, 423–35.
67. MacMillan, M.D., Cowell, V.M. (1985) Experimental candidiasis in the hamstercheek pouch. *Arch. Oral Biol.*, **30**, 249–55.
68. Magee, B.B., D'Souza, T.M., Magee, P.T. (1987) Strain and species identification by restriction fragment length polymorphism in the ribosomal DNA repeat of *Candida* species. *J. Bacteriol.*, **169**, 1639–43.
69. Magee, P.T., Rikkerink, H.A., Magee, B.B. (1988) Methods for the genetics and molecular biology of *Candida albicans*. *Analyt. Biochem.*, **175**, 361–72.
70. Magee, B.B., Koltin, J., Gorman, J.A., Magee, P.T. (1988) Assignment of cloned genes to the seven electrophoretically separated *Candida albicans* chromosomes. *Mol. Cell. Biol.*, **18**, 4721–6.
71. Matthews, R.C., Burnie, J.P., Tabaqchali, S. (1984) Immunoblot analysis of the serological response in systemic candidosis. *Lancet*, **ii**, 1415–8.
72. Matthews, R.C., Burnie, J.P., Tabaqchali, S. (1987) Isolation of immunodominant antigens from sera of patients with systemic candidiasis and characterization of serological response to *Candida albicans*. *J. Clin. Microbiol.*, **25**, 230–7.
73. Matthews, R.C., Burnie, J.P., Smith, D., *et al.* (1988) *Candida* and AIDS: evidence for protective antibody. *Lancet*, **ii**, 263–6.
74. Matthews, R.C. (1988) Immunotherapy of systemic candidiasis. *Serodiagn. Immunother.*, **2**, 1–6.
75. Matthews, R.C., Burnie, J.P. (1988) Diagnosis of systemic candidosis by an enzyme-linked dot immunobinding assay for a circulating immunodominant 47–kilodalton antigen. *J. Clin. Microbiol.*, **26**, 459–63.
76. Matthews, R.C., Burnie, J.P. (1989) Assessment of DNA fingerprinting for rapid identification of outbreaks of systemic candidiasis. *Br. Med. J.*, **298**, 354–7.
77. Matthews, R.C., Burnie, J.P. (1989) Cloning of a DNA sequence encoding a major fragment of the 47 kilodalton stress protein homologue of *Candida albicans*. *FEMS Microbiol. Letts.*, **60**, 25–30.
78. Matthews, R.C., Burnie, J.P., Fox, A., *et al.* (1989) The diagnostic and therapeutic potential of a monoclonal antibody to the 60 kilodalton nuclear antigen of *Candida albicans*. *Serodiagn. Immunother.*, **3**, 75–86.
79. Matthews, R.C., Burnie, J.P., Lee, W. (1991) The application of epitope mapping to the development of a new serological test for systemic candidosis. *J. Immunol. Methods*, **74**, 20–4.
80. Matthews, R.C., Burnie, J.P., Howat, D., Rowland, T., Walton, F. (1991) Autoantibody to heat shock protein 90 can mediate protection against systemic candidosis. *Immunology*, **73**, 20–4.

81. McCreight, M.C., Warnock, D.W. (1982) Enhanced differentiation of isolates of *Candida albicans* using a modified resistogram method. *Mykosen*, **25**, 589–98.
82. McNamara, M.P., Wiessner, J.H., Collins-Lech, C., Hahn, B.L., Sohnle, P.G. (1988) Neutrophil death as a defence mechanism against *Candida albicans* infections. *Lancet*, **ii**, 1163–5.
83. Minota, S., Camerson, B., Welch, W.J., Winfield, J.B. (1988) Autoantibodies to the constitutive 73-kD member of the hsp 70 family of heatshock proteins in systemic lupus erythematosus. *J. Exp. Med.*, **168**, 1475–80.
84. Minota, S., Koyasu, S., Jahara, I., Winfield, J. (1988) Autoantibodies to the heat-shock protein 90 in systemic lupus erythematous. *J. Clin. Invest.*, **81**, 106–9.
85. Morelli, R., Rosenburg, L.T. (1971) The role of complement in the phagocytosis of *Candida albicans* by mouse peripheral blood leucocytes. *J. Immunol.*, **107**, 476–80.
86. Mourad, S., Friedman, L. (1961) Pathogenicity of *Candida*. *J. Bacteriol.*, **81**, 550–6.
87. Mourad, S., Friedman, L. (1968) Passive immunization of mice against *Candida albicans*. *Sabouraudia*, **6**, 103–5.
88. Muller, H.L., Kirchoff, G. (1969) Serologische typen von *Candida albicans*. *Zentralbl. Bakteriol.*, **210**, 114–21.
89. Muller, S., Brian, J.P., Van Regenmortel, M.H.V. (1988) Presence of antibodies to ubiquitin during the autoimmune response associated with systemic lupus erythematosus. *Proc. Natl. Acad. Sci. U.S.A.*, **85**, 8176–80.
90. Munk, M.E., Schoel, B., Modrow, S., Karr, R.W., Young, R.A., Kaufman, S.H.E. (1989) T lymphocytes from healthy individuals with specificity to self-epitopes shared by the mycobacterial and human 65-kilodalton heat shock protein. *J. Immunol.*, **143**, 2844–9.
91. Neale, T.J., Muir, J.C., Drake, B. (1987) The immunochemical characterisation of circulating immune complex constitutents in *Candida albicans* osteomyelitis by isoelectric focusing, immunoblot and immunoprint. *Aust. N.Z. J. Med.*, **17**, 201–9.
92. Negroni, P. (1935) Variecion hacia el tipo R de *Mycotorula albicans*. *Rev. Soc. Argent. Biol.*, **11**, 449–53.
93. Nelson, R.D., Herron, M.J., McCormack, R.T., Gehrz, R.C. (1984) Two mechanisms of inhibition of human lymphocyte proliferation by soluble yeast mannan polysaccharide. *Infect. Immun.*, **43**, 1041–6.
94. Odds, F.C. (1988) Candidosis of the genitalia, in *Candida and Candidosis*, 2nd edn, Ballière Tindall, Oxford, pp. 124–35.
95. Odds, F.C. (1988) Morphogenesis of *Candida* with special reference to *C. albicans*, in *Candida and Candidosis*, 2nd edn, Baillière Tindall, Oxford, pp. 42–59.
96. Odds, F.C. (1988) Pathogenesis of candidosis, in *Candida and Candidosis*, 2nd edn, Bailliere Tindall, Oxford, pp. 252–78.
97. Odds, F.C. (1988) Isolation, identification and other laboratory aspects of

Candida, in *Candida and Candidosis*, 2nd edn, Bailliere Tindall, Oxford, pp. 60–7.

98. Odds, F.C., Abbott, A.B. (1980) A simple system for the presumptive identification of *Candida albicans* and differentiation of strains within the species. *Sabouraudia*, **18**, 301–18.
99. Odds, F.C., Abbott, A.B. (1983) Modification and extension of tests for differentiation of *Candida* species and strains. *Sabouraudia*, **21**, 79–81.
100. Olivo, P.D., McManus, E.J., Riggsby, W.S., Jones, J.M. (1987) Mitochondrial DNA polymorphism in *Candida albicans*. *J. Infect. Dis.*, **156**, 214–5.
101. Parker, J.C. Jr, McCloskey, J.T., Knauer, K.A. (1976) Pathobiologic features of human candidiasis. *Am. J. Clin. Pathol.*, **65**, 991–1000.
102. Pearsall, N.N., Adams, B.L., Bunni, R. (1978) Immunological responses to *Candida albicans*. III. Effect of passive transfer of lymphoid cells or serum on murine candidosis. *J. Immunol.*, **120**, 1176–80.
103. Pesti, M., Frenczy, L. (1982) Protoplast fusion hybrids of *Candida albicans* sterol mutants differing in nystatin resistance. *J. Gen. Microbiol.*, **128**, 123–8.
104. Phongpaichit, S., Mackenzie, D.W.R., Fraser, C. (1987) Strain differentiation of *Candida albicans* by morphotyping. *Epidemiol. Infect.*, **99**, 421–8.
105. Polonelli, L., Archibusacci, C., Sestito, M., Morace, G. (1983) Killer system: a simple method for differentiating *Candida albicans* strains. *J. Clin. Microbiol.*, **17**, 774–80.
106. Pope, L.M., Cole, G.T. (1982) Comparative studies of gastrointestinal colonization and systemic spread by *Candida albicans* and non-lethal yeast in the infant mouse. *Scanning Electron Microsc.*, **IV**, 1667–76.
107. Radin, D.R., Fong, T.L., Halls, J.M., Pontrelli, G.N. (1983) Monilial enteritis in acquired immunodeficiency syndrome. *Am. J. Roentgenol.*, **141**, 1289–90.
108. Ray, T.L., Wuepper, D. (1976) Activation of the alternative (properdin) pathway of complement by *Candida albicans* and related species. *J. Invest. Dermatol.*, **67**, 700–3.
109. Reichart, C.M., Kelly, V.L., Machert, A.M. (1985) *Pathologic Features of AIDS: Aetiology, Diagnosis, Treatment and Prevention*, Lippincott, Philadelphia, pp. 161–84.
110. Richardson, M.D., Smith, H. (1981) Production of germ tubes by virulent and attenuated strains of *Candida albicans*. *J. Infect. Dis.*, **144**, 565–9.
111. Richardson, M.D., Smith, H. (1981) Resistance of virulent and attenuated strains of *Candida albicans* to intracellular killing by human and mouse phagocytes. *J. Infect. Dis.*, **144**, 557–64.
112. Rogers, T.J., Balish, E. (1980) Immunity to *Candida albicans*. *Microbiol. Rev.*, **44**, 660–82.
113. Rogers, T.J., Balish, E., Manning, D.D.D. (1976) The role of thymus dependent cell-mediated immunity in resistance to experimental disseminated candidiasis. *J. Reticuloendothel. Soc.*, **20**, 291–8.
114. Reading, D.S., Hallberg, R.L., Myers, A.M. (1989) Characterization of the yeast hsp 60 gene coding for a mitochondrial assembly factor. *Nature*, **337**, 655–59.

115. Safranek, W.W., Buckley, H.R. (1990) Characterization of *Candida albicans* enolase, Poster Session, *2nd Conference on Candida and Candidiasis*, Philadelphia, USA.

116. Scheld, W.M., Calderone, R.A., Brodeur, J.P., Sande, M.A. (1983) Influence of preformed antibody on the pathogenesis of experimental *Candida albicans* endocarditis. *Infect. Immun.*, **40**, 950–5.

117. Scherer, S., Stevens, D.A. (1987) Application of DNA typing methods to the epidemiology and taxonomy of *Candida* species. *J. Clin. Microbiol.*, **25**, 675–9.

118. Scherer, S., Stevens, D.A. (1988) A *Candida albicans* dispersed, repeated gene family and its epidemiologic applications. *Proc. Natl. Acad. Sci. U.S.A.*, **85**, 1452–6.

119. Schreiber, B., Lyman, C.A., Gurevich, J., Needham, C.A. (1985) Proteolytic activity of *Candida albicans* and other yeasts. *Diagn. Microbiol. Infect. Dis.*, **3**, 1–5.

120. Selsted, M.A., Szklarek, D., Gang, T., Lehrer, R.I. (1985) Activity of rabbitl eukocyte peptides against *Candida albicans. Infect. Immun.*, **49**, 202–6.

121. Seligman, M., Chess, L., Fahey, J.L., *et al.* (1984) AIDS: an immunologic re-evaluation. *N. Engl. J. Med.*, **311**, 1286–292.

122. Slutsky, B., Buffo, J., Soll, D.R. (1985) High-frequency switching of colony morphology in *Candida albicans. Science*, **230**, 666–9.

123. Smith, R.A., Hitchcock, C.A., Evans, E.G.V., Lacey, C.J.N., Adams, D.J. (1989) The identification of *Candida albicans* strains by restriction fragment length polymorphism analysis of DNA. *J. Med. Vet. Mycol.*, **27**, 431–4.

124. Snider, W.D., Simpson, D.M., Nielson, S., Gold, J.W., Metroka, C.E., Posner, J.B. (1984) Neurological complications of acquired immune deficiency syndrome. *Ann. Neurol.*, **14**, 403–18.

125. Soles, P., Lim, L.Y., Louria, D.B. (1967) Active immunity in experimental candidiasis in mice. *Sabouraudia*, **5**, 315–22.

126. Soll, D.R., Langtimm, C.J., McDowell, J., Hicks, J., Galask, R. (1987) High-frequency switching in *Candida* strains isolated from vaginitis patients. *J. Clin. Microbiol.*, **25**, 1611–22.

127. Soll, D.R., Slutsky, B., MacKenzie, S., Langtimm, C., Staebell, M. (1987) Switching systems in *Candida albicans* and their possible role in oral candidiasis, in (eds I.C. Mackenzie, C.A. Squier, E. Dabelsteen), *Oral Mucosal Diseases: Biology, Etiology and Therapy*, Laegeforeningen Forlag, Copenhagen, pp. 52–9.

128. Soll, D.R., Galask, R., Joley, S., *et al.* (1989) Switching of *Candida albicans* during successive episodes of recurrent vaginitis. *J. Clin. Microbiol.*, **27**, 618–90.

129. Stone, H.H. (1974) Studies in the pathogenesis, diagnosis and treatment of *Candida* species in children. *J. Pediatr. Surg.*, **9**, 127–33.

130. Strockbine, N.A., Jargen, M.T., Zweibel, S.M., Buckley, H.R. (1984) Identification and molecular weight characterisation of antigens from *Candida albicans* that are recognised by human sera. *Infect. Immun.*, **43**, 715–21.

131. Strockbine, N.A., Jargen, M.T., Buckley, H.R. (1984) Production and characterisation of three monoclonal antibodies to *Candida albicans* proteins. *Infect. Immun.*, **43**, 1012–8.
132. Suzuki, T., Miyanae, Y., Ishida, I. (1991) Variation in colony morphology and chromosomal rearrangement in *Candida tropicalis* pK233. *J. Gen. Microbiol.*, **137**, 161–7.
133. Tsoulfa, G., Rook, G.A., Van Embden, JD, *et al.* (1989) Raised serum IgG and IgA antibodies to mycobacterial antigens in rheumatoid arthritis. *Ann. Rheum. Dis.*, **48**, 118–23.
134. Watson, J.D. (1989) Leprosy: understanding protective immunity. *Immunol. Today*, **10**, 218–21.
135. Weller, B.I, Simmons, P.D., Ivanyi, L. (1990) Identification of immunodominant antigens of *Candida albicans* in patients with superficial candidosis. *Clin. Immunol. Immunopathol.*, **54**, 347–53.
136. Whelan, W.L. (1987) The genetics of medically important fungi. *CRC Crit. Rev. Microbiol.*, **14**, 99–170.
137. Whelan, W.L., Kirsch, D.R., Kwon-Chung, K.J., Wahl, S.M., Smith, P.D. (1990) *Candida albicans* in patients with the acquired immunodeficiency syndrome: absence of a novel or hypervirulent strain. *J. Infect. Dis.*, **162**, 513–8.
138. Wills, J.W., Troutman, W.B., Riggsby, W.S. (1985) Circular mitochondrial genome of *Candida albicans* contains a large inverted duplication. *J. Bacteriol.*, **164**, 7–13.
139. Winner, H.I. (1956) Immunity in experimental moniliasis. *J. Pathol. Bacteriol.*, **71**, 234–37.
140. Yamazaki, M., Akaogi, K., Miwa, T., Imai, T., Soeda, E., Yokoyama, K. (1989) Nucleotide sequence of a full-length cDNA for 90 kDa heat-shock protein from human peripheral blood lymphocytes. *Nucleic Acids Res.*, **17**, 7108.
141. Zunino, S.J., Hudig, D. (1988) Interactions between human natural killer (NK) lymphocytes and yeast cells: human NK cells do not kill *Candida albicans*, although *C. albicans* blocks lysis of K562 cells. *Infect. Immun.*, **56**, 564–9.

10

Mycoplasmas and AIDS

SHYH-CHING LO, HITOSHI KOTANI and WENSI S. HU

10.1 INTRODUCTION

The pathogenicity or virulence of an agent depends on the precise conditions under which host and infectious agent interact. The infectious organism has to overcome the defensive responses of the host to entry and establish a favourable microenvironment within the host for its continuous survival and replication. Thus, in a broader term, all clinically significant infections are in a sense opportunistic infections. Individual conditions of such an interaction may determine if the infection will have a fulminant course, or be completely asymptomatic.

For years, patients belonging to the high risk groups for the development of AIDS (such as homosexuals, intravenous drug abusers and haemophiliacs) have been known to show different degrees of immune deficiency. They suffer repetitive infections and are more susceptible to many 'less virulent' organisms. Human immunodeficiency virus (HIV) is believed to be the primary cause of AIDS, producing profound immune deficiency in these patients.

Numerous studies in animals have shown that mycoplasmas are extraordinary pathogens, causing a wide spectrum of diseases. Man, however, has been considered to be the rare exception and is immune from any severe mycoplasmal disease. In the AIDS era, the previously unrecognized mycoplasmal diseases have become clinically significant. The rapid development of molecular biology and molecular diagnostic techniques have also begun to unveil the complexity and potential scope of mycoplasmal infections in human diseases.

Molecular and Cell Biology of Opportunistic Infections in AIDS
Edited by Steven Myint and Alan Cann
Published in 1992 by Chapman & Hall, London ISBN 0 412 45330 4

This chapter will cover general biological properties of mycoplasmas, diseases in animals and humans previously known to be caused by mycoplasmas, and the molecular biology of mycoplasmas. The focus of attention will be on recent studies of the AIDS-associated mycoplasma, *M. fermentans*. The studies include pathology, animal experiments, diagnostic assays and cultures of *M. fermentans* infections in patients with AIDS. Other species of mycoplasma which may be associated with AIDS disease are also discussed. We are just beginning to learn the complex disease process of mycoplasmal infections in AIDS.

10.2 GENERAL BIOLOGICAL PROPERTIES OF MYCOPLASMAS

Mycoplasmas, originally called pleuropneumonia-like organisms (PPLO) because of their first isolation from bovine pleuropneumonia [72], are the simplest forms of self-replicating micro-organisms. They are unique prokaryotes without a cell wall. Thus, they are resistant to cell wall active antibiotics such as penicillin. They are pleomorphic and intermediate in size between bacteria and viruses.

Mycoplasmas belong to the class *Mollicutes* (soft skin), division *Tenericutes* (wall-free prokaryotes). The class *Mollicutes* (Table 10.1) contains four orders; *Mycoplasmatales*, *Archoleplasmatales*, *Anaeroplasmatales* and *Spiroplasmatales*. There are six genera in the class. *Mycoplasma*, *Ureaplasma*, *Spiroplasma* and *Anaeroplasma* require sterols for growth. *Acholeplasma* and *Asteroplasma* do not require sterols. *Mycoplasma*, *Ureaplasma*, *Acholeplasma* and *Spiroplasma* are generally facultative anaerobes. *Anaeroplasma* and *Asteroplasma* are obligate anaerobes.

Electron microscopy reveals that mycoplasmas have a simple ultrastructure, with a plasma membrane, tightly packed ribosomes and genetic material (Figure 10.1). The genetic material has the typical properties of prokaryotes, consisting of circular double stranded DNA. The genome size is approximately one sixth that of *Escherichia coli* with low G + C content (24–35 mol %) in most species [81]. Therefore, their biosynthetic capabilities are limited and they require a nutritionally complex medium for *in vitro* growth. The rich medium components usually include beef heart infusion, peptone, glucose and yeast extract. In addition, *Mollicutes*, with the exception of *Acholeplasma*, need animal sera to provide lipoproteins and cholesterol for synthesis of the plasma membrane.

The host range of mycoplasmas includes plants, insects, fish, reptiles and mammals, including non-human primates and man. Mycoplasmas cause diseases of the respiratory tract, joints and urogenital tract in a variety of species of animals as well as in humans. Also, they are the major occult contaminants of cell cultures and virus stocks, producing artefactual results in various studies [2, 64].

Table 10.1 Taxonomy and properties of the class *Mollicutes*

Classification	*No. species*	*Genome size (MDa)*	*G + C of DNA (Mol%)*	*Cholesterol requirement*	*Host*
Division: *Tenericutes* (wall–free prokaryotes)					
Class: *Mollicutes*					
Order I: *Mycoplasmatales*					
Family I: *Mycoplasmataceae*					
Genus I: Mycoplasma	100	500	23–41	+	Humans, animals, insects, plants
Genus II: *Ureaplasma*	4	500	27–30	+	Humans, animals
Order II: *Acholeplasmatales*					
Family I: *Acholeplasmataceae*					
Genus I: *Acholeplasma*	11	1000	27–63	–	Animals, plants, insects
Order III: *Anaeroplasmatales*					
Family I: *Anaeroplasmataceae*					
Genus I: *Anaeroplasma*	4	1000	29–33	+	Rumen of cattle and sheep
Genus II: *Asteroplasma*	1	1000	40	–	
Order IV: *Spiroplasmatales*					
Family I: *Spiroplasmataceae*					
Genus I: *Spiroplasma*	9 (with many serogroups)	1000	25–31	+	Arthropods, insects, plants

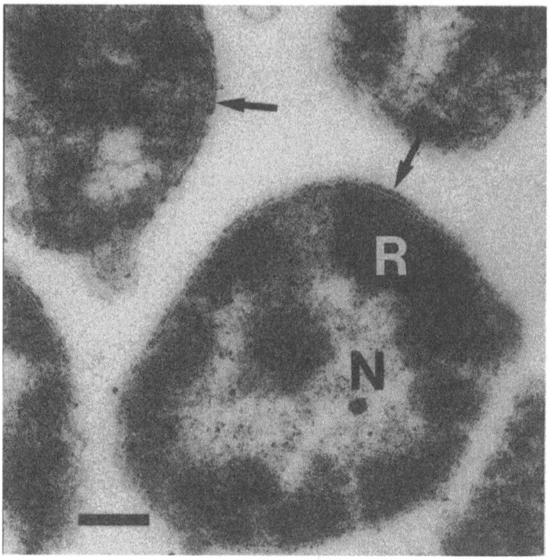

Figure 10.1 Electron photomicrography of mycoplasmas. *M. fermentans* (ATCC 19989) grown in SP-4 broth reveals trilaminar unit membrane (arrows), tightly packed ribosomal structure (R) and genetic material (N). B = 100 nm.

Mollicutes vary in shape from spherical (200–800 nm in diameter) to long branched filaments of variable length (up to 150 μm) depending on species, growth cycle, media and culture conditions. The genus *Spiroplasma* has a unique helical morphology. Most mycoplasmas divide by binary fission. Some mycoplasmas, however, elongate the filamentous forms and subsequently break up to form new spherical bodies [79].

Mycoplasmas are not usually motile. However, some *Mycoplasma* species such as *M. pneumoniae*, *M. genitalium*, *M. pulmonis*, *M. pirum* and *M. gallisepticum* possess specialized tip-like structures and exhibit gliding motility. The structures enable the organisms to attach to the mucosal surfaces of the respiratory or genital tracts [37, 39]. *Spiroplasma*, a helical organism without flagella or periplasmic filaments, demonstrates significant motility via a specialized protein [13, 111].

Optimum temperature for growth of most mycoplasmas is about 37 °C. They generally prefer aerobic conditions. However, primary cultures of human and animal mycoplasmas may need 10% CO_2 in the atmosphere.

Colonies of *Mollicutes* produce a characteristic fried egg appearance on agar media. Colonies show flat transparent surface growth surrounding a central granular area called the nipple. However, the appearance of colonies may vary depending on species and culture conditions. *Ureaplasmas* develop very small or tiny colonies on agar leading to the designation of 'T-strain'.

Most mycoplasmas utilize either glucose or arginine as their major energy source. Fermentative mycoplasmas catabolize glucose to lactic acid by the glycolytic pathways, producing an acidic pH shift in growth medium. Non-fermentative mycoplasmas hydrolyze arginine by a complex of three enzymes to convert arginine via ornithine to ammonia, producing an alkaline pH shift in growth medium. Few mycoplasmas utilize both glucose and arginine. *Ureaplasma* metabolizes neither substrate, but hydrolyzes urea to ammonia as its energy source [92, 93].

10.3 PATHOBIOLOGY OF MYCOPLASMAS

10.3.1 Mycoplasma diseases in animals

The first mycoplasma was isolated by Nocard and Roux in 1898 [72] from cattle with contagious bovine pleuropneumonia (CBPP). Since that time, a large number of mycoplasmas have been isolated from a wide variety of animals. Each animal mycoplasma adapts to a specific host and colonizes the mucosal surfaces of respiratory and urogenital tracts, eyes, digestive tract, mammary glands and joints. Most of the mycoplasmas are considered to be commensals and form part of the normal flora [79]. Some species of mycoplasmas, however, are undoubtedly pathogens and play important roles in causing a variety of diseases in laboratory and domestic animals. Mycoplasmas have been the cause of serious agricultural and economic losses in many parts of the world. Clinical manifestations of mycoplasmal infections reveal a wide spectrum of pathogenicity ranging from mild to severe. The diseases associated with mycoplasma infections include pneumonia, polyarthritis, keratoconjunctivitis, mastitis and genitourinary infections in domestic and laboratory animals. Table 10.2 summarizes the *Mollicutes* associated with a variety of diseases in animals.

The most serious mycoplasma disease in cattle is CBPP. The aetiological organism is *M. mycoides* subsp. *mycoides*, the first species of mycoplasma to be identified [72]. The disease is characterized by serious fibrinous interstitial pneumonia and pleurisy. The organism can be isolated from many different anatomical sites in cattle. CBPP was widespread in the past and still occurs in Africa and some parts of Asia, including India and China. *M. bovis* also causes calf pneumonia in many countries. This infection is commonly associated with bovine mastitis and joint manifestations. *M. bovigenitalium*, *M. canadense*, *M. californicum*, and *M. alkalescens* also cause mastitis in cattle [33]. Infections of *U. diversum* are frequently associated with calf pneumonia and urogenital tract infection such as vulvovaginitis [33].

Contagious caprine pleuropneumonia (CCPP) in goats is caused by both *M. mycoides* subsp. *capri* and subsp. *mycoides*, and contagious agalactia

Table 10.2 *Mollicutes* associated with diseases in animals

Animals	Mycoplasma *species*	*Disease*
Cattle	*M. alkalescens*	Arthritis, mastitis
	M. bovigenitalium	Arthritis, mastitis
	M. bovirhinis	Arthritis, mastitis
	M. bovis	Arthritis, mastitis, calf pneumonia
	M. bovoculi	Conjunctivitis
	M. californicum	Arthritis, mastitis
	M. canadense	Arthritis, mastitis
	M. dispar	Bronchialitis, alveolitis
	M. mycoides subsp. *mycoides*	Contagious bovine pleuropneumonia (CBPP)
	U. diversum	Calf pneumonia
Sheep and/or goat	*M. agalactiae*	Contagious agalactia, arthritis, pneumonitis
	M. capricolum	Mastitis, polyarthritis
	M. conjunctivae	Keratoconjunctivitis
	M. mycoides subsp. *capri*	Contagious caprine pleuropneumonia (CCPP)
	M. ovipneumonia	Pneumonia
Pig	*M. hyopneumoniae*	Enzootic pneumonia
	M. hyorhinis	Arthritis, pneumonia
	M. flocculare	Pneumonia
	M. hyosynoviae	Polyarthritis
Chicken	*M. anatis*	Sinusitis
	M. gallisepticum	Air sacculitis, sinusitis, arthritis
	M. synoviae	Synovitis
Turkey	*M. meleagridis*	Air sacculitis, arthritis, synovitis, perosis
Mouse and/or rat	*M. pulmonis*	Rhinitis, bronchopneumonia, salpingitis
	M. neurolyticum	Conjunctivitis, rolling disease (experimentally)
	M. arthritidis	Polyarthritis

in goats and sheep is caused by *M. agalactiae*. CCPP occurs in many countries of Africa and Asia, posing a major constraint on goat production because of its high mortality. *M. capricolum* may produce severe polyarthritis in goats and sheep. Contagious conjunctivitis is frequently associated with *M. conjunctivae* infection in many countries.

The economically most important porcine mycoplasma is *M. hyopneumoniae*. It is a primary agent of enzootic pneumonia of pigs (EPP) in many parts of the world. Infections with the mycoplasma produce characteristic chronic catarrhal pneumonia in young pigs. *M. hyorhinis* is a common inhabitant in the nasal cavity of pigs and is identified in pneumonic lesions with *M. hyopneumoniae*, suggesting that it may be a secondary invader in EPP lesions.

M. gallisepticum and *M. synoviae* cause respiratory and arthritic diseases in both chickens and turkeys. *M. meleagridis* produces similar disease in turkeys. These diseases are widespread in the poultry industry. In addition, *M. anatis* causes air-sacculitis in chickens and *M. iowae* causes low hatchability in chickens and turkeys.

M. pulmonis is the aetiological agent of a chronic murine respiratory disease called murine respiratory mycoplasmosis (MRM). This infection causes some of the most serious spontaneous diseases in laboratory mice and rats. *M. pulmonis* also causes genital diseases resulting in low birth rates in mouse breeding colonies. *M. arthritidis* in rats and *M. neurolyticum* in mice are also considered to be pathogens, producing arthritis and experimentally-induced rolling disease, respectively.

10.3.2 Mycoplasmal diseases in humans

Dienes and Edsall first isolated a human mycoplasma from a Bartholin's gland abscess in 1937 [24]. Since then 12 *Mycoplasma* species, *Ureaplasma urealyticum* and two *Acholeplasma* species have been isolated from humans (Table 10.3).

M. pneumoniae pneumonitis is the most thoroughly studied human mycoplasmosis. *M. pneumoniae* is a common cause of community acquired pneumonia ('atypical pneumonia') in children and young adults. A mysterious cyclic pattern of epidemics with intervals of 4 to 6 years has been noted. *M. pneumoniae* pneumonia is responsible for approximately 20% of all pneumonia cases [21]. The incubation period is about 2–3 weeks. Twenty-five percent of *M. pneumoniae* infections are asymptomatic. The symptomatic infection also has a favourable prognosis.

Descriptions of pathological changes are limited because human mycoplasmal infections are rarely associated with severe or fatal diseases. Pathogenicity of *M. pneumoniae* has been mainly studied in hamsters, guinea pigs and tracheal organ cultures [21]. In experimental animals, pulmonic lymphocytes infiltrate peribronchiolar and perivascular areas, to be replaced by an exudate of neutrophils and macrophages. Mycoplasmas trigger formation of prominent lymphofollicular sheets around the airways of sheep, goats and cattle. This disease pattern is commonly called cuffing pneumonia.

Non-respiratory complications of *M. pneumoniae* infection appear to be associated with some forms of autoimmune reaction [10]. The infected patients often develop autoantibodies, mainly IgM, such as cold agglutinins, antibodies to lymphocytes, brain, lung and smooth muscle, and rheumatoid factors. These complications result in haematological

Table 10.3 *Mollicutes* from humans

Mycoplasma *species*	*Site of isolation*	*Isolation frequency*	*Disease*
M. orale	Oropharynx	Frequent	None
M. salivarium	Oropharynx	Frequent	None
M. buccale	Oropharynx	Very rare	None
M. faucium	Oropharynx	Very rare	None
M. lipophilum	Oropharynx	Very rare	None
M. primatum	Genitourinary tract	Very rare	None
M. pirum	Blood	Very rare	Unknown
M. fermentans	Genitourinary tract, oral cavity (?), blood, bone marrow	Very rare	Systemic disease
M. genitalium	Genitourinary tract, respiratory tract	Very rare	Nongonococcal urethritis (?), pneumonia (?)
M. pneumoniae	Respiratory tract	Frequent	Pneumonia, tracheobronchitis
M. spermatophilum	Genitourinary tract	Very rare	Unknown
M. hominis	Genitourinary tract, extragenitourinary tract	Frequent Rare	Pelvic inflammatory disease Postpartum fever, pyelonephritis, arthritis, septicaemia, meningitis, peritonitis, neonatal pneumonia or meningitis
U. urealyticum	Genitourinary tract Joints	Frequent Rare	Nongonococcal urethritis, chorioamnionitis, neonatal pneumonia or meningitis Polyarthritis
A. laidlawii	Genitourinary tract	Very rare	None
A. oculi	Genitourinary tract	Very rare	None

(anaemia or thromboembolic phenomena), neurological (encephalitis, meningoencephalitis or other CNS diseases), cardiovascular (pericarditis or perimyocarditis), dermatological (skin manifestations or mucocutaneous lesions), musculoskeletal (arthralgia, myalgia or arthritis) and renal (glomerulonephritis) disease manifestations. The actual mechanisms of pathogenesis are still poorly understood.

U. urealyticum and *M. hominis* are found most frequently in the urogenital tract, mainly in the vagina of asymptomatic sexually mature women, with prevalence ranging from 40% to 80% and 21% to 50%, respectively [18, 102]. The finding of these organisms is much lower in the urethra of normal men. *M. hominis* and *U. urealyticum* are implicated in genitourinary tract infections in adults and a variety of diseases in newborn infants. The pathogenic significance of these mycoplasmas remains controversial. However, well-documented cases reveal mycoplasmal infections associated with disease of the urogenital tract in adults as well as in neonatal pneumonia, and meningitis, as shown in Table 10.3 [19]. *U. urealyticum* has been isolated from joints of immunocompromised hypogammaglobulinaemic patients with polyarthritis. Furthermore, *M. hominis* has been associated with postpartum fever, septicaemia, wound infections, CNS infections, joint infections and acute respiratory diseases [19, 59]. All these extragenital tract infections suggest the possibility of haematogenous dissemination of mycoplasmas in some infected patients.

M. genitalium was first isolated from urethral swabs of male patients with non-gonococcal urethritis in 1981 [104]. The organism is slow growing, fastidious, and consequently difficult to isolate. *M. genitalium* shares many antigenic and biological properties with *M. pneumoniae*. Serological studies for *M. genitalium* infections in women with pelvic inflammatory disease, as well as pathogenicity in humans, remain controversial [66]. However, in animal studies, inoculation of *M. genitalium* into oviducts of grivet monkeys and marmosets induces severe salpingitis [103]. In baboons, intracervical inoculations with the mycoplasma produces parametritis [103].

Recently, using a PCR assay, Jensen and colleagues showed that 12% of urogenital specimens from men (3/26) and 7% from women (5/76), originally submitted for *Chlamydia* cultures, tested positive for *M. genitalium* [44]. A high prevalence of *M. genitalium* infection in the urogenital tracts of asymptomatic heterosexuals (9%), homosexuals (21%), and heterosexual patients with non-gonococcal urethritis (22%) was also revealed in a separate report using genomic DNA as a probe [36]. If the report is confirmed, *M. genitalium* is a common human urogenital mycoplasma. However, this fastidious organism is apparently very difficult to isolate. In 1988, *M. genitalium*, along with *M. pneumoniae* was identified in throat swabs from patients with pneumonia [6]. The finding suggests that *M. genitalium* may also be a respiratory tract mycoplasma. Recently, Tully and Baseman also describe evidence of mixed *M. pneumoniae* and *M. genitalium* infections in the synovial fluid of a patient developing polyarthritis after having primary pneumoniae [105]. The present data support the possibility that *M. genitalium* may play a role in mycoplasmal pneumonia and in the development of extra-pulmonary complications.

10.3.3 Mycoplasms and the immune system

Surface infection of mucosal epithelial cells is a general feature of mycoplasmal infections. The infected host develops chronic, frequently subclinical, diseases of the respiratory or urogenital tracts. Typical examples of mycoplasmal infections of the respiratory mucosa are *M.*

pneumoniae in humans, *M. hyopneumoniae* in pigs, *M. gallisepticum* in chickens and *M. dispar* in cattle. On the other hand, some mycoplasmas appear to be more invasive. *M. bovis* in cattle, *M. agalactiae* in goats, *M. pulmonis* in rodents and *M. fermentans* in humans (discussed in detail later) are found associated with systemic forms of infection.

Natural infection with mycoplasmas in animals and in humans produce detectable levels of circulating antibodies. Humoral immunity is associated with production of specific antibodies against the infectious organism. Agglutination, growth inhibition, metabolic inhibition, immunofluroescence, and enzyme linked immunosorbent, immunoblot and Western blot assays can all be used to detect specific antibodies produced by infected hosts. However, antibody responses in mycoplasmal infections are often atypical. Production of non-specific, autoimmune-type antibodies and alteration of the immune function of the host as well as the mycoplasma's rapid phase variation or surface antigen diversification (discussed later) compromise the sensitivity and specificity of clinical serological assays.

Mycoplasmas are refractory to host immune defence systems such as phagocytosis and fixation of complement, often leading to persistent infections. Stimulation of local immune mechanisms by mycoplasmal infections on mucosal surfaces produces specific secretory antibodies. Although IgA is generally present, IgG is believed to be more important in defending against mycoplasmal infections.

Many mycoplasmas elude host immune responses and act as immunomodulators [22]. Mycoplasmas alter host cell functions by several different mechanisms. A number of *in vitro* studies have examined these mycoplasma–lymphoid cells interactions [22, 64, 89]. The suppressive effect of mycoplasmas on lymphocyte and macrophage functions includes capping and adherence of the organism to cells and acquisition of lymphocyte alloantigens such as T-cell differentiation antigens (Thy-1) and Class I antigens of the major histocompatibility complex (MHC). Mycoplasmal infections can also highly suppress the mitogenic effects of lymphocytes stimulated with lectins [22, 89].

Mycoplasmal components, mainly proteins, are also mitogens and polyclonal activators of both B- and T-lymphocytes from human, mouse, rat and guinea pigs [22]. *M. pulmonis*, *M. arginini*, *M. arthritidis*, *M. fermentans* and *M. pneumoniae* have been shown previously to have immune modulatory properties. The data suggest that mycoplasmas have the ability to activate normal lymphocytes, leading to an early inflammatory (lymphocytic) response before the development of a sensitized lymphocyte population. Mycoplasmal infection may also result in marked local cytokine production.

Mycoplasmas can apparently regulate gene expression in various types of cells. *M. hominis* suppresses specific immunoglobulin gene expression in

B-cells, causing significant reduction of the relative ratio of membrane to secretory μ heavy chain. Secretory IgM synthesis increases 50% to 100% following *M. hominis* infection [35]. Sitia and colleagues reported that *M. fermentans* induces murine B-lymphoma I.29 cells to differentiate into plasma cells, resulting in significantly increased levels of IgA secretion [94]. *M. arginini* also stimulates growth of B-cells and promotes immunoglobulin secretion [88]. Mycoplasma infections can also produce variable effects on the cellular expression of antigenic receptor sites and the affinity of a monoclonal antibody for its target on human colorectal cells [74].

Recent studies demonstrate that activated lymphocytes or macrophages release various cytokines. These cytokines may exert direct toxicity or enhance toxicity towards host cells [89]. Cytokines may also mediate many biological phenomena and the toxicity found in mycoplasma-infected cell cultures. Interferons (IFN), in addition to their antiviral effects, may have important functions in regulation of the host immune response, such as natural killer cell activities [11, 107]. Viable and even non-viable *M. arginini* induces class II MHC antigens on murine myeloma cells to levels of that induced by IFN, suggesting this organism could induce the production of IFN [100]. *M. pulmonis*, *M. hominis* and *M. pneumoniae* also induce Class II MHC antigens [99].

M. pulmonis membranes stimulate cells from mouse lymph node and spleen to produce interleukin-2 (IL-2) [49]. *M. pneumoniae*, on the other hand, induces granulocyte-macrophage colony-stimulating factor [17]. High molecular weight material from *M. fermentans* induces IL-6 in cultures of murine macrophages and human monocytes [77]. *M. orale* lysate stimulates human macrophages to produce tumour necrosis factor (TNF) which causes tumour cell lysis. In this regard, *M. gallisepticum*, *M. capricolum* and *A. laidlawii* also induce TNF production [20, 30].

Infections with *M. pneumoniae*, *M. orale*, *M. arginini*, *M. hyorhinis*, *M. salivarium*, *M. gallisepticum* or *A. laidlawii* in human lymphocyte cultures can induce high production of cytokines, including IL-1, IL-2, IL-6 and TNF-alpha [45]. The effectiveness of cytokine production induced by *Mycoplasma* in human lymphocyte cultures further suggests that infections with *Mycoplasma* may have a significant effect on the pathobiology and immune functions of the host.

10.4 MOLECULAR BIOLOGY OF MYCOPLASMAS

10.4.1 Genome and codon usage

Mollicutes, the smallest free-living organisms, have a circular double stranded DNA genome of 600–800 kilobase pairs (kbp) for *Mycoplasma* and *Ureaplasma* and 1400–1800 kbp for *Acholeplasma*, *Anaeroplasma*, *Asteroplasma* and *Spiroplasma* [8, 14, 71, 76, 80, 86, 110]. It is estimated that the smaller genomes of *Mycoplasma* and *Ureaplasma* could code about 500 genes [70]. *Mollicutes* is thought to derive phylogenetically by degenerative evolution from an ancestor of Gram-positive bacteria, such as *Clostridium* species [81]. The mycoplasmal DNA is highly A-T rich (61–78%) [63, 79, 81, 98] This fact suggests that mycoplasmas undergo selective pressure to replace their DNA G + C base pairs (bp) with A + T bp [75]. In other words, the codon usage in mycoplasmas appears to be preferentially A and T instead of G and C [69, 75]. Codon UGA, a universal termination codon, is used to code for tryptophan in many mycoplasmas [43, 112]. Because of this unusual genetic code usage, it results in premature termination of mycoplasmal genes when expressed in other eubacteria, such as *Escherichia coli*. This has hindered the progress of molecular genetic studies of mycoplasmas.

10.4.2 Cloning and expression of mycoplasma gene

To date, there is no effective method for manipulating recombinant DNA in mycoplasmas. Although plasmids have been described in *Spiroplasma* [1, 68, 73, 78, 83] and *M. mycoides* subsp. *mycoides* [9], they have not been developed like bacterial plasmids, which are used extensively as cloning vehicles. Two transposons, Tn916 and Tn4001, from Gram-positive bacteria have conjugative properties and a unique structure, enabling their introduction into mycoplasmas [25, 26, 60, 61]. The antibiotic resistance genes (serving as markers) of these transposons have previously been cloned in *E. coli*. The plasmids constructed in this way integrate into the *M. pulmonis* chromosome at a high frequency [60]. A two plasmid cloning system has been developed in *M. pulmonis* [62]. The first integrated plasmid, bearing the cloned antibiotic resistance gene of either Tn916 or Tn4001, was introduced into the *M. pulmonis* chromosome and served as a region of homology for a second plasmid cloning vector carrying the counter selectable marker and a segment of inserted foreign DNA [62]. This system could be applied theoretically to any *Mycoplasma* capable of transformation, homologous recombination and expression of antibiotic resistance markers. However, this system was found to be unable to express cloned foreign DNA [62].

As mentioned earlier, unusual codon usage in mycoplasmas results in poor expression of mycoplasmal genes in bacteria. It is possible to use oligonucleotide-directed site-specific mutagenesis to alter UGA codons to UGG, then the altered mycoplasmal genes can be fully expressed in bacterial expression systems. Expression of foreign genes in mycoplasmas remains difficult. It will be necessary to identify strong regulatory sequences such as promotor sequences for expression of genes in mycoplasmas. Recently, a genetic element resembling a prokaryotic insertion sequence (IS) was reported in the porcine mycoplasmas *M. hyopneumoniae* and *M. hyorhinis* [29]. We also reported the finding of a separate IS-like element in *M. fermentans* (incognitus strain) [40]. The presence of multiple copies of the IS-like elements in the genomes of these mycoplasmas suggests that they may be transposable elements. Development of mycoplasmal transposons will provide a powerful tool to study the organization and regulation of mycoplasmal genes. It should facilitate the understanding of the basic biology and molecular pathogenesis of these organisms.

10.4.3 Molecular pathogenesis

Most of the human and animal mycoplasmas are surface parasites. They adhere to and colonize epithelial cells of the respiratory and urogenital tracts. The property of adherence, named adhesion, depends on unique membrane proteins (adhesins) present on the mycoplasmal surface [82]. Most of the information on adhesins comes from studies of *M. pneumoniae*. By using monoclonal or monospecific antibodies, the P1 protein is identified as a major *M. pneumoniae* adhesin [5, 28, 38, 47]. Mutants lacking the P1 protein show reduced adherence capacity [47]. The gene encoding P1 protein has been molecularly cloned by two groups [41, 101]. Inamine *et al.* reported a P1 operon that is composed of the P1 gene flanked by two open reading frames (ORF), ORF-4 and ORF-6 [42]. The DNA sequence of the P1 gene, published by both groups, and its deduced amino acid sequence have been determined [41, 42, 101]. The peptide deduced from the nucleotide sequence has a molecular weight of about 170 kDa, consistent with data derived from biochemical characterization of the P1 protein [41, 42, 101]. Recently, both ORF-4 and ORF-6 in the P1 operon were expressed in *E. coli* [96]. Antisera raised against ORF-4 fusion proteins did not recognize *M. pneumoniae* antigens in Western blot analysis, but antisera directed against ORF-6 fusion proteins reacted with two *M. pneumoniae* proteins of 40 kDa and 90 kDa [96]. These two proteins are missing in *M. pneumoniae* B176, a non-cytadsorbing strain, which still contains the P1 protein but fails to colonize the lungs of hamsters [2, 96]. This finding suggests that several proteins, in addition to P1, take part in the complex process of adhesion. The gene products of ORF-6 may act as structural components or accessory factors in membrane morphogenesis [96].

Some mycoplasmas, similar to several bacterial species, show high-frequency changes in phenotype, which includes phase variation and antigenic diversification of surface macromolecules [15, 27, 65, 87, 91,

106]. Most of the information on phase variation comes from studies on the porcine pathogenic mycoplasma *M. hyorhinis* [15, 87] and the rodent pathogenic mycoplasma *M. pulmonis* [27, 109]. Both mycoplasmas show a unique phenomenon. There is an unusual surface antigen complex which forms multiple bands and a 'ladder' pattern in protein separation gels [87, 109]. In *M. hyorhinis*, this ladder pattern represents expression of diverse, lipid-modified and size-variant mycoplasmal surface proteins [87]. In *M. pulmonis*, this ladder pattern represents variable antigen 1 (V-1) which is composed of many subunits aggregated via hydrophobic interactions. These aggregates can at least be partially dissociated when exposed to harsh denaturing conditions, resulting in the characteristic ladder pattern [109]. The phenomenon of the ladder pattern, similar to that of streptococcal M proteins, is apparently controlled by complex regulatory systems at the genetic level [16]. The molecular mechanisms controlling the high frequency phenotypic changes and surface antigen diversification may play an important role in the pathogenesis of mycoplasmal diseases.

10.5 AIDS-ASSOCIATED MYCOPLASMAS

As discussed earlier, many mycoplasmas are known to cause severe and fatal diseases in animals. Infections in these animals are often systemic and associated with prominent alteration of immune function. However, other than atypical pneumonia (*Mycoplasma pneumoniae* pneumonia), some infections of the genitourinary system by the ureaplasmas or *M. hominis* and occasional reports of haematogenous invasion or dissemination, mycoplasmas have not been previously associated with any significant human diseases. This is mainly due to the fact that mycoplasmas are difficult to culture from the diseased tissues. Neither is the serological evidence convincing, but recent reports of infections with the AIDS-associated mycoplasmas have drawn widespread attention to the possible role of mycoplasmas in human diseases, including AIDS [51, 52, 53, 54, 55].

Detailed biochemical, antigenic and DNA analyses in these reports revealed a newly recognized, possibly pathogenic mycoplasma. This is thought to be a unique strain of the *M. fermentans* species, named incognitus [54, 90]. *M. fermentans* is a rarely isolated human urogenital mycoplasma. In the laboratory, the incognitus strain of *M. fermentans* produced fatal systemic infection in experimental monkeys [53]. These silver-leaf monkeys had little, late, or only a transient antibody response against the mycoplasma. The monkeys developed progressive weight loss and died 7–9 months after inoculation.

These infected animals revealed periodic antigenaemia during the course of the experiment. An additional monkey, kept as a control and killed 8 months after the last mycoplasma-inoculated monkey succumbed, showed neither an antibody response nor evidence of antigenaemia. Mycoplasma-specific DNA could be directly detected in necropsy tissues of all four monkeys inoculated with *M. fermentans* using the polymerase chain reaction (PCR) method. Mycoplasmal infection was identified in all four spleens, two of four livers, one of two kidneys, and all three brains tested from these four animals, but not in the tissues from the control monkey. The necropsy examination of the four mycoplasma-inoculated animals revealed no obvious opportunistic infections, acute inflammatory lesions, malignancies or cause of death other than mycoplasmal infection. Thus, it seems that the mycoplasma apparently caused a fatal systemic infection in these monkeys.

Infection with the organism has also been associated with previously healthy non-AIDS patients dying of an acute fatal disease [54]. These patients developed acute flu-like illnesses with persistent fever, lymphadenopathy, diarrhoea, pneumonia, and/or heart, liver and adrenal failure. They died in 1–7 weeks. These patients had no serological evidence of HIV infection, did not belong to any high risk groups for AIDS and could not be classified as AIDS patients according to CDC criteria. The clinical signs as well as laboratory and pathological studies of these patients, suggested an active infectious process, although no aetiological agent was found despite extensive infectious disease work-ups during their hospitalization. Post-mortem examinations showed histopathological lesions of fulminant necrosis involving the lymph nodes, spleen, lungs, liver, adrenal glands, heart and/or brain. No viral inclusion cells, bacteria, fungi or parasites could be identified in these tissues using special tissue stains. We reported that immunohistochemistry using rabbit antiserum raised against the incognitus strain revealed mycoplasma-specific antigens in these necrotizing lesions. *In situ* hybridization using a ^{35}S-labelled *M. fermentans*-specific DNA probe also detected the mycoplasmal genetic material in the areas of necrosis. Furthermore, mycoplasma-like particles closely resembling the incognitus strain were identified ultrastructurally in these histopathological lesions. The infection was highly concentrated at the advancing margins of these lesions. As controls, the subsequent tissue sections were also immunostained by a non-specific monoclonal antibody with the same isotype (IgG/k). No positive immunochemical reaction was identified. ^{32}P-labelled cloning vector DNA not containing the mycoplasmal DNA failed to hybridize to these tissues. Mycoplasma-like particles could be identified intracellularly and extracellularly in the necrotizing lesions [54]. This mycoplasma is apparently cytopathic and cytocidal *in vivo*.

The observation that infection by *Mycoplasma* can be intracellular is new and not consistent with previous dogma. The mycoplasmas have previously been thought to be strict extracellular pathogens. For this particular species of mycoplasma, *M. fermentans*, there is no specialized structure for attachment. It probably has other unique properties enabling cell invasion. The possible molecular mechanisms are currently being studied by many mycoplasmologists. At the 91st General Meeting of the American Society for Microbiology (1991) in Dallas, Texas, a report from Stadtlander and Cassell confirmed the unique property of *M. fermentans* (incognitus strain) of invasiveness and pathogenicity using primary tracheal ring organ cultures and also in a rat model following intranasal infections [97].

Many patients with AIDS also have systemic infections with the incognitus strain of *Mycoplasma* [7, 51, 55]. In one study, monoclonal antibodies (Mabs) were developed against antigens from a pure culture of the incognitus strain grown in modified SP-4 medium [54]. The Mabs reacted only with the incognitus strain and not with other species of human mycoplasmas, including the closely related prototype strain of *M. fermentans* (PG-18). In immunohistochemical analysis, the Mabs identified antigen specific for the incognitus strain in thymus, liver, spleen, lymph node, or brain from 22 of 32 patients with AIDS, as well as in two placentas from patients with AIDS [55]. In *in situ* hybridization assays, a ^{35}S-labelled DNA probe specific for *M. fermentans* also detected genetic material specific for the microbe in these infected tissues. Similarly, ultrastructural studies of target areas of tissues co-identified as highly positive for the antigens specific for the incognitus strain by immunohistochemistry, revealed characteristic structures of mycoplasma organisms (Figure 10.2).

Patients with AIDS often have significant dysfunction in multiple organ systems, including the immune system, haematopoietic system, brain, heart, liver, joints, gastrointestinal tract and kidney. In our studies, the functional deficit of various organ systems that show no other aetiological agent, appears to be more directly associated with hidden mycoplasmal infection. As a typical example, infection with *M. fermentans* in kidneys was shown to be associated with development of AIDS-associated nephropathy in HIV-positive AIDS patients [7]. In contrast, kidneys from the AIDS patients without clinicopathological diagnosis of AIDS-associated nephropathy did not reveal similar mycoplasmal infection in the parenchymal cells in casts of renal tubules [7].

Infection due to the incognitus strain is apparently systemic and shows little specific tissue tropism. The human immunodeficiency virus (HIV), a human retrovirus, is believed to be the causative agent of AIDS. This is primarily based on seroepidemiological studies of anti-HIV antibodies in patients with AIDS [12, 31, 32] and the virus tropism for $CD4^+$ lymphocytes [4, 46]. HIV produces cytopathic effects in $CD4^+$ cells and

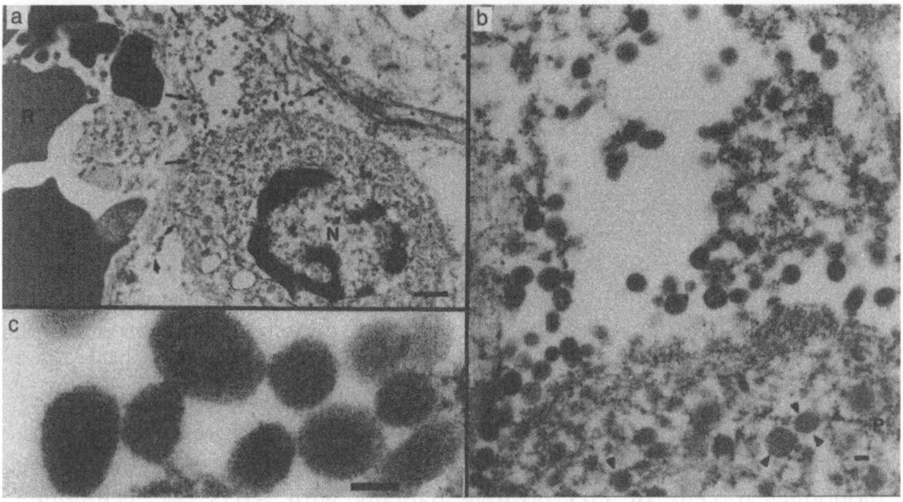

Figure 10.2 Electron microscopy of AIDS patient liver immunostained for incognitus strain specific antigens. (a) Electron micrograph of a periporal area of an AIDS liver with adjacent necrosis. Electron densed mycoplasma-like particles are seen in the surrounding area of a small vessel extracellularly. Nucleus of mononuclear lymphohistiocytic (N) and red blood cells (R) in the small vessel are shown. Bar = 500 nm. (b) Higher magnification of the mycoplasma-like organisms found in the empty extra-cellular space and lining along the outer surface of the lymphohistiocyte shown in A. Many intracellular particles (arrowheads) can also be identified and are difficult to differentiate with the extracellular particles. Polysomal structure (P) is shown. Bar = 100 nm. (c) Higher magnification of the mycoplasma-like microorganisms lining the outer surface of the lymphohistiocytes. Bar = 100 nm. The figure is partially adopted from [55].

cell death. On the other hand, *in vivo* studies of HIV infection in experimental animals have not produced specific illness other than transient lymphadenopathy. In chimpanzees, despite more than 8 years of persistent HIV viraemia, typical opportunistic infections or malignancies of clinical AIDS have not been seen. Furthermore, HIV infected patients often show a wide variation in times of disease incubation and rapidity of disease progression [50]. It is not known whether any specific infectious agent other than HIV can affect the complex pathogenesis of this disease. Our finding of a high prevalence of mycoplasmal infections in patients with AIDS has introduced the possibility that they could play a significant disease promoting role in HIV-infected patients.

Consistent with this possibility, co-infection with the incognitus strain of *Mycoplasma* was recently shown to greatly enhance the cell killing effects of HIV in $CD4^+$ human lymphocytes [57]. Syncytium formation of HIV-infected T-cells was essentially eliminated in the presence of *M. fermentans* (incognitus strain), despite prominent cell death. However, replication and production of HIV-1 particles continued during the co-infection. Furthermore, the supernatant from cultures co-infected with HIV-1 and the mycoplasma contained a factor that inhibited the standard reverse transcriptase enzyme assay. The modification of the biological properties of HIV-1 by co-infection with mycoplasmas may be involved clinically in the pathogenesis of acquired immunodeficiency syndrome (AIDS).

Montagnier and associates also reported the cell killing effect of HIV-1 could be significantly reduced by tetracycline antibiotics [48]. The tetracycline treated cultures continued to produce a high titre of HIV-1 in the absence of cytocidal effects. The authors suggested that a prokaryotic agent, most likely a mycoplasma, was involved with the cytocidal effect observed in the HIV-1 infected cultures. Indeed, in additional studies, they confirmed a mycoplasma as the hidden agent in the cultures [84]. Both *M. fermentans* and *M. pirum* have been isolated from peripheral blood mononuclear cells of HIV-positive patients with AIDS. *M. genitalium* specific DNA sequences have also been found in the blood of another AIDS patient, using a PCR assay. The HIV-1 associated pathogenicity in human lymphocytes was proposed to be a mycoplasma-dependent process. This controversial hypothesis is still being actively debated.

M. pirum has never been isolated directly from human clinical specimens. The organism was previously identified only in contaminated cell cultures. If the isolation from AIDS patients is confirmed, this organism with a tip-like structure is a human mycoplasma. More recently, Montagnier reported rabbit antiserum was raised against the site of adhesion of *M. pirum* to the cells. The adhesion protein is a high molecular weight membrane protein. In two other species of mycoplasma, *M. pneumoniae* and *M. genitalium*, this adhesion protein has been well characterized and analysed. The gene has been cloned and sequenced. An antibody against the specific adhesion site reduces the replication of some but not all strains of HIV-1 [67]. This inhibition of HIV infectivity appeared to be specific, the effect being blocked by absorption of the antibody with the adhesion site peptide. It was therefore concluded that mycoplasmas could help HIV-1 replication, in addition to enhancing HIV-1 cytopathogenicity. These in vitro observations and the isolation of *M. fermentans* and *M. pirum* from the blood of patients with AIDS have prompted Montagnier to propose that mycoplasmas are playing key role as co-factors in AIDS [85].

Our study has shown the pathogenesis of mycoplasmal infection per se is complicated and atypical. Histopathology of infected tissues from experimental animals and patients varied from no pathological change to fulminant necrosis, with or without an associated inflammatory reaction. We believe the unusual form of pathogenesis is due either to the mycoplasmal infection concomitantly causing damage to key components of the host's immune system, or the pathogen having special biological properties to elude immunosurveillance. In this context, although the role of *M. fermentans* infection and its precise relationship to AIDS or HIV infection is not clear, the evidence of systemic infection by direct identification of the pathogen in brain, liver, spleen or thymus of these patients should be considered clinically important.

In order to investigate further the possible pathogenic role of *Mycoplasma* in human diseases and to develop a productive culture system for isolating the apparently fastidious micro-organism from clinical specimens, we would need a highly sensitive and specific technique for detection. We have previously developed a polymerase chain reaction (PCR) assay based on the specific nucleotide (nt) sequence found at one terminus of the cloned DNA segment psb-2.2 from the incognitus strain of *M. fermentans*. Unfortunately, the reaction failed to detect some strains and clinical isolates of *M. fermentans* [51]. The targeted sequence is apparently not present in all the strains. A more sensitive PCR assay which is able to detect all the different strains and clinical isolates of *M. fermentans*, yet remaining highly selective or specific, is needed.

We have recently developed a highly selective PCR assay to detect *M. fermentans* with remarkable sensitivity. We have taken advantage of the presence of multiple copies of a unique insertion sequence (IS)-like genetic element in *M. fermentans* DNA [40]. The sequence of the targeted segment, as well as the primers for PCR assay, are located within the IS-like element (Figure 10.3). The assay detects all the different strains tested, including the new clinical isolates of *M. fermentans*, which our previous PCR assay failed to detect. Furthermore, the limitation of reaction sensitivity per assay for our current PCR is 0.1–1.0 fg *M. fermentans* DNA within a background of 1 μg of human DNA, compared with 1–10 fg of organism DNA in our previous PCR assay [108]. Thus, a molecular technique capable of selectively detecting essentially a single *M. fermentans* organism in a reaction is now available.

Using the synthetic oligonucleotide primer pair RW004 and RW005, which contain sequences within the IS-like genetic element specific for mycoplasmas of the *M. fermentans* species, 10 of 43 urine sediments obtained from 40 HIV-positive patients at different stages of AIDS disease tested positive [23]. *M. fermentans* was isolated and grown in cultures from three urine sediments obtained from two HIV-positive patients with AIDS. Fifty urine sediments from age-matched HIV-negative healthy controls tested negative for *M. fermentans* by PCR. *M. fermentans* was not isolated from any of the control urine samples. The preliminary results thus show a high prevalence of the AIDS-associated mycoplasma, *M. fermentans*, in the urine from HIV-positive patients but not from HIV-negative non-AIDS controls.

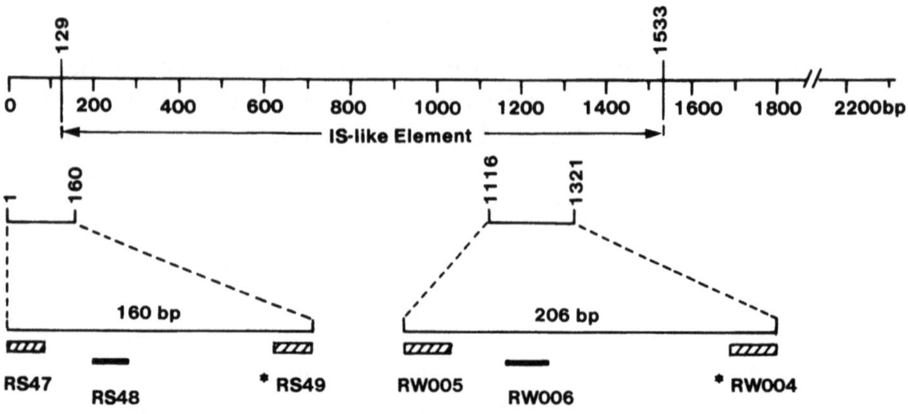

Synthetic Primer and Probe			Sequence (5′-3′)
RW	004	Primer (24 nt)	GGA CTA TTG TCT AAA CAA TTT CCC
RW	005	Primer (24 nt)	GGT TAT TCG ATT TCT AAA TCG CCT
RW	006	Probe (24 nt)	GCT GTG GCC ATT CTC TTC TAC GTT

Figure 10.3 Sequences of the synthetic oligonucleotide primer pair (RW004/RW005) and the verification probe (RW006) used for specific DNA amplification of *M. fermentans* and their positions in the IS-like element. The relative positions of the primer pair RS47/RS49 and probe RS48 used in our previous PCR assay [51] are also indicated.

The potential clinical significance arising from the identification of mycoplasmal infections is that they may be treatable with antibiotics. Quantitative assays of antimicrobial susceptibility of the incognitus strain of *Mycoplasma* in SP-4 broth cultures to representative antibiotics has revealed that the organism is not sensitive to erythromycin, the most commonly used antibiotic for human mycoplasmal infections [34, 56]. The *in vitro* testing shows that incognitus strain has varying degrees of sensitivity to the antibiotics, tetracycline, doxycycline, chloramphenicol, clindamycin, lincomycin and ciprofloxacin [34, 56]. Ciprofloxacin appears the most effective. Quinolones were also found to be the most effective antibiotics against many *Mycoplasma* species. Appropriate reservation and caution is always necessary in extrapolating *in vitro* biological results to *in vivo* clinical conditions.

Recently, we examined a 30-year-old non-AIDS male patient who had fever, cough, malaise, progressive weight loss, diarrhoea and extensive tissue necrosis involving the liver and spleen. *M. fermentans* (incognitus strain) was shown at the advancing margins of these necrotizing lesions. Following treatment with 300 mg doxycline/day for 6 weeks, he recovered fully. He currently has no fever or diarrhoea and his liver function tests have returned to normal. He has regained lost strength, and 14 kg of lost weight, and has remained disease free for more than one year [58]. Successful treatment by antibiotics in this previously healthy non-AIDS patient with fulminant *M. fermentans* infection may reflect a still intact immune system. Although mycoplasmal infections can be treated or suppressed, they are none the less extremely difficult to eradicate [95]. Mycoplasmal infections in patients with AIDS without a functional immune system can be very refractory to antibiotic treatments, even with a high dose and a long course of therapy. Under these circumstances, selection of resistant organisms becomes a major concern. In this context, the presence of IS-like elements may certainly provide an effective genetic tool for *M. fermentans* to rapidly develop antibiotic resistant traits. The true efficacy of antimicrobial treatments in immunocompromised AIDS patients infected by the mycoplasmas is awaiting systemic clinical trials.

10.6 CONCLUSION

Although much remains to be done, our studies and those of others, show that *M. fermentans* is found at a much higher frequency in HIV-positive AIDS patients than in HIV-negative controls. There are three general possibilities for the potential significance of infections with the 'AIDS-associated mycoplasma'. Firstly, the mycoplasma simply represents another opportunistic infection found in immunocompromised patients with AIDS. Secondly, infection with mycoplasmas may enhance pathogenicity of other human pathogens including HIV, a phenomenon demonstrable *in vitro*. Therefore, mycoplasmal infection may promote disease progression of AIDS. Thirdly, the organism itself is pathogenic in humans. This final possibility is supported by our findings in the animal experiment using non-human primates and the association between mycoplasmal infection and those previously healthy non-AIDS patients dying of an acute disease. Furthermore, this possibility is also suggested by the commonly seen alteration of host immune functions associated with mycoplasmal infections in animals.

Even if only another opportunistic infection in AIDS, mycoplasmal infection is still clinically significant. After all, opportunistic infections are the direct cause of death in more than 80% of patients with AIDS. Needless to say, the findings of mycoplasmal infection become more significant if the microbe is playing the disease promoting 'co-factor' role or itself is responsible for some aspects of AIDS pathogenesis. Viewing the apparently complex disease process of AIDS, we believe it is likely the organism may have a role in all of the three processes above. If so, disease associated with infection by this previously unrecognized prokaryotic agent may be potentially amenable to antibiotic therapy. Treatment of AIDS disease should then include treatment for mycoplasmal infection.

ACKNOWLEDGEMENTS

We thank Douglas J. Wear and Michael M. Hayes for reviewing this article and Ms Viola Penn for typing the manuscript.

REFERENCES

1. Archer, D.B., Best, J., Barber, C. (1981) Isolation and restriction mapping of a *Spiroplasma* plasmid. *J. Gen. Microbiol.*, **126**, 511–14.
2. Barile, M.F., Yoshida, C.H., Grabowski M.W., Harasawa, R., Razin, S. (1980) Parameters of *Mycoplasma pneumoniae* infection in Syrian hamster. *Infect. Immun.*, **56**, 2443–9.
3. Barile, M.F. (1981) Mycoplasma infection of cell cultures. *Isr. J. Med. Sci.*, **17**, 555–562.
4. Barre Sinoussi, F., Chermann, J.C., Rey, F., *et al.* (1983) Isolation of a T-lymphotropic retrovirus from a patient at risk for acquired immune deficiency syndrome (AIDS). *Science*, **220**, 868–70.
5. Baseman, J.B., Cole, R.M., Krause, D.C., Leith, D.K. (1982) Molecular basis for cytadsorption of *Mycoplasma pneumoniae. J. Bacteriol.*, **151**, 1514–22.
6. Baseman, J.B., Dallo, S.F., Tully, J.G., Rose, D.L. (1988) Isolation and characterization of *Mycoplasma genitalium* strains from the human respiratory tract. *J. Clin. Microbiol.*, **26**, 2266–9.
7. Bauer, F.A., Wear, D.J., Angritt, P., Lo, S-C. (1991) *Mycoplasma fermentans* (incognitus strain) infection in the kidneys of patients with acquired immunodeficiency syndrome and associated nephropathy: A light microscopic, immunohistochemical and ultrastructural study. *Hum. Pathol.*, **22**, 63–9.
8. Bautsch, W. (1988) Rapid physical mapping of the *Mycoplasma* mobile genome by two-dimensional field inversion gel electrophoresis techniques. *Nucleic Acids Res.*, **16**, 11461–7.
9. Bergemann, A.D., Finch, L.R. (1988) Isolation and restriction endonucleases: analysis of a mycoplasma plasmid. *Plasmid*, **19**, 68–70.
10. Biberfled, G. (1985) Infection sequelae and autoimmune reactions in *Mycoplasma pneumoniae* infection, in *The Mycoplasmas*, Vol. IV, (eds S. Razin, M.F. Barile), Academic Press, New York, pp. 293–308.
11. Birke, C., Peter, H.H., Langenberg, U., *et al.* (1981) Mycoplasma contamination in human tumor cell lines: Effect on interferon induction and susceptibility to natural killing. *J. Immunol.*, **127**, 94–8.
12. Blattner, W., Gallo, R.C., Temin, H.M. (1988) HIV causes AIDS. *Science*, **214**, 515–6.
13. Bove, J.M., Saillard, C. (1979) Cell biology of spiroplasmas, in *The Mycoplasmas*, Vol. III, (eds R.F. Whitcomb, J.G. Tully), Academic Press, New York, pp. 83–117.

14. Bove, J.M., Carle, P., Garnier, M., Laigret, F., Renaudin, J., Salliard, C. (1989) Molecular and cellular biology of *Spiroplasmas*, in *The Mycoplasmas*, Vol. V, (eds R.F. Whitcomb, J.G. Tully), Academic Press, New York, pp. 244–355.
15. Boyer, M.J., Wise, K.S. (1989) Lipid-modified surface protein antigens expressing size variation within the species *Mycoplasma hyorhinis*. *Infect. Immun.*, 57, 245–54.
16. Caparon M.G., Scott, J.R. (1987) Identification of a gene that regulates expression of M protein, the major virulence determinant of group A streptococci. *Proc. Natl. Acad. Sci. U.S.A.*, **84**, 8677–81.
17. Capobianchi, M.R., Lorino, G., Lun, M.T., Mancini, C., Di Marco, P., Dianzani, F. (1987) Membrane interactions involved in the induction of interferon-alpha by *Mycoplasma pneumoniae*. *Antiviral Res.*, **8**, 115–24.
18. Cassell, G.H., Cole, B.C. (1981) Mycoplasmas as agents of human disease. *N. Engl. J. Med.*, **304**, 80–9.
19. Cassell G.H., Davis, J.K., Waites, K.B., *et al.* (1987) Pathogenesis and significance of urogenital mycoplasmal infections. *Adv. Exp. Med. Biol.*, **224**, 93–115.
20. Chowdhury, M.I.H., Koyanagi, Y., Kobayashi, S., Yamamoto, N., Munakata, T., Arai, S. (1990) Mycoplasma and AIDS. *Lancet*, 336, 247–8.
21. Clyde, W.A., Jr. (1979) *Mycoplasma pneumoniae* infections of man, in *The Mycoplasmas*, Vol. II, (eds J.G. Tully, R.F. Whitcomb), Academic Press, New York, pp. 275–302.
22. Cole, B.C., Naot, Y., Stanbridge, E.J., Wise, K.S. (1985) Interaction of mycoplasmas and their products with lymphoid cells *in vitro*, in *The Mycoplasmas*, Vol. IV, (eds S. Razin, M.F. Barile), Academic Press, New York, pp. 203–57.
23. Dawson, M.S. Hayes, M.M., Wang, R.Y-H., *et al.* (1992) Detection and isolation of *Mycoplasma fermentans* from urine of HIV-1 infected patients. *Arch. Pathol. Lab. Med.* (in press).
24. Dienes, L., Edsall, G. (1937) Observations on the L-organisms of Klieneberger. *Proc. Exp. Biol. Med.*, **36**, 740–4.
25. Dybvig, K., Cassell, G.H. (1987) Transposition of Gram-positive transposon Tn916 in *Acholeplasma laidlawii* and *Mycoplasma pulmonis*. *Science*, **235**, 1392–4.
26. Dybvig, K., Alderete, J. (1988) Transformation of *Mycoplasma pulmonis* and *Mycoplasma hyorhinis*, Transposition of Tn916 and formation of cointegrate structures. *Plasmid*, 20, 33–41.
27. Dybvig, K., Simecka, J.W., Watson, H.L., Cassell, G.H. (1989) High-frequency variation in *Mycoplasma pulmonis* colony size. *J. Bacteriol.*, **171**, 5165–8.
28. Feldner, J., Gobel, U., Bredt, W. (1982) *Mycoplasma pneumoniae* adhesin localized to tip structure by monoclonal antibody. *Nature*, **298**, 765–7.
29. Ferrell, R.V., Heidari, M.B., Wise, K.S., McIntosh, M.A. (1989) A mycoplasma genetic element resembling prokaryotic insertion sequences. *Mol. Microbiol.*, 3, 957–67.

30. Gallily, R., Sher, T., Ben-Av, P., Loewenstein, J. (1989) Tumor necrosis factor as a mediator of *Mycoplasma orale*-induced tumor cell lysis by macrophages. *Cell. Immunol.*, **121**, 146–53.
31. Gallo, R.C., Wong-Staal, F. (1985) A human T-lymphotropic retrovirus (HTLV-III) as the cause of the acquired immunodeficiency syndrome. *Ann. Intern. Med.*, **103**, 679–89.
32. Gallo, R.C., Montagnier, L. (1988) AIDS in 1988. *Sci. Am.*, **259**, 41–8.
33. Gourlay, R.N., Howard, C.J. (1979). Bovine mycoplasma, in *The Mycoplasmas*, Vol. II, (eds M.F. Barile, S. Razin, J.G. Tully, R.F. Whitcomb), Academic Press, New York, pp. 50–91.
34. Hayes, M.M., Wear, D.J., Lo, S-C. (1991) *In vitro* antimicrobial susceptibility testing for the newly identified AIDS-associated mycoplasma: *Mycoplasma fermentans* (incognitus strain). *Arch. Pathol. Lab. Med.*, **115**, 464–6.
35. Hendershot, L., Levitt, D. (1985) Effects of mycoplasma contamination on immunoglobulin biosynthesis by human B lymphoblastoid cell lines. *Infect. Immun.*, **49**, 36–9.
36. Hooton, T.M., Roberts, P.L., Stamm, W.E., Roberts, M.C., Holmes, K.K., Kenny, G.E. (1988) Prevalence of *Mycoplasma genitalium* determined by DNA probe in men with urethritis. *Lancet*, **i**, 266–8.
37. Hu, P.C., Cole, R.M., Huang, Y.S., *et al.* (1982) *Mycoplasma pneumoniae* infection: Role of a surface protein in the attachment organelle. *Science*, **216**, 313–15.
38. Hu, P.C., Collier, A.M., Baseman, J.B. (1977) Surface parasitism by *Mycoplasma pneumoniae* of respiratory epithelium. *J. Exp. Med.*, **145**, 1328–41.
39. Hu, P.C., Schaper, U., Collier, A.M., *et al.* (1987) A *Mycoplasma genitalium* protein resembling the *Mycoplasma pneumoniae* attachment protein. *Infect. Immun.*, **55**, 1126–31.
40. Hu, W.S., Wang, R.Y-H., Liou, R-S., Shih, J.W-K., Lo, S-C. (1990) Identification of an insertion-sequence-like genetic element in the newly recognized human pathogen *Mycoplasma incognitus*. *Gene*, **93**, 67–72.
41. Inamine, J.M., Denny, T.P., Loechel, S., *et al.* (1988) Nucleotide sequence of the P1 attachment-protein gene of *Mycoplasma pneumoniae*. *Gene*, **64**, 217–29.
42. Inamine, J.M., Loechel, S., Hu, P-C. (1988) Analysis of the nucleotide sequence of the P1 operon of *Mycoplasma pneumoniae*. *Gene*, **73**, 175–83.
43. Inanime, J.M., Ho, K-C., Loechel, S., Hu, P-C. (1990) Evidence that UGA is read as a tryptophan codon rather than as a stop codon by *Mycoplasma pneumoniae*, *Mycoplasma genitalium*, and *Mycoplasma gallisepticum*. *J. Bacteriol.*, **172**, 504–6.
44. Jensen, J.S., Uqdum, S.A., Sondergard-Andersen, J., Vuust, J., Lind, K. (1991) Polymerase chain reaction for detection of *Mycoplasma genitalium* in clinical samples. *J. Clin. Microbiol.*, **29**, 46–50.
45. Kita, M. (1990) Induction of cytokines by mycoplasmas in human lymphocyte culture. *Tiss. Cult. Res. Comm.*, **94**, 64.

46. Klatzmann, D., Barre-Sinoussi, F., Nugeyre, M.T., *et al.* (1984) Selective tropism of lymphadenopathy associated virus (LAV) for helper-inducer T lymphocytes. *Science*, **225**, 59–62.
47. Krause, D.C., Leith, D.K., Wilson, R.M., Baseman, J.B. (1982) Identification of *Mycoplasma pneumoniae* proteins associated with hemadsorption and virulence. *Infect. Immun.*, **35**, 809–17.
48. Lemaitre, M., Guetard, D., Henin, Y., Montagnier, L., Zerial, A. (1990) Protective activity of tetracycline analogs against the cytopathic effect of the human immunodeficiency viruses in CEM cells. *Res. Virol.*, **141**, 5–16.
49. Levin, D., Gershon, H., Naot, Y. (1985) Production of interleukin-2 by rat lymph node cells stimulated by *Mycoplasma pulmonis* membranes. *J. Infect. Dis.*, **151**, 541–4.
50. Levy, J.A. (1988) Mysteries of HIV: challenges for therapy and prevention. *Nature*, **333**, 519–22.
51. Lo, S-C., Shih, J.W-K., Yang, N.Y., Ou, C.Y., Wang R.Y-H. (1989) A novel virus-like infectious agent in patients with AIDS. *Am. J. Trop. Med. Hyg.*, **40**, 213–28.
52. Lo, S-C., Shih, J.W-K., Newton, *et al.* (1989) Virus-like infectious agent (VLIA) is a novel pathogenic mycoplasma: *Mycoplasma incognitus*. *Am. J. Trop. Med. Hyg.*, **41**, 586–600.
53. Lo, S-C., Wang, R.Y-H., Newton, P.B., Yang, N.Y., Sonoda, M.A., Shih, J.W-K. (1989) Fatal infection of silvered leaf monkeys with a virus-like infectious agent (VLIA) derived from a patient with AIDS. *Am. J. Trop. Med. Hyg.*, **40**, 399–409.
54. Lo, S-C., Dawson, M.S., Newton, P.B., *et al.* (1989) Association of the virus-like infectious agent originally reported in patients with AIDS with acute fatal disease in previously healthy non-AIDS patients. *Am. J. Trop. Med. Hyg.*, **41**, 364–76.
55. Lo, S-C., Dawson, M.S., Wong, D.M., *et al.* (1989) Identification of *Mycoplasma incognitus* infection in patients with AIDS: An immunohistochemical *in situ* hybridization and ultrastructural study. *Am. J. Trop. Med. Hyg.*, **41**, 601–16.
56. Lo, S-C., Wear, D.G., Hayes, M.M. (1990) *In vitro* susceptibility testing for the newly identified pathogenic human mycoplasma, *M. incognitus*. *The VI International Conference on AIDS*. Abstract no. Th. b 536.
57. Lo, S-C., Tsai, S., Benish, J.R., Shih, J.W-K., Wear, D.J., Wong, D.M. (1991) Enhancement of HIV-1 cytocidal effects in $CD4^+$ lymphocytes by the AIDS-associated mycoplasma. *Science*, **251**, 1074–6.
58. Lo, S-C., Buchholz, C.L., Wear, D.J., Hohm, R.C., Marty, A.M. (1991) Histopathology and doxycycline treatment in a previously healthy non-AIDS patient systemically infected by *Mycoplasma fermentans* (incognitus strain). *Mod. Pathol.*, **6**, 750–4.
59. Madoff, S., Hooper, D.C. (1988) Nongenitourinary infections caused by *Mycoplasma hominis* in adults. *Rev. Infect. Dis.*, **10**, 602–13.
60. Mahairas, G.G., Minion, C. (1989) Transformation of *Mycoplasma pulmonis*: Demonstration of homologous recombination, introduction of

cloned genes, and preliminary description of an integrating shuttle system. *J. Bacteriol.*, **171**, 1775–80.
61. Mahairas, G.G., Minion, F.C. (1989) Random insertion of the gentamicin resistance transposon Tn4001 in *Mycoplasma pulmonis. Plasmid*, **21**, 43–7.
62. Mahairas, G.G., Jian, C., Minion, F.C. (1990) Development of a cloning system in *Mycoplasma pulmonis. Gene*, **93**, 61–6.
63. Maniloff, J., Morowitz, H.J. (1972) Cell biology of the mycoplasmas. *Bacteriol. Rev.*, **36**, 263–90.
64. McGarrity, G.J., Kokani, H. (1985) Cell culture mycoplasmas, in *The Mycoplasmas*, Vol. IV, (eds S. Razin, M.F. Barile, M.F.), Academic Press, New York, pp. 353–90.
65. Miller, J.F., Mekalanos, J.J., Falkow, S. (1989) Coordinate regulation and sensory transduction in the control of bacterial virulence. *Science*, **243**, 916–22.
66. Moller, B.R., Taylor-Robinson, D., Furr, P.M. (1984) Serological evidence implicating *Mycoplasma genitalium* in pelvic inflammatory disease. *Lancet*, **i**, 1102–3.
67. Montagnier, L., Berneman, D., Guetard, D., *et al.* (1990) Inhibition of HIV prototype strains infectivity by antibodies directed against a peptide sequence of mycoplasma. *C.R. Acad. Sci. Paris.*, **311**, 425–30.
68. Mouches, C., Barroso, G., Bove, J.M. (1983) Characterization and molecular cloning of *Escherichia coli* of plasmid from the *Mollicute, Spiroplasma citri. J. Bacteriol.*, **156**, 952–5.
69. Muto, A., Kawauchi, Y., Yamao, F., Osawa, S. (1984) Preferential use of A- and U-rich codons for *Mycoplasma capricolum* ribosomal proteins S8 and L6. *Nucleic Acids Res.*, **12**, 8209–17.
70. Muto, A. (1987) The genome structure of *Mycoplasma capricolum. Isr. J. Med. Sci.*, **23**, 334–41.
71. Neimark, H.C., Lange, C.S. (1990) Pulse-field electrophoresis indicates full-length mycoplasma chromosomes range widely in size. *Nucleic Acids Res.*, **18**, 5443–8.
72. Nocard, E., Roux, E.R. (1898) Le microbe de le peripneumoniae. *Ann. Int. Pasteur. (Paris)*, **12**, 240–62.
73. Nur, I., Glaser, G., Razin, S. (1986) Free and integrated plasmid DNA in *Spiroplasmas. Curr. Microbiol.*, **14**, 169–76.
74. Oredipe, O.A., Barth, R.F., Rotaru, J.H., Hinkle, G.H., Steplewski, S. (1990) Alterations in monoclonal antibody affinity and antigenic receptor site expression on mycoplasma-infected human colorectal cancer cells. *Proc. Soc. Exp. Biol. Med.*, **194**, 301–7.
75. Osawa, S., Jukes, T.H., Muto, S., Yamao, F., Ohama, T., Andachi, Y. (1987) Role of directional mutation pressure in the evolution of the eubacterial genetic code. *Cold Spring Harb. Symp. Quant. Biol.*, **52**, 777–89.
76. Poddar, S.K., Maniloff, J. (1989) Determination of microbial genome sizes by two-dimensional denaturing gradient gel electrophoresis. *Nucleic Acids Res.*, **17**, 2889–95.
77. Quentmeier, H., Schmitt, E., Kirkhoff, H., Grote, W., Muhlradt, P.F. (1990)

Mycoplasma fermentans-derived high-molecular-weight material induces interleukin-6 release in cultures of murine macrophages and human monocytes. *Infect. Immun.*, **58**, 1273–80.

78. Ranhand, J.M., Mitchell, W.O., Popkin, T.J., Cole, R.M. (1980) Covalently closed circular deoxyribonucleic acids in *Spiroplasmas. J. Bacteriol.*, **143**, 1194–9.
79. Razin, S. (1978). The *Mycoplasmas. Microbiol. Rev.*, **42**, 414–70.
80. Razin, S., Freundt, E.A. (1984). The Mycoplasmas, in *Bergey's Manual of Systemical Bacteriology*, Vol. I, (eds N.R. Kreig, J.G. Holt), Williams and Wilkins, Baltimore, pp. 740–92.
81. Razin, S. (1985) Molecular biology and genetics of mycoplasmas (*Mollicutes*). *Microbiol. Rev.*, **49**, 419–55.
82. Razin, S. (1985) Mycoplasma adherence, in *The Mycoplasmas*, Vol. IV, (eds S. Razin, M.F. Barile), Academic Press, New York, pp. 161–98.
83. Razin, S., Nur, I., Glaser, I. (1987) *Spiroplasma* plasmids. *Isr. J. Med. Sci.*, 23, 678–82.
84. Research News (1990) Mycoplasma in the AIDS spotlight. *Science*, **248**, 682–3.
85. Research News (1991) Montagnier pursues the mycoplasma-AIDS link. *Science*, **251**, 271.
86. Robinson, I.M., Freundt, E.A. (1987) Proposal for an amended classification of anaerobic mollicutes. *Int. J. Sys. Bacteriol.*, **37**, 78–81.
87. Rosengarten, R., Wise, K.S. (1990) Phenotypic switching in mycoplasma: phase variation of diverse surface lipoproteins. *Science*, **247**, 315–8.
88. Ruuth E., Lundgren, E. (1986) Enhancement of immunoglobulin secretion by the lymphokine-like activity of a *Mycoplasma arginini* strain. *Scand. J. Immunol.*, **23**, 575–80.
89. Ruuth E., Praz, F. (1989) Interactions between mycoplasmas and the immune system. *Immunol. Rev.*, **112**, 133–60.
90. Saillard, C., Carle, P., Bove, J.M., *et al.* (1990) Genetic and serologic relatedness between *Mycoplasma fermentans* strains and a mycoplasma recently identified in tissues of AIDS and non-AIDS patients. *Res. Virol.*, **141**, 385–95.
91. Seifert, H.S., So, M. (1988) Genetic mechanisms of bacterial antigenic variation. *Microbiol. Rev.*, **52**, 327–36.
92. Shepard, M.C., Lunceford, C.D. (1976) Differential agar medium (A7) for identification of *Ureaplasma urealyticum* (human T mycoplasmas) in primary cultures of clinical material. *J. Clin. Microbiol.*, **3**, 613–25.
93. Shepard, M.C., Masover, G.K. (1979) Special features of ureaplasmas, in *The Mycoplasmas*, Vol. I, (eds M.F. Barile, S. Razin), Academic Press, New York, pp. 451–94.
94. Sitia, R., Ribartelli, A., Deambrosis, S., Pozzi, D., Hammerling, U. (1985) Differentiation in the murine B cell lymphoma I. 29: inductive capacities of lipopolysaccharide and *Mycoplasma fermentans* products. *Eur. J. Immunol.*, **15**, 570–5.
95. Smith, C.B., Friedewald, W.T., Chanock, R.M. (1967) Shedding of *Mycoplasma pneumoniae* after tetracycline and erythromycin therapy. *N. Engl. J. Med.*, **276**, 1172–5.

96. Sperker, B., Hu, P., Herrmann, R. (1991) Identification of gene products of the P1 operon of *Mycoplasma pneumoniae. Mol. Microbiol.*, 5, 299–306.
97. Stadtlander, C.T.K-H. (1991) Respiratory epithelial cell interactions: *Mycoplasma fermentans* compared with other mycoplasmal species. *91st Gen. Meet. Am. Soc. Microbiol.*, Session 218.
98. Stanbridge, E.J., Reff, M.E. (1979) The molecular biology of mycoplasmas, in *The Mycoplasmas*, Vol. I, (eds M.F. Barile, S. Razin, J.G. Tully, R.F. Whitcomb), Academic Press, New York, pp. 157–82.
99. Stuart, P.M., Cassell, G.H., Woodward, J.G. (1989) Induction of class II MHC antigen expression in macrophages by mycoplasma species. *J. Immunol.*, **142**, 3392–9.
100. Stuart, P.M., Cassell G.H, Woodward, J.G. (1990) Mycoplasma: induction of class II MHC antigens, possible role in auto immunity. *Zenbralbl. Bakteriol.*, **Suppl. 20**, 570–7.
101. Su, C.J., Tryon, V.V., Baseman, J.B. (1987) Cloning and sequence analysis of cytadhesin P1 gene from *Mycoplasma pneumoniae. Infect. Immun.*, 55, 3023–9.
102. Taylor-Robinson, D., McCormack, W.M. (1980) The genital mycoplasmas. *N. Engl. J. Med.*, **302**, 1003–10.
103. Taylor-Robinson, D., Furr, P. M., Tully, J.G., Barile, M.F. (1987) Animal models of *Mycoplasma genitalium* urogenital infection. *Isr. J. Med. Sci.*, **23**, 561–4.
104. Tully, J.G., Cole, R.M., Taylor-Robinson, D., Rose, D.L. (1981) A newly discovered mycoplasma in the human urogenital tract. *Lancet*, i, 1288–91.
105. Tully, J.G., Baseman, J.B. (1991) *Mycoplasma. Lancet*, **337**, 1297.
106. Valdivieso-Garcia, A., Rosendal, S. (1982) Variation in colony size of *Mycoplasma mycoides* subspecies *mycoides* isolated from goats. *Vet. Rec.*, **110**, 470–1.
107. Wayner, E.A., Brooks, C.G. (1984) Induction of NKCF-like activity in mixed lymphocytes–tumor cell culture: direct involvement of mycoplasma infection of tumor cells. *J. Immunol.*, **132**, 2135–42.
108. Wang, R.Y-H., Hu, W.S., Dawson, M.S., Shih, J.W-K., Lo, S-C. (1992) Selective detection of *Mycoplasma fermentans* by the polymerase chain reaction using a nucleotide sequence within the insertion-sequence-like element. *J. Clin. Microbiol.*, **30**, 245–8.
109. Watson, H.L., McDaniel, L.S., Blalock, D.K., Fallon, M.T., Cassell, G.H. (1988) Heterogeneity among strains and a high rate of variation within strains of a major surface antigen of *Mycoplasma pulmonis. Infect. Immun.*, **56**, 1358–63.
110. Wenzel, R., Herrmann, R. (1989) Cloning of the complete *Mycoplasma pneumoniae* genome. *Nucleic Acids Res.*, **17**, 7029–043.
111. Williamson, D.L., Brink, P.R., Zieve, G.W. (1984) *Spiroplasma* fibrils. *Isr. J. Med. Sci.*, **20**, 830–5.
112. Yamao, F., Muto, A., Kawauchi, Y., *et al.* (1985) UGA is read as tryptophan in *Mycoplasma capricolum. Proc. Natl. Acad. Sci. U.S.A.*, **82**, 2306–9.

11

Herpesviruses

VINCENT C. EMERY, ALISON WEBSTER and PAUL D. GRIFFITHS

11.1 INTRODUCTION

The herpesviruses form a clinically important group of viruses resulting in overt disease in patients whose immune system is deranged. Most readers will be familiar with the manifestations of herpesvirus infections in transplant patients whose immune system is suppressed during the time of transplant and their effects in the newborn where the immune system is immature [94]). However, since the identification of HIV as the aetiological agent of AIDS, a new clinical situation has arisen where a set of patients undergo a progressive decline in immune function. Concomitant with this decline is the increasing likelihood of opportunistic herpesvirus infections. Manifestations of disease (Table 11.1) may be the result of reactivation of latent herpesvirus genomes or indeed reinfection with new strains of herpesviruses. This chapter aims to give an overview of the molecular organization of the herpesviruses, the *in vitro* and *in vivo* evidence for their role as viral co-factors for HIV disease progression, and the molecular methods used in the diagnosis and treatment of herpesvirus opportunistic infections.

11.2 MOLECULAR STRUCTURE AND ORGANIZATION

11.2.1 Morphology

The physical size of a mature herpesvirus particle varies between 250 nm and 300 nm. Herpesviruses have a characteristic morphology consisting of a toroidal core containing the genomic DNA suspended within a

Molecular and Cell Biology of Opportunistic Infections in AIDS
Edited by Steven Myint and Alan Cann
Published in 1992 by Chapman & Hall, London ISBN 0 412 45330 4

Table 11.1 Opportunistic diseases associated with the herpesviruses in AIDS patients

Herpesvirus	*Opportunistic disease*
HCMV	Retinitis
	Gastrointestinal ulceration
	Hepatitis
HSV–1, HSV–2	Oral–genital/rectal ulceration
	Encephalitis
EBV	B-cell lymphomas
	Oral hairy leukoplakia
	Lymphoid interstitial pneumonitis
VZV	Shingles
HHV–6	Not known

icosahedral capsid consisting of 162 capsomeres, approximately 100 nm in diameter. In turn, the capsid is surrounded by a tegument and, finally, an outer lipid envelope completes the architecture of the viral particle. The particle exhibits characteristic glycoprotein spikes projecting from the envelope. An electron micrograph of a typical herpesvirus particle is shown in Figure 11.1. At present the herpesviruses consist of six distinct types: herpes simplex virus type 1 (HSV-1), herpes simplex virus type 2 (HSV-2), varicella zoster virus (VZV), human cytomegalovirus (HCMV), Epstein–Barr virus (EBV) and human herpes virus 6 (HHV-6). In addition, a recent report [31] has identified a possible human herpes virus type 7 (HHV-7). The most studied of these viruses, with respect to the molecular details of infection and replication, is HSV.

11.2.2 Genomic structure

The genomic organization of the herpesviruses is complex and variable. However, common themes are apparent. The genome consists of a linear double stranded DNA molecule ranging from 120 to 230 kilobase pairs. Although the viral DNA is linear in the mature virion, it rapidly circularizes after release into the nuclei of infected cells. As a general rule the genome is relatively G + C rich, ranging from 42% in HHV-6 to 69% in HSV-2. The genomes of some of the viruses, such as EBV, contain internal repeated sequences, whereas in others such as HSV, HCMV and VZV, the genome is composed of two unique regions, the unique long (U_L) and unique short (U_S) which are bounded by internal and terminal repeat sequences. Thus, the genomes of HSV and HCMV can exist in four isomeric forms whilst that of VZV exists in two isomeric forms. Genomic arrangements of the human herpesviruses are shown in Figure 11.2.

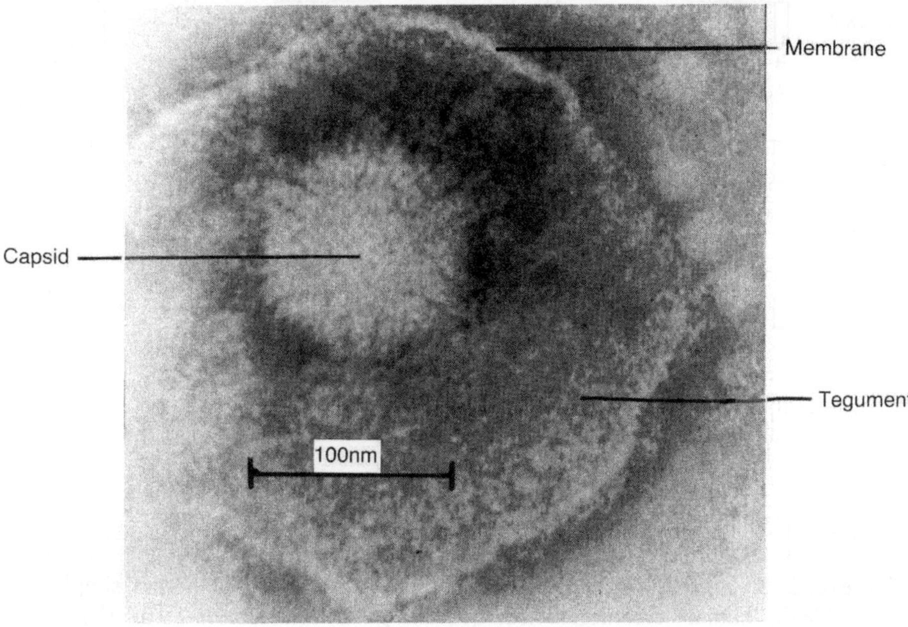

Figure 11.1 Electron micrograph of a typical herpesvirus (in this case cytomegalovirus). The viral outer membrane, tegument and capsid are indicated. (Micrograph prepared by Mr A. McDermott.)

The herpesviridae are frequently sub-divided into three sub-families according to biological parameters; the alpha-herpesvirinae, the beta-herpesvirinae and the gamma-herpesvirinae. The alpha-herpesvirinae are classified on the basis of their broad host range, relatively short replicative cycle, lytic nature, and their ability to establish latency in sensory ganglia. Thus HSV-1, HSV-2 and VZV fall into this category. The beta-herpesvirinae tend to have a restricted host range and *in vitro* infection proceeds slowly with infected cells exhibiting cytomegalic morphological changes. Unlike the alpha-herpesvirinae, a definitive site of viral latency has not been elucidated, although possible sites include the salivary glands, lymphoreticular cells, kidneys and many other tissues. The group is exemplified by HCMV and also provisionally contains HHV-6. An example of the gamma-herpesvirinae is EBV. *In vivo*, these viruses are specific for T- or B-lymphocytes and *in vitro*, replication occurs in lymphoblastoid cells. In lymphocytes, infection does not always produce infectious progeny and extrachromosomal latent virus is frequently detected in lymphoid tissue.

An underlying feature of all herpesviruses is their ability to establish a latent infection subsequent to the initial infection. At later stages in the

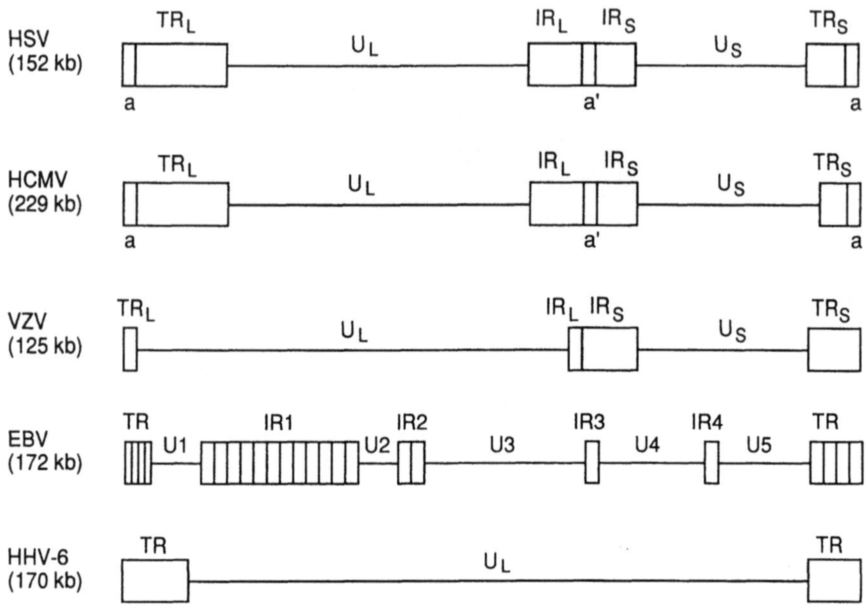

Figure 11.2 Stylized illustration of the genomic organization of herpes simplex virus (HSV), human cytomegalovirus (HCMV), varicella zoster virus (VZV), Epstein–Barr virus (EBV) and human herpesvirus type 6 (HHV-6). The genomic size in kilobase pairs is given for each virus. It should be noted that the sizes of the genomic segments are not drawn to scale. TR = terminal repeat, L = left, R = right, U = unique, L = long, S = short, IR = internal report.

life-span of the host, reactivation of this latent virus may occur leading to overt clinical disease. Examples of this phenomenon are primary infection with VZV causing chickenpox (varicella) with later reactivation manifesting as shingles (zoster), and the primary asymptomatic infection with HSV-1, reactivation of which manifests itself as common coldsores.

11.2.3 Polymorphism of herpesvirus genomes

Investigation of the micropolymorphisms of human herpesvirus genomes has occurred in the case of HSV-1, HSV-2, HCMV and, to a lesser extent, EBV. The genome of these viruses appears to be stable both in cell culture and in infected individuals. However, restriction fragment length polymorphism (RFLP) studies of HSV have shown that each individual HSV isolate is different except when the isolates are derived from epidemiologically related individuals [39, 82]. A similar situation exists with HCMV [18]. In contrast, the genome of VZV appears to be remarkably conserved between epidemiologically unrelated sources [40, 64]. With the exception of VZV, these RFLP analyses provide an important epidemiological tool in which to study the transmission of viruses within populations. Thus, it has been shown that multiple strains of HCMV can infect the same HIV-infected individual [19]. A logical extension of such analyses is to perform PCR and DNA sequencing of the amplified fragments of the herpesvirus genomes in order to study microheterogeneity of key regions of the genome. Such analysis has been effected on a major neutralizing epitope within the glycoprotein molecule of HCMV and nucleotide variations leading to amino acid changes within this epitope have been identified [15, 21]. The biological significance of these variations has yet to be ascertained.

11.3 GENE EXPRESSION AND VIRAL REPLICATION

Gene expression within the herpesviruses can be broadly classified into three major groups, immediate early, early and late (an alternative nomenclature is alpha, beta and gamma, respectively). The expression of these genes is co-ordinately regulated and sequentially ordered in a cascade fashion [43, 44]. The most studied herpesvirus with respect to gene expression and replication is HSV-1 and most of this section will refer to HSV as the representative member of the group. The gross organization of gene expression and viral replication is similar between all the herpesviruses even though the fine details may vary considerably.

The HSV α genes consist of five members of ICP (infected cell protein), 0, 4, 22, 27 and 47. The α genes do not require prior viral protein synthesis for expression and all possess a *cis* regulatory sequence element 5′ NCGyATGnTAATGArATTCyTTGnGGG 3′ (commonly known as 'TAATGARAT') in one or more copies within the 400 nucleotides (nt) upstream of the transcription initiation site [54, 62]. During the immediate early phase of gene expression two other genes are transcribed, the latency associated transcript (LAT) which is antisense to the α0 gene [90] and a ORI_sRNA_1 which initiates within the inverted repeat sequence of the U_s and is antisense to the α22 and α47 genes [45]. The majority of the α genes are involved in gene regulation, that is trans-activation of the β and γ gene promoters.

The α genes are induced by a virion structural protein known as α0 trans-inducing factor (α-TIF) [72]. The virion contains between 500 and 1000 copies of α-TIF [41]. The precise mechanism of transactivation of the a genes by α-TIF has not been elucidated but the *cis* acting sequence within the genes binds α cellular transcription factor OTF-1 [55, 56] and α-TIF appears to interact with this transcription factor-α gene promoter DNA complex to yield a full α gene transcriptional activation [33, 69].

As expected, β genes have an absolute requirement for α gene products for their synthesis; ICP4 is the major effector of transactivation of the β genes. HSV specifies many proteins required for the efficient production of deoxynucleotide pools and for DNA synthesis. These include a DNA polymerase [12, 74], DNase [49], dUTPase [11, 95], ribonucleotide reductase, uracil-DNA glycosylase, a major DNA binding protein and a thymidine kinase [47]. All these gene products are directly or indirectly involved in viral replication and must be synthesized prior to the onset of replication. It is interesting that not all of this repertoire of genes has been identified in the other herpesviruses, for example, HCMV does not possess a thymidine kinase. Viral DNA synthesis occurs in the nucleus and satisfies the tenets of the rolling circle model for DNA replication. It has been shown that only a small proportion of input DNA (virion DNA) is replicated [46].

While DNA replication is proceeding the late or γ gene products are synthesized ready for the encapsidation and envelopment of the viral DNA. These genes include eight viral glycoproteins gB, gC, gD, gE, gG, gH, gI and gJ, and structural proteins present in the capsid and tegument [88]. Glycoproteins gB and gD are involved in binding to and fusion with target cell membranes whereas gC binds to the C3b component of complement and gE and gI specify a molecule that binds to the Fc portion of IgG [48]. The Fc receptor is present on the mature virion. The replicated DNA is packaged into preformed capsids [34] following cleavage of the concatameric replicated virion DNA [8, 9]. The capsids containing DNA are then enveloped at the nuclear membrane in discrete patches containing the envelope glycoproteins [22, 85]. Recently the mechanism by which the tegument is acquired has been elucidated for HHV-6 and involves invagination of the cytoplasm into the nucleus as a 'tegusome' [80]. The enveloped capsids accumulate in the space between the inner and outer lamellae of the nuclear membrane and in the cysternae of the endoplasmic reticulum. The viral particles egress from the infected cell through the endoplasmic reticulum.

One result of α-herpesvirinae infection is the death of the infected cell. Following infection, a disaggregation of host polyribosomes occurs together with cessation of host cell DNA synthesis, degradation of host mRNA and consequent cessation of host cell protein synthesis [29, 67]. These effects are probably mediated by virion structural gene products rather than α gene products. The host cell metabolic pathways are further inhibited following expression of the viral β and γ gene products [43]. These effects include cessation of host ribosomal RNA synthesis and chromosome fragmentation. Insertion of the viral glycoproteins in the cellular membranes may cause morphological changes in the cell, resulting, finally, in the destruction of the cell architecture [81].

11.3.1 Viral latency

All the herpesviruses have the propensity to remain latent. At this juncture it is worth noting that the *Oxford English Dictionary* definition of latent is 'existing but not developed or manifest'. The molecular basis for latency is not well defined and is exacerbated by the paucity of experimental systems which allow establishment of the latent state and reactivation of latent genomes to be investigated. The sites of latency for the herpesviruses have been elucidated for HSV, VZV (neurones of the sensory and autonomic ganglia) and EBV (B-lymphocytes) [38]. In these two examples, the biochemical nature of the cell type differs substantially in that EBV is latent in a cell that is capable of dividing whereas HSV is dormant in neuronal cells that do not divide. Thus, it is likely that the individual herpesviruses use distinct mechanisms to establish, maintain and possibly reactivate from the latent state. The herpesvirus genomes appear to exist in an episomal state in the latently infected cells.

Initial experiments with HSV provided a possible mechanism for latency through the latency associated transcript (LAT) RNA species. The latter is transcribed in an antisense orientation to the α0 product and was found to be transcribed at low levels during lytic infection but at high levels in latently infected cells. Unfortunately, more recent data using HSV-1 variants that do not synthesize the LAT but can establish latency, argue that the LAT is involved in the reactivation of latent virus rather than its establishment and maintenance. A more attractive thesis for the establishment of the latent state has been provided by the elegant work of Latchman and colleagues [80, 93] which has identified a neuronal cell specific repressor that can bind to the octomer related (OCT-1) motifs in the α promoters [50]. The repressor factor has been detected in immortalized neurones from rat dorsal root ganglia, the natural site of latent HSV infection *in vivo*. Conceivably, the modulation of this factor by nerve growth or stress offers a mechanism for the reactivation of latent virus.

11.4 MOLECULAR MECHANISMS OF PATHOGENESIS

Apart from the obvious role of loss of immune function, there are several distinct molecular mechanisms by which herpesvirus pathogenicity may be enhanced in patients with HIV infection.

11.4.1 Enhanced switch from latent to productive infection

Individuals with HIV infection may have a series of herpesvirus genomes maintained in the latent form. Clinical latency may obviously be maintained by effective immune surveillance which declines as the patient progresses towards AIDS. Reciprocal enhancement of viral expression has been demonstrated for CMV in monocyte and T-cell lines co-infected with CMV and HIV [84]. Ho *et al.* [42] have shown that transfection with HIV *tat*, or infection with HIV-1 upregulates CMV IE gene expression and replication. No reciprocal enhancement of gene expression has been shown for HSV [76] or EBV [71]. Alternatively, production of transcriptional factors such as NFKB by the the herpesvirus may allow increased expression of HIV. Even if the herpesvirus and HIV do not infect the same cell they may nevertheless interact through a bystander effect *via* release of cytokines or by antigen presentation to immune cells harbouring latent herpesvirus genomes such as EBV and CMV.

11.4.2 Generation of genetic diversity

Given the large size of herpesvirus genomes, it is likely that genetic variants arise during replication. These variants might have novel features facilitating their escape from normal immune surveillance and so cause disease. Mutant viruses with pathogenic potential may be controlled by an intact immune system but as the immune response declines, 'less fit' but 'more damaging' mutants may become predominant. A consequence of this phenomenon might be the selection of strains resistant to antiviral drugs, a subject which is discussed below.

It is also possible that intertypic genetic recombination may occur between strains, as has been reported for CMV in transplant patients [14].

11.4.3 Stimulation of host immune responses

The immune response of AIDS patients is not absent but deranged. Aberrant immune responses might cause some of the manifestations of HIV disease, such as auto-immune haemolytic anaemia. Since EBV produces polyclonal B-cell activation it has the potential to drive the appearance of such pre-existing auto-antibodies [86]. Interestingly, the B-cells immortalized by EBV are not those with cytoplasmic immunoglobulin [4] which suggests that chronic stimulation by EBV may be necessary for polyclonal B-cell activation. It is also interesting to note that expression of 8 EBV encoded latent proteins can protect B-cells from apoptosis, the programmed cell death which normally controls the expression of a full repertoire of immunoreactive memory [37]. Thus, expression of such latent genes could potentially prevent deletion of B-cell clones reactive with 'self' antigens. It remains to be determined whether the mechanisms controlling emergence of B-cells expressing EBV latent genes are impaired in HIV-infected individuals.

Alternatively, herpesviruses might cause the presentation *de novo* to the immune system of host antigens leading to induction of auto-immunity. This could be indirect, through the induction of interferon production or could be direct in the case of CMV, which encodes its own class 1 HLA-like gene [7].

11.4.4 Down regulation of host immune responses

To avoid immune surveillance, herpesviruses may down regulate host proteins. For example, B-cells immortalized by EBV have decreased expression of cell surface adhesion molecules LFA-3 and ICAM-1, which normally facilitate recognition by EBV-specific T-lymphocytes [37].

Recently, the HLA class I-like gene product of CMV has been shown to bind to β2 microglobulin and inhibit the cell surface display of mature class I HLA molecules when both genes were expressed in vaccinia virus vectors [10]. Such a mechanism, if proven to be operational when CMV infects cells *in vivo*, would parallel the situation in adenovirus infected cells where the display of mature class I molecules is decreased by viral products that interfere with post-translational modifications or the transcriptional activity of the class I heavy chain.

11.4.5 Interaction with host cell DNA

Those herpesviruses which integrate into host cell chromosomes might mediate insertional mutagenesis. Alternatively, a specific association, such as that between EBV and translocation of the *myc* oncogene from chromosome 8 to 14, might ensue. In either case, if the altered cell had improved growth capabilities, then clonal expansion could lead to tumour formation. Some of the non-Hodgkin's lymphomas found in AIDS patients have been shown to contain EBV genomes, although no rearrangement of *myc* was found [52].

11.5 HERPESVIRUSES AS COFACTORS FOR HIV DISEASE PROGRESSION

Infection with HIV is followed by the development of symptomatic immunodeficiency usually after a long, but variable, interval. This is, to some extent, combapable to integration of HIV proviral DNA into the host cell genome, in which form the virus may remain latent for prolonged periods and from which it may subsequently reactivate. A number of *in vitro* studies have suggested that the herpesviruses may act as activators of HIV gene expression. However, complementary *in vivo* studies have not proved conclusive.

11.5.1 *In vitro* studies

Both viral and cellular proteins interact with sequences within the HIV LTR to modulate expression of the genome. These include the HIV transactivating protein (*tat*), and a variety of cellular transcriptional factors such as NFkB, SP1, and NFAT-1. The ability of various heterologous viral proteins to transactivate the HIV LTR has been studied using cells co-transfected with plasmids containing the HIV-1 LTR linked to the reporter gene chloramphenicol acetyl transferase (CAT), and plasmids containing the viral gene(s) of interest. Using this system, a wide range of viruses, including the herpesviruses HSV-1 and HSV-2, CMV, HHV-6 and EBV [23, 26, 32, 51, 60, 66, 70, 78, 87] have all been reported to transactivate the HIV LTR. Immediate early (IE) gene products have been shown to mediate these effects [1, 23, 75]. The cellular transcription factor NFϰB appears to play a central role in the activation by HSV and HHV-6, whereas in the case of CMV, a factor LBP-1 mediates the effect [5, 26, 35].

Molecular interaction between HIV and herpesviruses may be a two way process [77]. This may be a mechanism by which opportunistic viral infections develop, independently of a decline in immune surveillance. Induction of cellular transcription factors may also result from external stimuli such as mitogens or cytokines. The cytokine tumour necrosis factor alpha (TNF-alpha) has been shown to increase HIV expression in a chronically infected T-cell line [30]. Cytokine production can be stimulated by viral infections, and therefore this may be an alternative mechanism by which viruses upregulate HIV. Challenge of PBMCs with inactivated viral antigens results in the production of soluble mediators, in particular TNF-α which are capable of upregulating HIV expression in chronically infected T-cell lines [16]. CMV, EBV and HIV stimulated cells produced cytokines which upregulate HIV, whereas HSV-1, HSV-2, VZV and HHV-6 were ineffective in this model [17].

Antibody-coated HIV is able to enter cells *via* interaction with the receptor for the Fc component of IgG on the cell surface [91]. CMV and HSV induce the expression of these receptors on the surface of infected cells, which could then be exploited by HIV. McKeating and colleagues [65] have demonstrated the uptake of antibody-coated HIV into CMV-infected fibroblasts, cells which are non CD4–bearing, and which are not normally permissive for HIV. Uptake could be blocked by normal human serum, but not by soluble CD4, or by Leu 3a, an anti-CD4 antibody, confirming that uptake is independent of CD4.

Infection with a cofactor virus may also facilitate HIV uptake by inducing transcription of the cellular gene encoding the HIV receptor, the CD4 molecule. HHV-6 infection has been reported to induce expression of the CD4 molecule on mature $CD8^{+}CD4^{-}$ lymphocytes, so rendering the cells susceptible to HIV infection [61]. In contrast, prior infection with CMV or HSV did not produce this effect.

Phenotypic mixing of HIV with other viruses may result in the formation of pseudotypes with altered target cell tropism. This has been reported to occur in T-cell lines co-infected with HIV and HSV [96]. Transformation of cord blood B-lymphocytes as a result of EBV infection has been shown to render these cells susceptible to HIV infection [71].

11.5.3 Epidemiological studies

It is important to place the *in vitro* data summarized in the previous section in the context of epidemiological studies. In general, such *in vivo* studies have not satisfactorily confirmed or refuted the clinical relevance of the experimental results described above, possibly due to other host factors such as age or genetic background.

The cofactor effect of CMV cannot be studied amongst cohorts of HIV-infected homosexuals, since CMV infection is virtually universal in this group as a result of high risk sexual behaviour and is a marker for increased exposure to HIV. One study has used a cohort of HIV-infected haemophiliacs to determine the influence of CMV on HIV disease progression. The results of this study showed that CMV-positive patients are significantly more likely to have progressed to AIDS, compared with CMV-negative patients, after controlling for age [92]. Other studies who have addressed the issue of CMV as a cofactor have not confirmed these findings. However, the majority of these studies can be criticized for experimental design and unknown dates of seroconversion in a number of the patients analysed.

HHV-6 infection is common amongst the general population, being acquired during the first few years of life. The progression to AIDS has been analysed in a group of homosexual men with lymphadenopathy, whose dates of HIV seroconversion were not known, 66% of whom were reported to be seropositive for HHV-6 [89]; 30% of these patients progressed to AIDS over 5 years. No association was found between HHV-6 and progression to AIDS.

Studies to date suggest that EBV is not a cofactor in HIV disease progression [57]. The proportion of HIV-infected homosexuals who have elevated IgA antibody titres to EBV viral capsid antigen increases with the stage of HIV disease [63], but this may be a consequence, rather than a cause, of the progressive immune deficiency.

11.6 MOLECULAR APPROACHES TO DIAGNOSIS

11.6.1 Serology

Now that the entire genomes of several human herpesviruses have been sequenced (EBV, HSV-1, VZV, CMV), it might be expected that improved immunoassays for detecting IgG antibodies would be available based upon expression of particular virus encoded proteins or synthetic peptides of the same. For example, the DNAse, thymidine kinase and gp340 membrane antigen of EBV have been expressed in recombinant baculovirus or bovine papillomavirus systems [59]. Antibodies of IgA class, reactive with the thymidine kinase, were shown to be predictive of nasopharyngeal carcinoma. Likewise, Gorgievski *et al.* [36] showed that IgA antibodies against diffuse early antigen, DNA binding protein and viral capsid antigen of EBV, could be detected in acute cases of infectious mononucleosis. For CMV, fusion proteins containing portions of a DNA binding protein and the major virion phosphoprotein have been shown to be recognized by sera from patients with CMV infection [79]. Synthetic peptides have been used to map linear epitopes on the same phosphoprotein [68].

Such reagents have the advantage of being free from contaminating host proteins associated with preparations made in cell culture (especially in the case of the cell-associated VZV and CMV). However, only limited steps have been taken to identify immunodominant epitopes which could be used to detect all individuals seropositive for each herpesvirus. Furthermore, the existing crude assays have proved to be fairly robust in clinical practice, with serum being used at high dilution to avoid cross reaction with nonspecific antibodies, and so it may be some time before they are replaced in routine practice.

11.6.2 Detection of virus strains resistant to antiviral agents

The appearance of herpesviruses resistant to the current range of antiviral agents has increased with their long term use in AIDS patients [13]. For example, HSV resistant to acyclovir [28] and CMV resistant to ganciclovir [25, 27] or ganciclovir and foscarnet [53], as well as VZV resistant to acyclovir [58] have been described. It is not clear whether this phenomenon is merely another manifestation of declining immune function, so that viruses with impaired replicative capacity are not eliminated, or is a consequence of HIV enhancing herpesvirus replication, or whether there is no relationship with herpesviruses. Resistant strains are currently identified by propagating the virus in cell culture in the presence of the antiviral agent. Strains isolated in this way are then subjected to sequence analysis to determine the mechanism by which the virus has escaped from antiviral control. Such an approach is facilitated by the fact that the majority of currently used anti-herpetic agents are DNA polymerase inhibitors, and hence mutations, yielding resistance to the drug, are likely to reside in either the herpesvirus DNA polymerase itself or in associated enzymes necessary for drug anabolism. However, now that humanized monoclonal antibodies which can neutralize HSV-1 and CMV are entering clinical trials, the molecular basis for the generation of herpesviruses resistant to these agents will require detailed analysis of other regions of the genome.

The identification of the genetic loci for resistance has been elucidated most rigorously for HSV. Thus, HSV strains resistant to acyclovir have mutations in either the thymidine kinase (TK) or DNA polymerase genes [20]. There are three phenotypes:

1. 'TK minus' mutants where the functional TK protein is lost by point mutation, insertion or deletion, leading to stop codon production and a truncated protein;
2. 'TK altered' mutants where the active site of the enzyme is mutated to exclude acyclovir as a substrate but still accepts thymidine;
3. 'DNA polymerase' mutants where the active site of the enzyme is mutated and does not accept acyclovir triphosphate as a substrate but still accepts guanosine triphosphate.

In general, the 'TK minus' mutants have low pathogenic potential [20] but may well produce disease in some individuals with AIDS [28].

11.6.3 Gene probes for diagnosis

The technique of *in situ* hybridization has been used to detect EBV genomes in particular clinical conditions. For example, Permeen *et al.* [73] used a non-radioactive probe to detect EBV DNA in cases of nasopharyngeal carcinoma. Schulten *et al.* [84] used *in situ* hybridization, Southern blot hybridization and the polymerase chain reaction for EBV DNA in cases of oral hairy leucoplakia. To avoid taking biopsies, De Souza *et al.* [24] have used filter and cytospin *in situ* hybridization on cells swabbed from the surfaces of the tongue lesions, and reported good sensitivity. Bashir *et al.* [6] studied central nervous system lymphomas and reported that EBV could be detected using a biotinylated EBV probe.

11.6.4 Genomic amplification

Since the herpesviruses can establish a latent infection, results obtained with PCR may detect a latent herpesvirus genome rather than an active infection. Thus, all studies using PCR, or other amplification methods, must evaluate this possibility in properly conducted clinical trials.

PCR has been used for the diagnosis of herpes simplex encephalitis in patients with normal immunity [3, 83] and may be expected to be useful in AIDS patients also; the technique has been used to detect EBV genomes in lymphoma biopsies from two AIDS patients [2].

11.7 CONCLUSIONS AND FUTURE PROSPECTS

The high prevalence of herpesviruses in the general population testifies to their success as human pathogens. The cycle of latency and reactivation within a host undoubtedly contributes towards this success. In the context of AIDS, these viruses take on a more sinister role, where the normal balance between pathogen and host is perturbed by a declining immune system. Herpesvirus infections in these patients may be persistent and life-threatening, may involve multiple strains and may result in the development of tumours or the emergence of virus strains resistant to antiviral agents [13].

Medicine is facing important challenges in the diagnosis and treatment of herpesvirus infections in HIV-infected patients. Molecular biology will undoubtedly make an increasing impact in both of these areas. The application of PCR to the diagnosis of herpesvirus infections needs to be assessed in well controlled clinical studies in AIDS patients. Amplification of mRNA transcripts may prove to be a powerful tool in distinguishing active from latent herpesvirus infections and is an area likely to receive attention in future. The requirement for only small amounts of material may obviate the need for invasive and sometimes hazardous biopsies. Prospective screening and early diagnosis may offer therapeutic advantages in these patients, but it is essential that these techniques are tested rigorously in comparison with currently available diagnostic methods and correlated with adequate follow-up data. Sequencing of virus isolates may yield important information on the epidemiology and pathogenic potential of various strains. Longitudinal studies of virus isolates, when linked to clinical data on immune function, may provide insight into the processes involved in the development of opportunistic infections and neoplasms.

REFERENCES

1. Albrecht, M.A., DeLuca, N.A., Byrne, R.A., Schaffer, P.A., Hammer, S.M. (1989) The herpes simplex virus immediate-early protein, ICP4, is required to potentiate replication of human immunodeficiency virus in $CD4^+$ lymphocytes. *J. Virol.*, **63**, 1861–8.
2. Ambinder, R.F., Lambe, B.C., Mann, R.B., *et al.* (1990) Oligonucleotides for polymerase chain reaction amplification and hybridization detection of Epstein–Barr virus DNA in clinical specimens. *Mol. Cell. Probes*, **4**, 397–407.
3. Aurelius, E., Johansson, B., Skoldenberg, B., Staland, A., Forsgren, M. (1990) Rapid diagnosis of herpes simplex encephalitis by nested polymerase chain reaction assay of cerebrospinal fluid. *Lancet*, 337, 189–92.
4. Azim, T., Crawford, D.H. (1988) Lymphocytes activated by the Epstein–Barr virus to produce immunoglobulin do not express CD23 or become immortalized. *Int. J. Cancer*, **42**, 23–8.

5. Barry, P.A., Pratt-Lowe, E., Peterlin, B.M., Luciw, P.A. (1990) Cytomegalovirus activates transcription directed by the long terminal repeat of human immunodeficiency virus type 1. *J. Virol.*, **64**, 2932–40.
6. Bashir, R.M., Hochberg, F.H., Harris, N.L., Purtilo, D. (1990) Variable expression of Epstein–Barr virus genome as demonstrated by *in situ* hybridization in central nervous system lymphomas in immunocompromised patients. *Mod. Pathol.*, 3, 429–34.
7. Beck, S., Barrell, B.G. (1988) Human cytomegalovirus encodes a glycoprotein homologous to MHC class-I antigens. *Nature*, **331**, 269–72.
8. Ben-Porat, T., Tokazewski, S. (1977) Replication of herpesvirus DNA. II. Sedimentation characteristics of newly synthesized DNA. *Virology*, **79**, 292–301.
9. Braun, D.K., Batterson, W., Roizman, B. (1984) Identification and genetic mapping of a herpes simplex virus capsid protein which binds DNA. *J. Virol.*, 50, 645–8.
10. Browne, H., Smith, G., Beck, S., Minson, A.C. (1990) A complex between the MHC class I homologue encoded by human cytomegalovirus and beta 2 microglobulin. *Nature*, **347**, 770–2.
11. Caradonna, S.J., Chang, Y.C. (1981) Induction of Uracil-DNA-glycosylase and dUTP hydrolase in herpes simplex virus infected human cells. *J. Biol. Chem.*, **256**, 9834–7.
12. Chartrand, P., Crumpacker, C.S., Schaffer, P.A., Wilkie, N.M. (1980) Physical and genetic analysis of the herpes simplex virus DNA polymerase locus. *Virology*, **103**, 311–25.
13. Chavanet, P., Malet, J., Waldner, A., *et al.* (1990) A double blind randomized placebo trial on very high doses of acyclovir in weakly symptomatic HIV-patients. *Cancer Detect. Prev.*, **14**, 669–73.
14. Chou, S.W. (1989) Reactivation and recombination of multiple cytomegalovirus strains from individual organ donors. *J. Infect. Dis.*, **160**, 11–5.
15. Chou, S., Dennison, K.M. (1991) Analysis of interstrain variation in cytomegalovirus glycoprotein B sequences encoding neutralisation-related epitopes. *J. Infect. Dis.*, **163**, 1229–34.
16. Clouse, K.A., Robbins, P.B., Fernie, B., Ostrove, J.M., Fauci, A.S. (1989) Viral antigen stimulation of the production of human monokines capable of regulating HIV-1 expression. *J. Immunol.*, **143**, 470–5.
17. Clouse, K.A., Powell, D., Washington, I., *et al.* (1989) Monokine regulation of human immunodeficiency virus-1 expression in a chronically infected human T cell clone. *J. Immunol.*, **142**, 431–8.
18. Colimon, R., Somogyi, T., Bertrand, C., Michelson, S. (1985) Demonstration of the colinearity of human cytomegalovirus genomes and construction of restriction maps of unknown isolates using cloned subgenomic fragments. *J. Gen. Virol.*, **66**, 2183–198.
19. Collier, A.C., Chandler, S.H., Handsfield, H.H., Corey, L., McDougall, J.K. (1990) Identification of multiple strains of cytomegalovirus in homosexual men. *J. Infect. Dis.*, **159**, 123–6.

20. Collins, P., Darby, G. (1991) Laboratory studies of herpes simplex virus strains resistant to acyclovir. *Rev. Med. Virol.*, **1**, 19–28.
21. Darlington, J., Super, M., Patel, K., Grundy, J.E., Griffiths, P.D., Emery V.C. (1991) Use of the polymerase chain reaction to analyse sequence variation within a major neutralising epitope of glycoprotein B (gp58) in clinical isolates of human cytomegalovirus. *J. Gen. Virol.*, **72**, 1985–9.
22. Darlington, R.W., Moss, L.H. III (1969) The envelope of herpesvirus. *Prog. Med. Virol.*, **11**, 16–45.
23. Davis, M., Kenney, S.C., Kamine, J., Pagano, J.S., Huang, E.S. (1987) Immediate–early gene region of human cytomegalovirus trans-activates the promoter of human immunodeficiency virus. *Proc. Natl. Acad. Sci. U.S.A.*, **84**, 8642–6.
24. De Souza, Y.G., Freese, U.K., Greenspan, D., Greenspan, J.S. (1990) Diagnosis of Epstein–Barr virus infection in hairy leukoplakia by using nucleic acid hybridization and noninvasive techniques. *J. Clin. Microbiol.*, **28**, 2775–8.
25. Drew, W.L., Miner, R.C., Busch, D.F., *et al.* (1991) Prevalence of resistance in patients receiving ganciclovir for serious cytomegalovirus infection. *J. Infect. Dis.*, **163**, 716–9.
26. Ensoli, B., Lusso, P., Schachter, F., *et al.* (1989) Human herpes virus-6 increases HIV-1 expression in coinfected T cells via nuclear factors binding to the HIV-1 enhancer. *EMBO J.*, **8**, 3019–27.
27. Erice, A., Chou, S., Biron, K.K., Stanat, S.C., Balfour, H.H. Jr., Jordan, M.C. (1989) Progressive disease due to ganciclovir-resistant cytomegalovirus in immunocompromised patients. *N. Engl. J. Med.*, **320**, 289–93.
28. Erlich, K.S., Mills, J., Chatis, P., *et al.* (1989) Acyclovir-resistant herpes simplex virus infections in patients with the acquired immunodeficiency syndrome. *N. Engl. J. Med.*, **320**, 293–6.
29. Fenwick, M.L., Walker, M.J. (1978) Suppression of the synthesis of cellular macromolecules by herpes simplex virus. *J. Gen. Virol.*, **41**, 37–51.
30. Folks, T.M., Justement, J., Kinter, A., Dinarello, C.A., Fauci, A.S. (1987) Cytokine-induced expression of HIV-1 in a chronically infected promonocytic cell line. *Science*, **238**, 800–2.
31. Fraenkel, L.D., Gaur, S., Tsolia, M., Scudder, R., Howell, R., Kesarwala, H. (1990) Cytomegalovirus infection in children with AIDS. *Rev. Inf. Dis.*, **12** Suppl 7, S820–6.
32. Gendelman, H.E., Phelps, W., Feigenbaum, L., *et al.* (1986) Trans-activation of the human immunodeficiency virus long terminal repeat sequence by DNA viruses. *Proc. Natl. Acad. Sci. U.S.A.*, **83**, 9759–63.
33. Gerster, T., Roeder, R.G. (1988) Herpesvirus trans-activating protein interacts with transcription factor OTF-1 and other cellular proteins. *Proc. Natl. Acad. Sci. U.S.A.*, **85**, 6247–51.
34. Gibson, W., Roizman, B. (1972) Proteins specified by herpes simplex virus. VIII. Characterization and composition of multiple capsid forms of subtypes 1 and 2. *J. Virol.*, **10**, 1044–52.
35. Gimble, J.M., Duh, E., Ostrove, J.M., Gendelman, H.E., Max, E.E., Rabson, A.B. (1988) Activation of the human immunodeficiency virus long terminal

repeat by herpes simplex virus type 1 is associated with induction of a nuclear factor that binds to the NF-kappa B/core enhancer sequence. *J. Virol.*, **62**, 4104–12.

36. Gorgievski Hrisoho, M., Hinderer, W., Nebel Schickel, H., *et al.* (1990) Serodiagnosis of infectious mononucleosis by using recombinant Epstein–Barr virus antigens and enzyme-linked immunosorbent assay technology. *J. Clin. Microbiol.*, **28**, 2305–11.
37. Gregory, C.D., Murray, R.J., Edwards, C.F., Rickinson, A.B. (1988) Downregulation of cell adhesion molecules LFA-3 and ICAM-1 in Epstein–Barr virus-positive Burkitt's lymphoma underlies tumour cell escape from virus-specific T cell surveillance. *J. Exp. Med.*, **167**, 1811–24.
38. Gregory, C.D., Dive, C., Henderson, S., *et al.* (1991) Activation of Epstein–Barr virus latent genes protects human B cells from death by apoptosis. *Nature*, **349**, 612–4.
39. Hammer, S.M., Buchman, T.G., D'Angelo, L.J., Karchmer, A.W., Roizman, B., Hirsch, M.S. (1980) Temporal cluster of herpes simplex encephalitis, investigation by restriction endonuclease cleavage of viral DNA. *J. Infect. Dis.*, **141**, 436–40.
40. Hayakawa, Y., Yamamoto, T., Yamanishi, K., Takahashi, M. (1986) Analysis of varicella zoster virus DNAs of clinical isolates by endonuclease Hpa I. *J. Gen. Virol.*, **67**, 1817–29.
41. Heine, J.W., Hariess, R.W., Cassai, F., Roizmann, B. (1974) Protein herpes simplex virus. XII. The virion polypeptide of type 1 strain. *J. Virol.*, **14**, 640–51.
42. Ho, W.Z., Harouse, J.M., Rando, R.F., Gonczol, E., Srinivasan, A., Plotkin, S.A. (1990) Reciprocal enhancement of gene expression and viral replication between human cytomegalovirus and human immunodeficiency virus type 1. *J. Gen. Virol.*, **71**, 97–103.
43. Honess, R.W., Roizman, B. (1974) Regulation of herpesvirus macromolecular synthesis I. Cascade regulation of the synthesis of three groups of viral proteins. *J. Virol.*, **14**, 8–19.
44. Honess, R.W., Roizman, B. (1975) Regulation of herpesvirus macro-molecular synthesis, sequential transition of polypeptide synthesis requires functional viral polypeptides. *Proc. Natl. Acad. Sci. U.S.A.*, **72**, 1276–80.
45. Hubenthal-Voss, J., Roizman, B. (1988) The properties of two 5′ co-terminal RNAs transcribed part way and across the S component origin of DNA synthesis of the herpes simplex virus 1 genome. *Proc. Natl. Acad. Sci. U.S.A.*, **85**, 8454–8.
46. Jacob, R.J., Roizman, B. (1977) Anatomy of herpes simplex virus DNA. VIII. Properties of the replicating DNA. *J. Virol.*, **23**, 394–411.
47. Jamieson, A.T., Subak-Sharpe, J.H. (1974) Biochemical studies on the herpes simplex virus-specified deoxypyrimidine kinase activity. *J. Gen. Virol.*, **24**, 481–92.
48. Johnson, D.C., Frame, M.C., Ligas, M.W., Cross, A.M., Stow, N.D. (1988) Herpes simplex virus immunoglobulin G Fc receptor activity depends on a complex of two viral glycoproteins gE and gI. *J. Virol.*, **62**, 1347–54.

49. Keir, H.M., Gold, E. (1963) Deoxyribonucleic acid nucleotidyltransferase and deoxyribonuclease from cultured cells infected with herpes simplex virus. *Biochim. Biophys. Acta*, 72, 263–76.
50. Kemp, L.M., Dent, C.L., Latchman, D.S. (1990) Octamer motif mediates transcriptional repression of HSV immediate-early genes and octamer-containing cellular promoters in neuronal cells. *Neurology*, **4**, 215–22.
51. Kenney, S., Kamine, J., Markovitz, D., Fenrick, R., Pagano, J. (1988) An Epstein–Barr virus immediate early gene product trans-activates gene expression from the human immunodeficiency virus long terminal repeat. *Proc. Natl. Acad. Sci. U.S.A.*, **85**, 1652–6.
52. Knowles, D.M., Inghirami, G., Ubriaco, A., Dalla Favera, R. (1989) Molecular genetic analysis of three AIDS-associated neoplasms of uncertain lineage demonstrates their B-cell derivation and the possible pathogenetic role of the Epstein–Barr virus. *Blood*, 73, 792–9.
53. Knox, K.K., Drobyski, W., Carrigan, D. (1991) Cytomegalovirus isolate resistant to ganciclovir and foscarnet from a marrow transplant patient. *Lancet*, **337**, 1292–3.
54. Kristie, T.M., Roizman, B. (1984) Separation of sequences defining basal expression from those conferring a gene recognition within the regulatory domains of herpes simplex virus 1α genes. *Proc. Natl. Acad. Sci. U.S.A.*, **81**, 4065–9.
55. Kristie, T.M., Roizman, B. (1987) Host cell proteins bind to the *cis*-acting site required for virion-mediated induction of herpes simplex virus 1 α genes. *Proc. Natl. Acad. Sci. U.S.A.*, **84**, 71–5.
56. Kristie, T.M., Roizman, B. (1988) Differentiation and DNA contact points of the host proteins binding at the *cis*-site for the virion mediated induction of genes of herpes simplex virus 1. *J. Virol.*, **62**, 1145–57.
57. Lang, D.J., Kovacs, A.A., Zaia, J.A., *et al.* (1989) Seroepidemiologic studies of cytomegalovirus and Epstein–Barr virus infections in relation to human immunodeficiency virus type 1 infection in selected recipient populations. *J. AIDS*, **2**, 540–9.
58. Linnemann, C.C., Jr., Biron, K.K., Hoppenjans, W.G., Solinger, A.M. (1990) Emergence of acyclovir-resistant varicella zoster virus in an AIDS patient on prolonged acyclovir therapy. *Aids*, **4**, 577–9.
59. Littler, E., Baylis, S.A., Zeng, Y., Conway, M.J., Mackett, M., Arrand, J.R. (1991) Diagnosis of nasopharyngeal carcinoma by means of recombinant Epstein–Barr virus proteins. *Lancet*, **337**, 685–9.
60. Lusso, P., Ensoli, B., Markham, P.D., *et al.* (1989) Productive dual infection of human $CD4^+$ T lymphocytes by HIV-1 and HHV-6. *Nature*, **337**, 370–3.
61. Lusso, P., DeMaria, A., Malnati, M., *et al.* (1991) Induction of CD4 and susceptibility to HIV-1 infection in human $CD8^+$ T lymphocytes by human herpesvirus 6. *Nature*, **349**, 533.
62. Mackem, S., Roizman, B. (1982) Structural features of the α gene 4, 0 and 27 promoter-regulatory sequences which confer a regulation on chimeric thymidine kinase genes. *J. Virol.*, **44**, 939–49.
63. Margalith, M., Sarov, B., Sarov, I., *et al.* (1990) Serum IgG and IgA antibodies

specific to Epstein–Barr virus capsid antigen in a longitudinal study of human immunodeficiency virus infection and disease progression in homosexual men. *AIDS Res. Hum. Ret.*, **6**, 607–16.

64. Martin, J.H., Dohner, D.E., Wellinghoff, W.J., Gelb, L.D. (1982) Restriction endonuclease analysis of varicella-zoster virus and wild-type DNAs. *J. Med. Virol.*, **9**, 69–76.
65. McKeating, J.A., Griffiths, P.D., Weiss, R.A. (1990) HIV susceptibility conferred to human fibroblasts by cytomegalovirus-induced Fc receptor. *Nature*, **343**, 659–61.
66. Mosca, J.D., Bednarik, D.P., Raj, N.B.K., *et al.* (1987) Herpes simplex virus type-1 can reactivate transcription of human immunodeficiency virus. *Nature*, **325**, 67–70.
67. Nishioka, Y., Silverstein, S. (1977) Degradation of cellular mRNA during infection by herpes simplex virus. *Proc. Natl. Acad. Sci. U.S.A.*, **74**, 2370–4.
68. Novak, J., Pavel, S., Krchnak, V., Hamsikova, E., Zavadova, H., Roubal, J. (1991) Mapping of serologically relevant regions of human cytomegalovirus phosphoprotein pp150 using synthetic peptides. *J. Gen. Virol.*, **72**, 1409–13.
69. O'Hare, P., Goding, C.R. (1988) Herpes simplex virus regulatory elements and the immunoglobulin octamer domain binds a common factor and are both targets for virion transactivation. *Cell*, **52**, 435–45.
70. Ostrove, J.M., Leonard, J., Weck, K.E., Rabson, A.B., Gendelman, H.E. (1987) Activation of the human immunodeficiency virus by herpes simplex virus type 1. *J. Virol.*, **61**, 3726–32.
71. Pagano, J.S., Kenney, S., Markovitz, D., Kamine, J. (1988) Epstein–Barr virus and interactions with human retroviruses. *J. Virol. Methods*, **21**, 229–39.
72. Pellett, P.E., McKnight, J.L.C., Jenkins, F.J., Roizman, B. (1985) Nucleotide sequence and predicted amino acid sequence of a protein encoded in a small herpes simplex virus DNA fragment capable of trans-inducing α genes. *Proc. Natl. Acad. Sci. U.S.A.*, **82**, 5870–4.
73. Permeen, A.M., Sam, C.K., Pathmanathan, R., Prasad, U., Wolf, H. (1990) Detection of Epstein–Barr virus DNA in nasopharyngeal carcinoma using a non-radioactive digoxigenin-labelled probe. *J. Virol. Methods*, **27**, 261–7.
74. Powell, K., Purifoy, D. (1977) Nonstructural proteins of herpes simplex virus. I. Purification of the induced DNA polymerase. *J. Virol.*, **24**, 618–26.
75. Quinlivan, E.B., Holley-Guthrie, E., Mar, E.C., Smith, M.S., Kennedy, S. (1990) The Epstein–Barr virus BRLF1 immediate–early gene product transactivates the human immunodeficiency virus type 1 long terminal repeat by a mechanism which is enhancer independent. *J. Virol.*, **64**, 1817–20.
76. Rando, R.E., Pellett, P.E., Luciw, P.A., Bohan, C.A., Srinivasan, A. (1987) Transactivation of the human immunodeficiency virus by herpesviruses. *Oncogene*, **1**, 13–8.
77. Rando, R.F., Srinivasan, A., Feingold, J., Gonczol, E., Plotkin, S. (1990) Characterization of multiple molecular interactions between human cytomegalovirus (HCMV) and human immunodeficiency virus type 1 (HIV-1). *Virology*, **176**, 87–97.
78. Rice, A.B., Matthews, M.B. (1988) Trans-activation of the human immunodeficiency virus long terminal repeat sequences, expressed in an adenovirus

vector, by the adenovirus E1A 13S protein. *Proc. Natl. Acad. Sci. U.S.A.*, **85**, 4200–4.

79. Ripalti, A., Landini, M.P., Mocarski, E.S., La Placa, M. (1989) Identification and preliminary use of recombinant lambda gt11 fusion proteins in human cytomegalovirus diagnosis. *J. Gen. Virol.*, 70, 1247–51.
80. Roffman, E., Albert, J.P., Goff, J.P., Frenkel, N. (1990) Putative site for the acquisition of human herpesvirus-6 virion tegument. *J. Virol.*, **64**, 6308–13.
81. Roizman, B., Tognon, M. (1983) Restriction endonuclease patterns of herpes simplex virus DNA , Application to diagnosis and molecular epidemiology. *Curr. Topics Microbiol. Immunol.*, **104**, 275–86.
82. Roizman, B., Furlong, D. (1974) The replication of herpesviruses, in *Comprehensive Virology*, Vol. 3, (eds H. Fraenkel-Conrat, R.R. Wagner), Plenum Press, New York, pp. 229–403.
83. Rowley, A.H., Whitley, R.J., Lakeman, F.D., Wolinsky, S.M. (1990) Rapid detection of herpes simplex virus DNA in cerebrospinal fluid of patients with herpes simplex encephalitis. *Lancet*, **335**, 440–1.
84. Schulten, E.A., Snijders, P.J., ten Kate, R.W., *et al.* (1991) Oral hairy leukoplakia in HIV infection: a diagnostic pitfall. *Oral Surg. Oral Med. Oral Pathol.*, **71**, 32–7.
85. Schwartz, J., Roizman, B. (1969) Concerning the egress of herpes simplex virus from infected cells: electron and light microscope observations. *Virology*, **38**, 42–9.
86. Seigneurin, J.M., Guilbert, B., Bourgeat, M.J., Avrameas, S. (1988) Polyspecific natural antibodies and autoantibodies secreted by human lymphocytes immortalized with Epstein–Barr virus. *Blood*, **71**, 581–5.
87. Skolnik, P.R., Kosloff, B.R., Hirsh, M.S. (1988) Bidirectional interactions between human immunodeficiency virus type 1 and cytomegalovirus. *J. Infect. Dis.*, **157**, 508–14.
88. Spear, P.G. (1984) Glycoproteins specified by herpes simplex virus, in *The Herpesviruses*, Vol. 3, (ed. B. Roizman), Plenum Press, New York, p. 315.
89. Spira, T.J., Bozeman, L.H., Sanderlin, K.C., *et al.* (1991) Lack of correlation between human herpesvirus-6 infection and the course of human immunodeficiency virus infection. *J. Infect. Dis.*, **161**, 567–70.
90. Stevens, J.G., Wagner, E.K., Devi-Rao, G.B., Cook, M.L., Feldman, L.T. (1987) RNA complementary to a herpesvirus a gene mRNA is prominent in latently infected neurons. *Science*, **235**, 1056–9.
91. Takeda, A., Tuazon, C.U., Ennis, F.A. (1988) Antibody-enhanced infection by HIV-1 via Fc receptor-mediated entry. *Science*, **242**, 580–3.
92. Webster, A., Lee, C.A., Cook, D.G., *et al.* (1989) Cytomegalovirus infection and progression towards AIDS in haemophiliacs with human immunodeficiency virus. *Lancet*, **ii**, 63–6.
93. Wheatley, S.C., Dent, C.L., Wood, J.N., Latchman, D.S. (1991) A cellular factor binding to the TAATGARAT DNA sequence prevents the expression of the HSV immediate–early genes following infection of nonpermissive cell lines derived from dorsal root ganglion neurons. *Exp. Cell Res.*, **194**, 78–82.
94. Whitley, R.J., Goldsmith, S., Grann, J. (1990) Herpesviruses in the

immunocompromised host, in *Control of Virus Diseases*, (eds N.J. Dimmock, P.D. Griffiths, C.R. Madely), Symposium 45, Society for General Microbiology, Cambridge University Press, pp. 315–40.

95. Wohlrab, F., Francke, B. (1980) Deoxypyrimidine triphosphatase activity specific for cells infected with herpes simplex virus type 1. *Proc. Natl. Acad. Sci. U.S.A.*, 77, 1872–6.
96. Zhu, Z., Chen, S.S.L., Huang, A.S. (1990) Phenotypic mixing between human immunodeficiency virus and vesicular stomatitis virus or herpes simplex virus. *J. AIDS*, 3, 215–9.

12

Future prospects

STEVEN H. MYINT, ALAN J. CANN and DAVID A.J. TYRRELL

Previous chapters have reviewed the cellular and molecular biology of a number of bacterial, viral and fungal diseases that represent the majority of the opportunistic infections occuring in HIV-positive patients. There is a great deal of interest in these agents and it is apparent that the techniques of modern biology provide an understanding of the pathogenesis of these infections that was not possible previously. This applies to pathogens and diseases that have been known for a long time as well as those that are newly recognized. An example of the former is salmonellosis: the transposon mutagenesis experiments described in Chapter 3 define virulence factors in *Salmonella.* Characterizing these and their mechanism of action will greatly assist in the development of future vaccines. Similarly, the identification of variants of *Entamoeba histolytica* is described in Chapter 5, and this will help greatly to understand the epidemiology and pathology of the disease and hence promote future treatment strategies. 'Newer' pathogens such as *Pneumocystis carinii* could only be investigated in detail by using modern molecular biological approaches and the results have enabled us to reclass this organism as a member of the fungi, rather than as a protozoon, and to identify a likely host receptor.

The new methodologies have also enhanced the study of the biology of these organisms *in vivo*, both in laboratory models and in patients. An example is the use of novel methods of detecting the organisms. Gene detection methods are described throughout this book, and in some cases have supplanted more traditional techniques. The use of gene amplification

Molecular and Cell Biology of Opportunistic Infections in AIDS
Edited by Steven Myint and Alan Cann
Published in 1992 by Chapman & Hall, London ISBN 0 412 45330 4

by the polymerase chain reaction, and allied techniques, holds particular promise for the future.

Despite the wealth of new information presented in this book, much more research needs to be undertaken. As so often happens, one question is answered but the answer suggests many more questions! Several spring to mind. What is the precise mechanism for the failure of BCG vaccine in India? Will strain analysis of field isolates provide the answer? How does *Cryptosporidium parvum* cause diarrhoea? If we knew this could we devise an effective treatment? If *Pneumocystis carinii* is genetically a fungus, does it behave biologically as one? Would modifications of current antifungal drugs give us improved treatment regimens? Are cytomegalovirus and mycoplasmal species co-factors in the progression of HIV disease? If so, by what mechanism? This is important since blocking this might enable us to prevent disease progression in the millions of people worldwide who will be infected with HIV in spite of all the effort put into health education and the development of vaccines.

The first chapter refers to many more infections than those detailed in subsequent chapters. It is evident from the literature that there is some work under way on the molecular and cellular biology of most of these other organisms, such as *Cryptococcus neoformans*, but it is still in its infancy. As the number of cases of AIDS increases throughout the world these infections will no longer be rare and we will need to understand their biology in order to manage them effectively. The prevalence of specific infections will increase and new pathogens will be recognized. It appears that there is a particular lack of research, and presumably funding, on the parasitic and fungal disorders. This may reflect their current minor role as pathogens in the developed world, relative to bacteria and viruses, but it is likely that their importance will increase and we should start to study them now, using molecular approaches.

So what of the future? It is hoped that, in the not too distant future, effective vaccines and more effective treatment directed against HIV will become available; large sums are being spent to this end. But until this time, and thereafter, the number of opportunistic infections, which after all are the final causes of mortality, will continue to grow. Efforts to understand and combat them should be increased. Furthermore, immunosuppression and hence opportunistic infection will also increase for other reasons, organ transplantation for example.

Although this book is a scientific compilation we feel we should end with a statement of policy: research on the basic biology of opportunistic infection must be developed further.

Index

MIX
Papier aus verantwortungsvollen Quellen
Paper from responsible sources
FSC® C105338

FSC
www.fsc.org

If you have any concerns about our products,
you can contact us on
ProductSafety@springernature.com

In case Publisher is established outside the EU,
the EU authorized representative is:
Springer Nature Customer Service Center GmbH
Europaplatz 3, 69115 Heidelberg, Germany

Printed by Libri Plureos GmbH
in Hamburg, Germany